T0143244

The Comparative Approach in Evolutionary
Anthropology and Biology

The **Comparative Approach** in **Evolutionary Anthropology** and **Biology**

Charles L. Nunn

THE UNIVERSITY OF CHICAGO PRESS | CHICAGO AND LONDON

Charles L. Nunn is associate professor in the Department of Human Evolutionary Biology at Harvard University. He is coeditor of *Evolution of Sleep: Phylogenetic and Functional Perspectives* and coauthor of *Infectious Diseases in Primates: Behavior, Ecology and Evolution.*

The University of Chicago Press, Chicago 60637
The University of Chicago Press, Ltd., London
© 2011 by The University of Chicago
All rights reserved. Published 2011.
Printed in the United States of America

20 19 18 17 16 15 14 13 12 11   1 2 3 4 5

ISBN-13: 978-0-226-60898-3 (cloth)
ISBN-13: 978-0-226-60899-0 (paper)
ISBN-10: 0-226-60898-0 (cloth)
ISBN-10: 0-226-60899-9 (paper)

Library of Congress Cataloging-Publication Data
Nunn, Charles L.
    The comparative approach in evolutionary anthropology and biology / Charles L. Nunn.
        p. cm.
    Includes bibliographical references and index.
    ISBN-13: 978-0-226-60898-3 (cloth : alk. paper)
        ISBN-10: 0-226-60898-0 (cloth : alk. paper)
        ISBN-13: 978-0-226-60899-0 (pbk. : alk. paper)
        ISBN-10: 0-226-60899-9 (pbk. : alk. paper)
    1. Evolution (Biology) I. Title.
    QH366.2.N87 2011
    599.93′8—dc22
                                                    2010052466

♾ This paper meets the requirements of ANSI/NISO Z39.48-1992 (Permanence of Paper).

Contents

Preface   vii

1   The Importance of Comparison   1

2   Basic Phylogenetic Concepts and "Tree Thinking"   20

3   Reconstructing Ancestral States for Discrete Traits   52

4   Reconstructing Ancestral States for Quantitative Traits   79

5   Modeling Evolutionary Change   98

6   Correlated Evolution and Testing Adaptive Hypotheses   126

7   Comparative Methods to Detect Correlated Evolutionary Change   148

8   Using Trees to Study Biological and Cultural Diversification   180

9   Size, Allometry, and Phylogeny   202

10   Human Cultural Traits and Linguistic Evolution   227

11   Behavior, Ecology, and Conservation of Biological and Cultural Diversity   255

12   Investigating Evolutionary Singularities   280

13   Developing a Comparative Database and Targeting Future Data Collection   299

14   Conclusions and Future Directions   317

References   323
Index   365

Preface

When I set out to write this book, I never anticipated how much I would learn and how greatly it would alter my research program. I was amazed to see how widely the comparative approach is applied in evolutionary anthropology and to discover that the comparative method means different things to people in different fields of anthropology. I also was impressed with the wide diversity of phylogenetic methods that are being used to investigate fundamental questions in evolutionary anthropology—including in emerging fields such as evolutionary archeology—and with the rapid development of new methods in biology to study phylogenetic patterns. I hope that this book captures some of the excitement I felt as I discovered these fields of study and that it showcases the phylogenetic research being conducted across evolutionary anthropology more generally. I also hope that it will make phylogenetic approaches more accessible to evolutionary anthropologists and biologists.

This book was conceived during three intellectually rewarding years of research at the Max Planck Institute for Evolutionary Anthropology in Leipzig, Germany. The institute houses five departments: linguistics, comparative psychology, evolutionary genetics, primatology, and human evolution. During this time, I witnessed collaborations blossom among researchers, and I was inspired to find commonalities among the different departments housed by the institute. One of my goals in writing this book is to increase communication among the subfields of evolutionary anthropology. By probing the common comparative foundations of the different subfields that I was exposed to in Leipzig—and by considering the differences in how the comparative method is applied in these fields—I believe that we can create a more integrative and productive approach to understanding the factors that have shaped our evolutionary history.

While the book was conceived in Leipzig, the first pages were written shortly after a job interview at Harvard University, where discussions with a number of my future colleagues inspired me to start writing. My subsequent interactions with faculty, postdoctoral researchers, and graduate students propelled me along new paths of discovery. In addition, I was fortunate to use the manuscript as a reading in an undergraduate class titled "Comparison and Adaptation in Primate Evolutionary Biology," which resulted in new insights from many different perspectives.

I had to make many difficult decisions when considering the methods and examples to include in this book. I apologize to those who might feel left out, and I acknowledge that my choices probably were biased toward examples that appealed most to my research interests. The literature is filled with many wonderful examples, but space was limited, and so I had to prioritize some topics over others and aim for key examples rather than exhaustive overviews.

Many colleagues provided comments and discussion on the material covered in this book. I would like to thank in particular Christian Arnold, Isabella Capellini, Luke Matthews, Terry Ord, and Liam Revell for reading a majority of chapters (in some cases, the entire book). Enrico Rezende and an anonymous reviewer also provided extremely valuable comments on the first version of the text. I also benefited greatly from discussion and comments on the manuscript from Quentin Atkinson, Christophe Boesch, Monique Borgerhoff Mulder, Carrie Brezine, Dave Collar, Michael Cysouw, Laura Fortunato, Mathias Franz, Ted Garland, Alex Georgiev, Russell Gray, Anja Hayen, Kate Jones, Jason Kamilar, Dan Lieberman, Patrik Lindenfors, Roger Mundry, Brian O'Meara, Kevin Omland, Chris Organ, Sheila Patek, David Pilbeam, Brian Preston, Andrew Ritchie, Richard Smith, Jamie Tehrani, Claudio Tennie, Linda Vigilant, and Søren Wichmann.

More generally, I learned much of what I know about comparative methods and their applications from discussions with Robert Barton, Monique Borgerhoff Mulder, Joe Felsenstein, Kate Jones, Ted Garland, John Gittleman, and Carel van Schaik.

I also wish to thank a number of people who contributed to the production of the book itself. I thank Christie Henry at the University of Chicago Press for her support of this project and Abby Collier and Mary Gehl for help with production. For assistance with figures and other aspects of the text, I thank Christian Arnold, Carrie Brezine, Judith Butler-Vincent, Sebastian Geidel, Peter Maciej, Christof Neumann, Joanna Rifkin, Karen Santospago, and Serena Zhao. A number of colleagues also provided data, information, or images for the figures, including Quentin Atkinson, Marcel Cardillo, Alexandre Diniz Filho, Michel Genoud, Kevin Omland, Ulrich Reichard, Katerina Rexová and Peter Ungar. Christian Arnold and Luke Matthews contributed numerous entries to the AnthroTree website that accompanies this book. I also thank the Dean of the Faculty of Arts and Sciences at Harvard for a grant from the "Publishing Fund" that supported redrawing some figures.

Last, and most importantly, I would like to acknowledge my family for their pivotal role in many aspects of this book. My wife, Sheila Patek, was a helpful sounding board on many aspects of this project, including coming up with the name "AnthroTree" and encouraging me to write the book. My daughter Sylvia taught me to look at the world with fresh eyes, and it was always a delight to return home to her after a long day of writing, especially on days when we found time to go to the Leipzig Zoo, the Botanical Gardens at Berkeley, or the museums of Boston. And I am grateful to my parents, Jack and Janet, for exposing me to ideas that led me down an academic path.

1 The Importance of Comparison

Comparison is fundamental to evolutionary anthropology. When anthropologists attempt to understand the new fossils from Flores, for example, they compare these remains to other fossils and extant humans (Brown et al. 2004; Lieberman 2005; Tocheri et al. 2007), and when psychologists study chimpanzee cognition, they often compare chimpanzee performance on cognitive tasks to the performance of human children on the same tasks (Whiten et al. 1996). Linguists expend tremendous effort describing the grammar and vocabulary of different languages; interesting results also emerge when efforts of multiple linguists are synthesized in a comparative study, for example, to construct language trees (Campbell 2004; Atkinson and Gray 2005; Dunn et al. 2005; McMahon and McMahon 2005). Similarly, the comparative method has a long history in primatology (Crook and Gartlan 1966; Milton and May 1976; Clutton-Brock and Harvey 1977). Primatologists and evolutionary morphologists continue to produce pioneering comparative studies in nonhuman primates, including research on sexual dimorphism, brain size, and social systems. When geneticists create evolutionary trees of primates, they also use data in a comparative context by identifying shared derived characters under specific models of evolution. Paleontologists use similar approaches to reconstruct the phylogeny of extinct hominins using morphological characteristics (e.g., Strait et al. 1997).

These examples demonstrate the central role that comparison plays in evolutionary anthropology. Comparison provides a way to draw general inferences about the evolution of traits and therefore has long been the cornerstone of efforts to understand biological and cultural diversity (Darwin 1859; Murdock and White 1969; Ridley 1983; Brooks and McLennan 1991; Harvey and Pagel 1991; Mace and Pagel 1994). Individual studies of fossilized remains, living species, or human populations are the essential units of analysis in a comparative study; bringing these elements into a broader comparative framework allows the puzzle pieces to fall into place, thus creating a way to test adaptive hypotheses and generate new ones.

Here, *comparative study* refers to studies of multiple populations or species, including studies of geographic variation at a continental or regional scale. In many cases, a comparative study stands on the shoulders of numerous, painstaking studies of individual populations or species; a single scientist would find it difficult—even impossible—to collect data at such a broad scale. *Evolutionary anthropology* lies at the interface of biology and anthropology; it aims to understand human behavior, ecology, and evolution. This includes studies of hominin fossils, language diversification, and the genetics of human populations. Evolutionary anthropologists also investigate cognitive abilities in different species of primates (or other animals), and they study primate morphology and behavioral ecology. Thus, evolutionary anthropology broadly encompasses studies of human evolution, comparative psychology, primatology, functional morphology, linguistics, human behavioral ecology, and comparative anatomy.

The overarching aims of this book are twofold: first, to explore the comparative foundations of evolutionary anthropology and biology in past and present research, and second, to point the way toward new uses of the comparative approach in the future, especially in terms of applying new phylogeny-based methods to address fundamental evolutionary questions. In thinking about the past, present, and future of comparative approaches, the past gives us a sense of how comparative approaches have played a central role in evolutionary anthropology and biology, even before rigorous phylogenetic approaches were possible. In the present, phylogeny-based methods are being applied with greater frequency, and with these methods researchers are addressing both new and old questions. As for the future, new methods are on the horizon, as are new comparative questions that have arisen through previous discoveries.

Thus, this is partly a "how-to" book to increase the use of new phylogeny-based comparative approaches in future anthropological and biological research (chapters 2–8), and partly an exploration of how comparative approaches have been applied to investigate biological and cultural variation (chapters 1 and 9–13). For the methodological sections in the early chapters, I provide an introduction to the methods and how they have been applied rather than giving a detailed statistical examination of the methods, and I focus on the methods that are likely to be of greatest use to evolutionary anthropologists and biologists rather than providing a comprehensive overview of all available methods. My target audiences are anthropologists and biologists who wish to become users of phylogenetic comparative methods and those who want to understand more about the burgeoning literatures that use these methods to test specific hypotheses. I generally (but not exclusively) focus on organismal or cultural traits rather than genetics. Links to computer programs and worked examples are provided on AnthroTree, a website designed around the book (http://www.anthrotree.info). Even though many of the analyses in the book were done with older programs, the website focuses heavily on using R (R Development Core Team 2009), which is emerging as the most flexible and powerful way to run comparative analyses. This site will be updated as new computer programs become available.

While this book is focused primarily on evolutionary anthropology, I tried to write in a way that will also appeal to organismal biologists interested in applying the comparative method. I have done this for several reasons. First, there is a need for an entry-level book on phylogenetic methods in comparative biology, and I hope this book helps fill that opening. In addition, I wish to share with biologists the incredible diversity of comparative questions that are being addressed in evolutionary anthropology. Last, by identifying key questions of general importance to human evolution, I aim to facilitate greater collaboration between evolutionary biologists and anthropologists.

The Need for This Book

Given that comparative approaches are so widely used, and given that there is little controversy over the value of comparison, why do we need this book? Indeed, one could argue that a book focused on comparative methods downplays the importance of other valuable approaches, such as experimental approaches that can uncover causality more directly. Certainly we should make use of experiments when possible. For many of the questions that evolutionary anthropologists and biologists address, however, experiments are neither feasible nor ethical. Even when experiments are possible, comparative approaches are essential for understanding evolutionary phenomena and thus should be used in tandem with experiments. One field in particular—animal cognition—provides opportunities to synthesize experiments on different primate species into a broader evolutionary context (E. MacLean, B. Hare, L. Matthews, and C. Nunn, unpublished data). By giving equivalent cognitive tasks to different species, it becomes possible to assess the roles of ecological and social factors in favoring particular types of cognition and to reconstruct the evolutionary history of primate cognitive abilities (see also Haun et al. 2006a, b).

More generally, a book on this topic is valuable and overdue for at least four reasons. First, *innovative phylogenetic approaches have fueled new discoveries in evolutionary biology, and these methods have yet to be applied to their full potential in evolutionary anthropology.* Many "nonphylogenetic" comparative studies in the past have yielded fruitful insights. However, phylogeny-based methods enable evolutionary anthropologists and biologists to investigate trait evolution more directly, to identify the drivers of speciation and extinction, and to study coevolutionary dynamics of genes, languages, and cultural traits. Thus, phylogenetic comparative methods represent an increasingly important toolkit for evolutionary anthropologists. The first part of this book reviews and synthesizes many advances in phylogenetic methods from biology and provides selected examples from evolutionary anthropology. The second part of the book reviews in greater depth how these approaches have been (or could be) applied to questions in different fields of evolutionary anthropology; thus, the latter chapters reinforce many concepts and approaches that are covered in the earlier, methodologically focused chapters.

Second, *evolutionary anthropology is increasingly interdisciplinary and informatics-driven, and many of these research projects have a strong comparative component.* This follows trends in other sciences, including biology, which are propelled by an appreciation that greater integration among fields can produce fundamentally new insights, and by greater availability of funding for large-scale collaborative research. In an interdisciplinary environment, it becomes essential to understand research in other fields, for example as a way to form new collaborative networks aimed at uncovering links between genes and languages (Diamond and Bellwood 2003) or associations between primate behavior and morphology (Plavcan et al. 1995, 1997b). This book links diverse fields of evolutionary anthropology through the common lens of comparative methods.

Third, *evolutionary anthropology is largely a nonexperimental, historical science; phylogeny-based methods provide a way to examine independent evolutionary origins and thus overcome problems of statistical non-independence in such data.* Many of the most enduring questions in evolutionary anthropology concern empirical patterns of evolution that are unsuited for experimental approaches. Examples include the spatial distribution of nonhuman primates, the spread of modern humans to all corners of the earth, and the evolution of human behavioral, morphological and life history traits in different populations. With their long life spans, most nonhuman primates are poorly suited for experimental manipulations of environmental factors thought to influence genetically determined traits, and the same is true of most questions involving human evolution. Relative to other fields, this means that evolutionary anthropologists rely to a greater extent on observational data to test hypotheses. Importantly in this context, observational data suffer from spatial and historical dependencies—the bane of statistical methods, and the source of many uncertainties in evolutionary anthropology. Given the high costs of acquiring biological and anthropological data from remote locations, it is imperative that we make the best use of these data by correctly applying methods that control for statistical nonindependence while also maximizing statistical power.

Finally, *from an intellectual standpoint, it is important to understand the epistemological basis to our science.* Given the importance of comparison to all subfields of evolutionary anthropology, the time is right to critically assess these methods and how they have increased our knowledge of primate and human evolution. What are the strengths and weaknesses of different comparative approaches? Are there phylogenetic methods that could be useful for evolutionary anthropology but have yet to be applied to study human and primate evolution? Conversely, can methods in evolutionary anthropology be useful for other fields in both the biological and social sciences? Can theoretical models, including simulations, provide new insights to the drivers of broad-scale patterns and by doing so lead to methodological advances to study these processes empirically?

Five Examples from Across Evolutionary Anthropology

In what follows, I briefly review five examples of how the comparative method is used in evolutionary anthropology. I focus on the past and present uses of the comparative method, and I hope to whet your appetite for learning about how phylogeny-based comparative approaches can be used in the future. This is also a good time to make an important caveat relevant to the entire book: given the large number of examples of comparative research in evolutionary anthropology and space limitations in writing this book, it would be impossible to provide a comprehensive summary of all the different comparative studies that have been undertaken (and it is unclear if such a vast litany of examples would be useful in any case). Instead, I focused on particular examples to illustrate different methods and approaches, often drawing on my own research interests in the evolution of behavioral, ecological, and cultural traits.

1. Sperm competition. A suitable example to start with involves the links between primate behavior and morphology. Most studies of mammalian sexual selection prior to about 1980 focused on female choice or direct competition among males, such as males competing for females. For example, Clutton-Brock et al. (1977) investigated the links between body mass dimorphism and group composition in primates. Consistent with the idea that increasing competition for mates selects for larger male body mass, they found that dimorphism increases in lineages characterized by a greater number of females per male and larger body mass (see also Clutton-Brock and Harvey 1977; Mitani et al. 1996a; Plavcan and Van Schaik 1997b; Lindenfors and Tullberg 1998). These findings suggest that primate males experience intense competition to monopolize access to females. In many primate species, however, females mate with multiple males during their time of fertility (Hrdy and Whitten 1987; Van Schaik et al. 1999). For example, Barbary macaque (*Macaca sylvanus*) females mate with up to ten males per day (!) during estrus (Taub 1980). This might reflect that males are unable to monopolize access to females when they become fertile, resulting in a failure of effective competition. In addition to overt physical competition, however, competition among males might also occur *after* mating, that is, within the female reproductive tract, with the sperm from different males competing to fertilize the egg of a female (Parker 1970). This situation is known as *sperm competition*, and it should result in selection for males to produce larger numbers of sperm and sperm that can more quickly reach a fertilizable egg.

Harcourt et al. (1981) used comparative data on primates to investigate a critical prediction of sperm competition. They predicted that in species characterized by greater female mating promiscuity, males should have larger testes mass (see also Short 1979). As testes mass shows a positive association with body mass, it is also necessary to control for body mass in such a comparison. Using data on thirty-three species of primates, they found compelling support for their prediction. Species in which females mate with multiple males have larger *relative* testes mass, measured, for example, as positive residuals from a regression of testes mass on body mass (figure 1.1). Harcourt et al. found that humans fall very close to the regression line (see arrow in figure 1.1), which is inconsistent with lifelong monogamy (under pure monogamy, they should fall further below the line). In a follow-up study, Harcourt et al. (1995) showed that mating system effects were significant after controlling for phylogenetic relationships among the primate species in their sample,

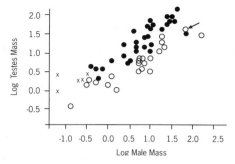

Figure 1.1. Testes mass in relation to body mass and mating system in primates. Filled dots represent species in which females typically mate with multiple males, open circles represent species in which females typically mate with single males, and x's indicate species for which mating patterns are uncertain (mainly small-bodied species with dispersed social systems). The arrow points to *Homo sapiens*. Data on testes mass and male mass come from Harcourt et al. (1995), while data on mating patterns are from several sources (Harcourt et al. 1995; Nunn 1999a; Van Schaik et al. 1999).

and that results were independent of effects of seasonality, which might also favor larger testes if males copulate at a higher rate in a shorter breeding season. By building on this pioneering comparative research on primates, sperm competition is now known to be pervasive across many groups of animals (Smith 1984b; Birkhead and Moller 1992, 1998).

2. Comparative cognition. Given that humans are characterized by sophisticated tool use, language, and social learning, evolutionary anthropologists are also interested in uncovering the factors that influence primate cognitive abilities. One way to probe cognition comparatively is to design an experimental task and then challenge individuals of different species to solve this task in a way that reveals how they learn new skills. Is it a case of imitation—"monkey see, monkey do"—or does individual learning play a larger role?

As an example, one study investigated social learning in the context of a food-processing task with an artificial "fruit" (Whiten et al. 1996). A whole series of such "fruits" have been generated over the years; what they have in common is that an experimental subject can open a box containing rewards in different ways. The investigators used this apparatus to assess whether apes imitate knowledgeable "model" individuals; in this case, humans who opened the box using only one of the solutions. They also tested young children on this task (using candy rather than fruit). The investigators aimed to assess whether the experimental ape subjects copied the same method that the model used—evidence for imitation—or whether they used another solution to open the box—evidence for other types of social learning. This is an important question, because previous research indicated that social learning differs in humans and other apes. Nonhuman apes were thought to be more likely to use emulation or stimulus enhancement with trial-and-error learning to solve tasks; humans were thought to imitate the actions of the model more precisely.

The investigators found that chimpanzees tended to duplicate the actions of the model they witnessed, which was taken to be evidence for imitation. The authors also compared the chimpanzees to young children and found that humans copy actions to a greater degree than chimpanzees (Whiten et al. 1996). Thus, a comparative approach was essential for assessing the relative amount of imitation found in nonhuman apes; only by having human subjects of different ages was it possible to investigate these more subtle effects. In subsequent research, the initial interpretation of the data in terms of chimpanzee imitation has been questioned; thus, it is still hotly debated whether great apes can and do learn observationally in the same way that humans do, sparking many other comparative studies that use experimentally generated data (e.g., Whiten et al. 2004; Call et al. 2005; Horner and Whiten 2005; Tennie et al. 2006, 2009; Herrmann et al. 2007).

Other studies in comparative psychology take a broader comparative approach in which tens or even hundreds of species are compared, typically using nonexperimental data. Reader and Laland (2002), for example, compiled a unique data set of 533 cases of innovation, 445 cases of social learning, and 607 cases of tool use. The authors defined innovation as the discovery of novel solutions to environmental or social problems, including new gestures or alarm calls, novel play behavior, or use of new sleep sites (Reader and Laland 2001). In total, their data set covered 116 different primate species. They found that social learning correlated positively with measures of brain size (the "executive brain ratio"; figure 1.2), as predicted by hypotheses suggesting that these behaviors select for larger brains. Their results remained significant after controlling for phylogeny and in other tests involving brain size and measures of tool use and innovation frequency. The major conclusion, then, is that larger brains provide advantages for social learning, innovation, and tool use and that these factors represent a package of selective pressures favoring "adaptive complex variable strategies" (Reader and Laland 2002, 4,438). Thus, investment in a bigger brain may provide a means to survive environmental and social challenges through the invention of novel solutions.

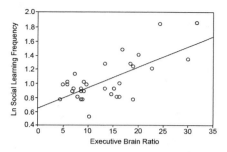

Figure 1.2. Comparative studies of brain size and social learning frequency in primates. Data show evolutionary change in executive brain ratio and evolutionary change in social learning frequency. "Executive brain ratio" is defined as the sum of the neocortex and striatum divided by the brain stem, and social learning frequency was corrected for research effort, based on the number of articles found on different species in the literature searched. (Figure redrawn from Reader and Laland 2002.)

3. Human cultural diversity. Comparative approaches are also essential in understanding patterns of human diversity at regional and even global scales. Research by Ruth Mace, Mark Pagel and their colleagues has shown how comparative research can generate new insights to cross-cultural variation (Pagel 2000b; Pagel and Mace 2004; Mace and Holden 2005). While many anthropologists have applied comparative approaches to investigate cultural diversity, an important niche that Mace and Pagel fill is to inject stronger methodological approaches to cross-cultural comparisons and to also address questions of broad ecological importance—including testing whether comparative patterns found in biological systems also hold for human populations.

For example, Mace and Pagel (1995) investigated the geographical distribution of Native American societies at the time of European contact. Specifically, they tested whether human language groups exhibited a latitudinal gradient in diversity, with the number of language groups declining from the equator to the poles—a pattern that is well-known in biological systems (Rosenzweig 1995), including North American mammals (Pagel et al. 1991). As shown in figure 1.3a, Mace and Pagel found support for this prediction in their study of language diversity: the number of language groups declines towards the poles, with six times as many languages spoken in the southern latitudes, compared to the most northern latitudes.

Mace and Pagel (1995) also investigated how the geographic ranges of languages change with latitude and found that latitudinal extent—an estimate of geographical range—increases as one moves from the equator to the poles in North America (figure 1.3b). Again, this pattern is well known in biological

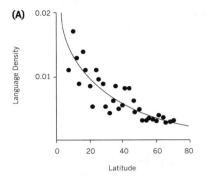

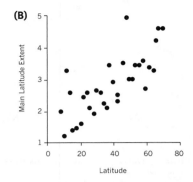

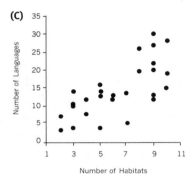

Figure 1.3. Language diversity and latitude. Plots show results for association between (a) latitude and language diversity (i.e., density per kilometer line of latitude), (b) latitude and mean latitude extent (i.e. a measure of a language's "geographic range"), and (c) number of habitats and number of languages. The results in (c) remain significant when controlling for both latitude and kilometers of terrestrial habitat at each latitude. (Redrawn from Mace and Pagel 1995.)

systems, where it is documented so consistently that it is called "Rapoport's Rule" (Rapoport 1982; Stevens 1989). Finally, Mace and Pagel investigated whether habitat diversity per area correlates with language "richness" under the expectation that more niches provide an opportunity for more human groups to coexist. Figure 1.3c shows that this prediction also was supported, and the effect remained significant when controlling for latitude and variation in the width of North America.

Another fascinating example is provided by Daniel Nettle's research on language diversity (1999a). In one study, for example, Nettle (1999c) investigated whether the number of languages is a simple linear function of the time since humans entered a region (see also Sutherland 2003). He focused in particular on a puzzling observation about languages in the Americas: this area is widely accepted as the last to be colonized by humans, but it has greater higher-level linguistic diversity (language "stocks") than other regions, such as Africa, New Guinea, Australia, and Eurasia. Thus, if we assume that language diversity accumulates at a constant rate, the large number of language stocks in the Americas indicates that the New World should have been settled *earlier* than archeological evidence suggests (Nichols 1990).

Nettle (1999c) provided evidence against the constant rate assumption and the accumulation of language stocks geographically. Specifically, he found no tendency for the number of language stocks to increase with time since founding, and fewer languages per stock exist in the Americas than in other regions. Nettle proposed instead that the number of language stocks reaches a peak early in an adaptive radiation of human populations, as in the peopling of the Americas, and then declines due to extinction of stock lineages (see also Dixon 1997). Under this scenario, a larger number of language stocks could indicate more recent human colonization rather than more ancient settlement.

A discussion of comparative research on human diversity would be incomplete without mentioning the innovative and highly synthetic work of Jared Diamond (e.g., Diamond 1997; Diamond and Robinson 2010). To take just one example, Diamond and Bellwood (2003) reviewed the role of farming in the spread of languages through demic expansion (i.e., the spread of a group of people through population expansion). They evaluated evidence for the hypothesis that farming leads to the dispersal of people (see also Ammerman and Cavalli-Sforza 1984; Sokal et al. 1991). If correct, this hypothesis predicts correspondences between five major sets of human traits—genes, languages, archaeological remains, domesticated plants and animals, and skeletal morphology—a list that reveals the interdisciplinary nature of evolutionary anthropology when conducted at a global scale. Evidence supporting the farming hypothesis includes the finding that major language families are found in close association with agricultural homelands, such as the Fertile Crescent. Also supported—although perhaps not as strongly—is the expectation that east-west expansions of languages are more common than north-south expansions. This clever prediction derives from reasoning that equivalent lines of latitude are more likely to have similar ecological conditions than equivalent lines of longitude; this should favor expansion of farming technology and domesticated animals along lines of latitude (east-west) rather than longitudinally (north-south).

4. Trapped in a tree: Pinworms and primate evolutionary history. What could be more "comparative" than a study of different primate species? One possibility is to compare primate radiation to the radiations of other organisms, including their parasites. A number of studies have done just that; in fact, some of

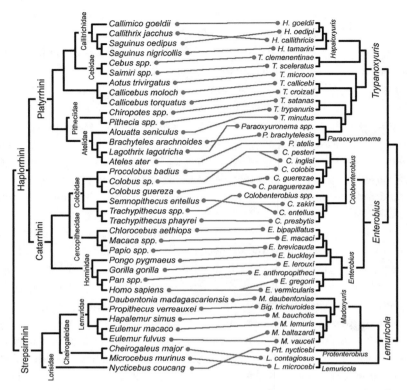

Figure 1.4. Co-speciation of primates and their pinworms. Primate phylogeny (Purvis 1995) is shown on the left, pinworm phylogeny on the right. The general congruence in the structure of the two phylogenies provides evidence for co-speciation. (Redrawn from Hugot 1999.)

these studies are considered classics in evolutionary parasitology, including research on primate pinworms (Brooks and Glen 1982). Pinworms are nematode parasites that infect the gastrointestinal tracts of a wide range of host species, including primates. They tend to be very host specific, meaning that a single species of pinworm infects only one or a few species of primates. A very simple question, then, is to ask whether the pinworm lineages tend to cospeciate with primate hosts, meaning that when a lineage of hosts splits, the parasites also show a corresponding split at the same point in evolutionary history, with each parasite lineage following one of the host lineages. In other words, one set of evolutionary lineages—in this case parasites—tends to "track" another set of lineages—in this case primates (Page and Charleston 1998).

This tracking pattern is found in the case of primates and their pinworms. For example, figure 1.4 shows a recent assessment of the hypothesis of cospeciation. The tree for primates comes from Purvis (1995), and the pinworm phylogeny was constructed by Hugot (1999) using forty-five morphological characters. While the associations are not perfect, general congruence is seen between the host and parasite phylogenies; statistical tests confirmed a significant association in reconstructed co-divergence scenarios for prosimians and New World monkeys. Cases of incongruence can be the result of a variety of factors, including parasite extinctions, speciation events, and transfers of the parasite to new hosts (i.e., host shifts). These different scenarios can be investigated with phylogenetic methods (Charleston 1998; Page and Charleston 1998; Page 2003).

Coevolutionary approaches also can be applied in novel ways, including to questions that are considerably less dismal than debilitating parasitic infections. For example, a recent study investigated patterns of genetic variation in yeast used to make alcoholic beverages and bread, which can reveal the origins of these practices and how they spread around the world (Legras et al. 2007). As with studies of parasites that track hosts, yeasts should track human migrations; as the proverbial grapevine spreads, so should the yeast strains used by humans to make wine, and the same goes for beer, bread, and other dietary items that use fermentation. In what might be called a "global pub crawl," Legras et al. obtained data on 651 yeast strains from 56 sources distributed from around the world, including Belgian ales, fermented milk from Morocco, rice wines from Asia, and, of course, fine European wines. From these data, they found that yeast strains tend to cluster according to type of method used, with, for example, sake strains, palm wines, and beers tending to represent distinct clades on a phylogenetic tree of the yeast. The authors also found evidence for isolation by distance, which indicates local differentiation of yeast populations, and among wines, they found that the phylogenetic distributions of yeast strains follow known migration routes.

These examples reveal how comparative studies can be used to examine the interdependent evolution of two lineages, that is, coevolution. Many opportunities exist to apply this coevolutionary perspective to questions in evolutionary anthropology, for example, if we view cultural traits as "agents" that coevolve with genetically based organismal lineages, as in studies of gene-culture coevolution or dual-inheritance theory (Boyd and Richerson 1985; Durham 1991; Feldman and Laland 1996), or in testing whether cultural traits exhibit similar evolutionary histories (Tehrani and Collard 2009).

5. Genes, geography, and languages. Luigi Cavalli-Sforza is another pioneer in applying comparative approaches to anthropological questions. Among many

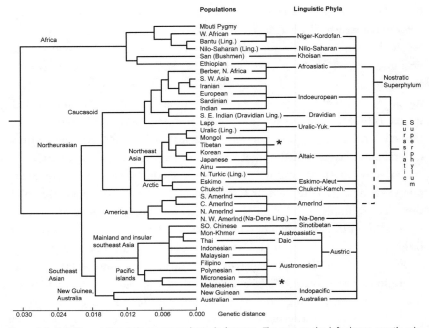

Figure 1.5. Genetic and linguistic correspondence in humans. The tree on the left shows genetic relationships among humans based on 120 allele frequencies. The tree on the right represents relationships among major linguistic groups (phyla). (Redrawn from Cavalli-Sforza et al. 1988.)

examples that can be drawn from his remarkable lifework, he undertook massive analyses of gene frequencies and languages to investigate the movement of humans and the relationships among different human populations (Cavalli-Sforza et al. 1994; Cavalli-Sforza 2000). His research laid a foundation that is still followed today. In one widely cited paper, for example, Cavalli-Sforza et al. (1988) found a correspondence between genetic trees and linguistic trees at a global scale, with major linguistic groups (phyla) tending to correspond to one of six genetic groupings (figure 1.5). Some incongruence can be seen in this figure, for example with Afroasiatic and Uralic languages that are not represented by true evolutionary groups, and it should be noted that criticisms of their methods and interpretations have been raised (e.g., Bateman et al. 1990). New phylogenetic approaches could be used to provide new insights to these patterns by studying congruence among trees using the co-evolutionary approaches described in the previous example, and of course new data are shedding further light on historical relationships among human groups (Li et al. 2008).

Other studies have addressed similar questions. For example, Barbujani and Sokal (1990) identified genetic disruptions, or "boundaries," in Europe based on gene frequencies. Of thirty-three boundaries that they discovered, thirty-one coincided with clear linguistic boundaries, often between different language families (in fifteen cases) and thus representing relatively deep splits. In twenty-two of the genetic boundaries, obvious physical boundaries were involved, such as mountains or oceans. The genetic boundaries did not always coincide with political boundaries, but they did often demarcate zones of contact from distantly related groups of people in European history. Overall, these results suggest that language differences have actively modified patterns of gene flow. Importantly, however, when looking at a more global, geographically based sample, it appears that discrete genetic continuities are replaced by more gradual patterns (Serre and Pääbo 2004).

These studies attempted to understand the association between genes and the distribution of languages. Another approach is to look more directly at correlated evolution of a particular gene and characteristics of the language itself. Recent work along these lines is provided by Dediu and Ladd (2007). These authors examined the distribution of two genes—*ASPM* and *Microcephalin*—in relation to whether languages exhibit "tonal distinctions," as found especially in parts of Asia and sub-Saharan Africa. In some languages, for example, the same word can have different meanings depending on voice pitch. Based on a series of statistical tests that controlled for historical and geographical nonindependence, the authors showed that derived versions of this gene are associated with non-tonal languages. As both of these genes are involved in brain development, the authors proposed that the genes produce some as yet unknown cognitive bias for use (or nonuse) of tonal languages. Further research is needed to assess whether the correlation reflects an underlying causal relationship.

This discussion would be incomplete without mentioning the work on mitochondrial "Eve" (Cann et al. 1987), which linked genetic and geographical data. The authors examined the mitochondrial DNA from 147 individuals, including 34 Asians, 46 Caucasians, 20 people of sub-Saharan African descent, and 47 individuals from New Guinea and Australia. Their phylogenetic analysis of these samples revealed that Africa is most likely to be the ancestral source of the human mitochondrial gene pool, with a common ancestor existing 140,000 to 290,000 years ago. Subsequent research confirmed these general conclusions using methods to deal with uncertainties in the original data and analysis (Vigilant et al. 1991; Ingman et al. 2000).

In summary, many exciting studies have already laid the groundwork for linking genetics and languages—from groundbreaking studies of linguistic, genetic, and geographic patterns by Cavalli-Sforza and others to more recent attempts to wed cultural, linguistic, and genetic data, as proposed in various forms by diverse research groups working today (Diamond 1997; Nettle 1999a; Mace and Holden 2005; Dediu and Ladd 2007).

The AnthroTree Website

An important component of this book is a freely accessible website called AnthroTree (http://www.anthrotree.info). Anthrotree provides data used in some examples in the book. In addition, it provides links to one or more computer programs and R packages that can be used to run the analyses. These links and the material on the website will be updated as new approaches and programs become available. Throughout the book, the reader will find reference to the site as "AnthroTree x.y," where "x.y" refers to the chapter and example in that chapter, respectively. Updates to the book—for example, in case clarification is needed or new methods become available—will also be provided on the AnthroTree website. These updates and clarifications can be found at the beginning of examples for each chapter (i.e., AnthroTree x.0). As a starting point, AnthroTree 1.1 provides an overview of the main programs that are used for the examples on the website.

Structure of this Book

As noted above, this book has two main parts. Chapters 2–8 give a broad overview of phylogenetic comparative methods and select examples from evolutionary anthropology to illustrate particular methods. Chapters 9–13 provide further methodological approaches and examples, but with an organization that is meant to convey the diversity of the fields of research in evolutionary anthropology (rather than the diversity of methods). Thus, the latter chapters of the book significantly expand on the questions asked in the first set of chapters, and they have some repetition with the first chapters in terms of methods that are used. This repetition is intentional, as I expect it will be useful for those who are just learning the methods to reinforce the key concepts and show how the approaches apply to real examples. To help readers make the links across chapters, at the end of each chapter I provide "pointers" to later chapters that contain related methods or concepts. In addition, here I provide a brief synopsis of the material covered in each chapter.

A central focus of this book concerns how evolutionary history—in the form of a phylogeny, a linguistic tree, or a gene network—provides essential information for most questions that are of interest to evolutionary anthropologists. Chapter 2 covers basic phylogenetic concepts and emphasizes the importance of "tree thinking." The chapter covers issues involving basic terminology, inferring phylogeny using different methods, and dealing with uncertainty in estimating both tree structure (topology) and the ages of ancestral nodes. Chapter 2 also provides an overview of how biological methods are being applied to study language evolution.

Chapters 3 and 4 discuss methods for reconstructing the values of ancestral states on a phylogenetic tree. Chapter 3 focuses on mapping the evolution of discrete traits, such as activity period in primates (e.g., nocturnal vs. diurnal) or the presence-absence of dowry in studies of human populations. Chapter 4 considers methods for reconstructing the evolution of continuously varying characters, such as body mass or group size. Chapter 4 also covers basic issues in reconstructing "protolanguages" in linguistics.

Chapter 5 covers modeling frameworks that can be used to investigate the evolution of continuous traits, including models of "Brownian motion" and stabilizing selection. This chapter also covers an important set of methods used to test whether traits show phylogenetic signal, that is, whether more closely related species exhibit more similar trait values. Diagnostic statistics presented in chapter 5 provide an essential tool kit for assessing trait evolution in a model-based framework, and some of these statistics play an important role in later chapters.

Chapter 6 discusses correlated trait evolution as a means for testing adaptive hypotheses. The *repeated* evolutionary origin of a trait with some other trait or environmental factor offers the strongest comparative test of an adaptive hypothesis, especially when this is combined with studies of the genetic underpinnings and selection on the trait. The chapter also reviews a central issue in comparative studies, namely that species-level data are not independent from one another in a statistical sense.

Chapter 7 presents three methods for investigating the correlated evolution of traits. Two of these methods can be used to study correlated change in continuous characters: independent contrasts (Felsenstein 1985b; Garland et al. 1992) and generalized least squares (Martins and Hansen 1997; Pagel 1999a; Garland and Ives 2000). The final method examines whether discrete traits show correlated evolution (Pagel 1994a). These methods represent only a subset of methods developed by comparative biologists to study correlated evolution (Maddison 1990; Harvey and Pagel 1991; Martins and Hansen 1996; Garland et al. 2005). However, they are among the most commonly used methods, and they offer the greatest flexibility for understanding patterns of correlated evolution in future studies.

Chapter 8 shifts the focus to consider how an evolutionary tree can reveal new insights to biological and cultural diversification. More specifically, here we are interested in the factors that influence rates of speciation and extinction, as revealed by the phylogeny itself. An exciting set of methods is now available to test hypotheses using data on the "shape" of phylogenetic trees, and to estimate rates of speciation and extinction even in the absence of a fossil record.

The remaining chapters in the book focus on specific fields of study within evolutionary anthropology while also amplifying, extending, and illustrating methods presented earlier in the book. Chapter 9 covers the important issue of investigating and controlling for body mass in comparative studies. This chapter also reviews how we can use knowledge of correlated evolution in extant species to make inferences about body mass in extinct organisms in a statistically rigorous way. Similar principles apply to reconstructing other characters in the fossil record.

Chapter 10 investigates approaches for studying human cultural traits, including language. An important issue here is *Galton's problem*, which concerns the nonindependence that arises when cultural traits are shared through common descent, or through diffusion from neighboring societies. The latter issue of horizontal trait transmission is potentially problematic for applying phylogenetic comparative methods, and it is interesting to investigate as a factor that may account for observed trait variation.

Chapter 11 considers the use of the comparative method to study behavior and ecology, both in nonhuman primates and in humans. The comparative method has played an important role in such studies, as in the example of sperm competition given above. In addition, new phylogeny-based methods can be used to study community ecology in a phylogenetic context, and to address questions of conservation concern in primates and in humans (e.g., the extinction of languages or cultural diversity).

Chapter 12 considers a particularly thorny problem for applying comparative methods to study human evolution. Many of the traits that probably played a role in making humans successful have evolved rarely (or only once) and are shared by all humans, including language, bipedal locomotion, and tool use. These "singularity problems" pose special challenges for inferring adaptation with the comparative method (Pagel 1994b). This chapter considers how we can make stronger statistical inferences in the context of single evolutionary origins.

Chapter 13 addresses a practical issue that is central to all comparative studies: how should one organize the data in a way that it is accessible and easy to manage and with each record in the database linked to a citation for that record? A number of tools are available for creating relational databases that can be shared securely with collaborators over the Internet or easily served to the wider public on a website. In terms of building databases, we can also use phylogenetic information to target species, populations or languages that need further study, or that could serve as additions to comparative databases to make more powerful inferences. I review this concept of "phylogenetic targeting" (Arnold 2008; Arnold and Nunn 2010). Chapter 14 provides an overview and synthesis and also discusses new directions for applying the comparative method to the evolutionary anthropology of the future.

A final comment concerns jargon. Because one of the aims of this book is to communicate recent advances in phylogenetic methodology among researchers in different fields of evolutionary anthropology and across biology more generally, I intentionally avoided using terms that might be confusing or unclear to researchers in different fields. For example, rather than using more formal taxonomic names, such as hominoids or cercopithecines, I generally use the terms apes and Old World monkeys, as researchers across fields of evolutionary anthropology and biology are likely to be more familiar with the latter terms than the former. Similarly, phylogenetic terminology is often intimidating to those working outside of phylogenetics, and some concepts have different terms in different subfields of evolutionary anthropology (especially in linguistics). Thus, aficionados of scientific and taxonomic terminology might be disappointed, but such a step helps to communicate ideas among researchers in different fields of evolutionary anthropology, and to share the results with researchers in other fields.

2 Basic Phylogenetic Concepts and "Tree Thinking"

One of the most significant breakthroughs in biology was the realization that lineages of organisms can be represented with a branching tree structure. Because languages, societies, and organisms all descend in a hierarchical fashion from ancestors, trees are a fundamental way of understanding the connections among the individual data points in a comparative study. Geography is also important, especially for situations in which adjacent locations have similar ecological conditions, or when traits can spread horizontally, such as "loanwords" among languages.

A principal theme of this book is that knowing the historical relationships among the organisms or populations in a comparative dataset can generate a richer understanding of variation. While this point might seem obvious, many anthropological, primatological, and linguistic studies fail to make use of phylogenetic information. At an even more basic level, a remarkable number of biologists and evolutionary anthropologists fail to appreciate the most basic concepts about phylogenies. Many readers of this book are likely to assume that they already know how to read a phylogeny and so might be tempted to skip this chapter. Before doing so, I recommend taking the "Tree-Thinking Challenge," published as part of a *Science* article on the need for increased phylogenetic perspective in biological research (Baum et al. 2005) and linked through AnthroTree (AnthroTree 2.1). It is possible to climb the "phylogenetics learning curve" simply by taking the test and reading the brief explanations to the questions.

In this book, I am concerned primarily with making use of phylogenetic information in comparative research rather than with creating phylogenies. Few recent books, most dating from the 1990s, have been devoted to using phylogenetic trees in comparative research (Harvey and Pagel 1991; Harvey et al. 1995; Martins 1996a). This is surprising, given the publication of many new and more powerful approaches for investigating comparative data since the publication of these books. In contrast, many recent books have been devoted

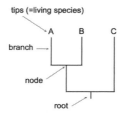

Figure 2.1. Nodes and branches on a phylogenetic tree. Nodes refer to the points at which two (or more) lineages diverge, while branches reflect the lineages. An undated tree topology is shown here. On a *dated phylogeny*, either the branches or the nodes are assigned ages or amounts of character change. Tips are usually living species, as in this case, but can also represent extinct lineages.

to building phylogenies (Page and Holmes 1998; Nei and Kumar 2000; Felsenstein 2004; Hall 2008). Thus, I refer readers seeking more information to the books on phylogenetics just cited, and I limit coverage of tree-building techniques and concepts to the basics in this chapter. I also use jargon associated with phylogenetics on a "need-to-know basis" relative to the topics covered later in the book, as a full listing of terms is not essential (and phylogeneticists have created a dense jungle of terms).

This chapter starts with the basics of how to read a phylogeny and then moves on to a brief overview of four major approaches to reconstructing phylogenies: distance-matrix methods, parsimony, maximum likelihood, and Bayesian approaches. I close the chapter by discussing linguistic phylogenies, as these provide the historical context for many studies in cross-cultural research (see chapter 10).

The Basics

Put simply, a *phylogenetic tree* shows the relationships among a set of species. It displays these relationships as a hierarchical set of nested taxa that can indicate either the relative or absolute times at which two or more species last shared a common ancestor. The building blocks of a phylogeny are the nodes and branches. The nodes represent speciation events in which a lineage separated into two or more descendent lineages, while the branches represent the lineages. The standard format for a tree is to have the tips of the tree—representing species—pointing toward the top of the page, with the ancestral "root" of the tree at the bottom (figure 2.1). In the literature, however, one will find phylogenies with the "fork tines" of the phylogeny pointing up, down, or sideways. In this book I present trees in the standard position upward or to the right, with the right position generally reserved for larger trees that fit best on the long edge of a page.

Let us take a real-world example from the phylogeny of apes, based on a recent inference of phylogeny across all mammals. On the tree in figure 2.2a, one can see that the two species of chimpanzees (*Pan troglodytes* and *Pan paniscus*) shared a common ancestor more recently than either did with *Homo sapiens*. These two species are therefore more closely related to one another than either is to humans or to any other species in this phylogeny. We can change the order in which the two species of *Pan* appear on the tree—for example, by moving *P. troglodytes* to the left of *P. paniscus*—and this will not alter the basic relationships among any of the species. In other words, trees can be rotated around nodes without changing the information that is represented by the tree itself, which means that the ordering of the tips is less important than their nested structure (see also Baum et al. 2005).

While many phylogenetic lineages are denoted as branching dichotomously, that is, into two lineages, phylogenies can also contain a *polytomy*, where more than two lineages emanate from a single node. An example is shown in figure 2.2a, where five lineages of gibbons (*Hylobates*) are represented as descending from a single node on the tree. Such a situation could reflect that these five species all radiated more or less simultaneously, as might happen when a continuous habitat rapidly breaks up into multiple isolated habitats—known as refugia—and a once widespread species thus finds itself fragmented into separate populations. This process can result in distinct species as reproductive capability declines among the individuals in different populations, resulting in a *hard polytomy*. In this case, the polytomy reflects the actual branching history of a single lineage into three or more new lineages at the same time.

More often, however, polytomies reflect uncertainty about the true branching order of a group of organisms. It could be that speciation events happened so closely in time that the available morphological or molecular evidence cannot discern the evolutionary relationships, or different genes may have conflicting evolutionary histories (e.g., Maddison 1997), resulting in ambiguity at a particular node. When a polytomy reflects uncertainty due to conflict in the data or general lack of knowledge, it is referred to as a *soft polytomy*, with the polytomy thus reflecting ignorance in our knowledge of the true evolutionary history for a group of organisms.

Many trees represent only the branching pattern, that is, a tree topology, rather than the exact age of the splits on the tree. Such a topology is often

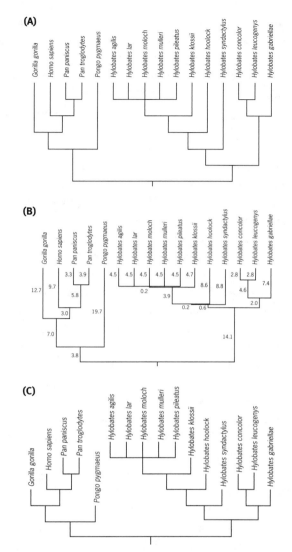

Figure 2.2. Ape phylogeny. This phylogenetic reconstruction was taken from a recent inference of mammalian phylogeny (Bininda-Emonds et al. 2007). In (a), an undated phylogeny (cladogram) is given, while (b) shows a dated phylogeny, with branch lengths indicated as numbers to the left of each branch and time in units of millions of years. Panels (a) and (b) show ultrametric trees. In (c), however, all branches were set to be equal in length—a common assumption in comparative analyses—resulting in a non-ultrametric tree. Note the "polytomy" among the gibbons in the middle of the list of species names, with five branches all emanating from a single node.

called a *cladogram*. Without dates, we can make inferences about the relative relatedness of nested taxa on the tree, but not about the timing of the splits. In figure 2.2a, we can say that *P. troglodytes* and *P. paniscus* are more closely related than either is to *H. sapiens*, and that *Hylobates concolor* is more closely related to *H. leucogenys* than either is to *H. gabriellae*. These pairs of more closely related species are called *sister species*, and they are connected by only a single node on the tree (i.e., the closest possible connection). From this undated tree topology, however, we cannot say if the two species of *Pan* shared a common ancestor more recently in time than did *H. concolor* and *H. leucogenys*.

Other trees give details on branch lengths in what is called a *dated phylogeny*, or a *chronogram*. When we add branch lengths to our ape phylogeny and represent branch lengths as proportional to time (figure 2.2b), we see that *H. concolor* and *H. leucogenys* shared a common ancestor more recently (2.8 million years ago) than did the two species of *Pan* (3.9 million years ago). Thus, nodal dates and branch lengths can give insights that simple topologies cannot. A *phylogram* is similar to a dated phylogeny but shows amounts of character changes along branches rather than amounts of time; in such cases, the tips of the tree usually do not line up to give the same age for all living species in the clade (i.e., the tree is not *ultrametric*). One type of character change commonly represented on phylogram branches is the amount or molecular change in DNA sequence, which is typically reported as the number of nucleotide substitutions per base pair site. In the past, before branch length information was commonly available, many comparative studies made the assumption that all branches were equal in length, which also produces a non-ultrametric tree (figure 2.2c). Now, however, a variety of methods have been developed to place dates on phylogenies (see Sanderson 2002; Thorne and Kishino 2002; Felsenstein 2004). Hence, the assumption of equal branch lengths should be justified through diagnostic statistics (see chapter 5).

In this book, I use the term *dated phylogeny* to refer to trees with divergence dates, and a *tree with branch lengths* to refer more generally to cases in which either absolute (a dated phylogeny) or relative branches (a phylogram) are given. This distinction is important for phylogenetic comparative methods because some methods require a dated tree, while other methods require branch lengths, but they need not be given in units of absolute time. As we will see in later chapters, it is possible to test the assumptions of comparative methods for a given trait (or traits) and to scale the branches to best meet these assumptions.

An important issue in reading trees is often misunderstood: a phylogeny based on living (extant) species does not represent the many extinct lineages that failed to produce lineages that are alive today. In other words, clades might have radiated and even persisted over a relatively long proportion of the history of the group but then went extinct, and these extinct lineages will not be seen on the phylogeny. An exception to this occurs for many trees that focus on fossil lineages, as is common in studies of hominin phylogeny (Strait et al. 2007). When the phylogeny is based on extant species, we should thus remember that the actual number of speciation events is in fact likely to be much higher than the number of nodes on the tree, because many *ghost lineages* have probably occurred and gone extinct. An important point, then, is that the rate of splitting reflects the rate of *diversification*, rather than speciation sensu stricto. Phylogenetic methods provide a means to estimate speciation and extinction rates from a dated tree, even when the fossil record itself is largely nonexistent (see chapter 8 and Nee 2006; Maddison et al. 2007).

How Does One Infer a Phylogenetic Tree?

As with many methodological questions, phylogenetics can be a contentious topic. One could easily write an entire book about phylogenetic methods— and of course people have (e.g., Page and Holmes 1998; Nei and Kumar 2000; Felsenstein 2004; Hall 2008). Interested readers may wish to refer to these resources for more information (particularly Felsenstein 2004). Here, I provide a brief introduction to some of the key concepts and methods.

Phylogenetic data. Phylogenies have been created from varied types of data. In studies of primate phylogeny, for example, biologists have used morphological and molecular traits to infer phylogeny, and fossils have often played a major role in this endeavor (e.g., Gingerich 1984; Goodman et al. 1994; Disotell 1996; Shoshani et al. 1996; Ruvolo 1997). When considering what kind of data to use to infer phylogeny, it is important to obtain characters that are as independent from one another as possible. This is usually assumed to be the case for genetic data, but it is more challenging to justify this assumption for morphological characters where the integrated nature of functional traits could make it difficult to identify independent characters (Cheverud 1995; Lieberman 1999; McCollum 1999; Ackermann and Cheverud 2004a).

In addition to genes and morphology, other types of data have been used to infer primate phylogeny, including characters related to vocal, visual, and olfactory communication (Macedonia and Stanger 1994); developmental traits

(Yoder 1992); and primate social organization (Di Fiore and Rendall 1994). While behavioral traits are often viewed as less useful for phylogenetic reconstruction and raise questions about how to identify homologous characters (Lauder 1986; Brooks and McLennan 1991; Wenzel 1992; Gittleman and Decker 1994; Robson-Brown 1999), a recent review found that this was not the case: behavioral traits exhibit as much "phylogenetic signal" as other types of data, including morphology and genes (Rendall and Di Fiore 2007). As will be discussed later in this chapter, researchers also have developed ways of combining multiple phylogenetic hypotheses for subsets of species into a "supertree" that includes all species in a group of organisms (Purvis 1995; Purvis and Webster 1999; Bininda-Emonds et al. 2007). Until recently, only morphology has been available to infer relationships among fossil taxa, but technical advances in extracting DNA from fossil material (i.e., ancient DNA) are providing new insights to evolution (Krings et al. 1997; Yoder et al. 1999; Hofreiter et al. 2001). And of course, data on different kinds of language characteristics are commonly used to infer the historical relationships among human populations (Gray and Jordan 2000; Gray and Atkinson 2003; McMahon and McMahon 2005).

Key concepts. A central philosophical issue surrounds most attempts to reconstruct phylogeny. Because evolutionary trees reflect "descent with modification" (Darwin 1859), the most powerful way to build these trees is to group taxa according to the characters that are modified from ancestral states and shared with other taxa (Hennig 1966). In other words, phylogeny should be based on the concept of *shared derived characters*, or *synapomorphies* (figure 2.3). These are traits of organisms that exhibit *homology*, that is, they are shared through common descent and thus reflect evolutionary history. In contrast to homology, *homoplasy* involves multiple origins of a character state on the tree where not all of the characters of a particular state can be attributed to descent from a common ancestor. Generating a tree from similarity in morphological, genetic, linguistic, or behavioral traits is insufficient, as many similarly appearing organisms are not in fact most closely related (see Stewart 1993). A good example of this in primatology would involve the relationships between chimpanzees, gorillas, and humans (see figure 2.2). While chimpanzees and gorillas might appear more closely related at first glance—they are large, hairy, knuckle-walking apes—in fact chimpanzees and humans are more closely related.

The critical methodological issue concerns how we identify these shared derived traits. How are we to make sense of huge amounts of data on living organisms, including data on behavior, morphology, genes, or physiol-

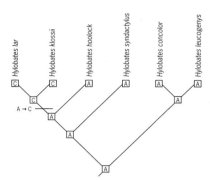

Figure 2.3. Similarity due to common descent. Shared derived characters (synapomorphies) provide evidence for common descent. In this figure, nucleotide codes are shown for one position in a hypothetical gene. In this case, the C nucleotide (cytosine) is a shared derived character that unites *Hylobates lar* and *H. klossii*. The A nucleotide (adenine) is ancestral and shared by four species in the tree.

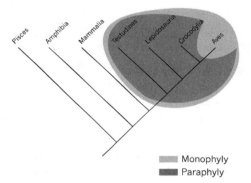

Figure 2.4. Monophyletic groups. A *monophyletic group* (shown in light gray) includes all descendants from a common ancestor. A *paraphyletic group* (shown in dark gray) fails to include all descendants of a most recent common ancestor, and would be illustrated by the grouping "reptiles." A *polyphyletic group* does not contain the most recent common ancestor of a group, such as "warm-blooded animals" that groups Aves with Mammalia (not shown).

ogy? And what about fossil taxa, which not only have data relevant for inferring some of these characters, but also provide information on the time span over which the lineages existed? Can we use data on organisms that have co-evolved with the primates or humans that interest us, such as symbiotic gut microorganisms or parasites? In terms of naming taxa, phylogenetic systematists aim to construct *monophyletic* groups, which are groups of organisms that have a single common ancestor and include all descendants from that ancestor (figure 2.4).

Many practical issues must be addressed before deciding on a method of phylogenetic reconstruction. With molecular sequences, for example, the user

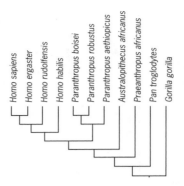

Figure 2.5. Hominin phylogeny. Phylogeny of hominins and two ape outgroups (from Strait et al. 1997). The taxa are given names based on nomenclatural revisions by the authors.

identifies bases that are homologous; this involves aligning the sequences. Although algorithms exist for gene sequence alignments (e.g., Gusfield 1997), it is often essential to check the alignments visually, and molecular phylogeneticists often make minor changes to the alignment based on their visual assessments of the similarity of two or more sequences. In the case of protein coding regions, for example, it is possible to assess alignment accuracy by checking the amino acid translation for gross anomalies that likely reflect alignment errors. One must similarly make careful decisions concerning how morphological characters are defined, and as noted above, it is important to choose characters that are largely independent of other characters. To build a tree of languages, a set of "meanings" must be chosen, such as words for "table" or "arm" in different languages. For these meanings, words are compiled for each language, and rules followed for determining whether the terms are homologous in different languages.

Data for phylogenetic analyses of extinct organisms come from fossil material, and of course the phylogeny of extinct hominins has received special scrutiny in evolutionary anthropology (e.g., Delson et al. 1977; Chamberlain and Wood 1987; Lieberman et al. 1996; Lieberman 1999; Strait and Grine 2004; Strait et al. 2007). One study, for example, used craniodental characters to investigate the evolutionary relationships among nine species of hominins (Strait et al. 1997). Across eight analyses that differed in terms of the characters used, character coding, and evolutionary assumptions, the authors found some consistent patterns (figure 2.5), for example, that the robust australopithecines group together.

A point that is baffling to many users of phylogenies involves the root-

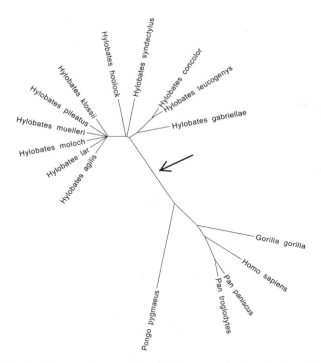

Figure 2.6. Unrooted phylogenetic tree showing relationships among apes. Compare this to figure 2.2, where the trees are rooted along the branch indicated by the arrow (which thus creates another node on the tree).

ing of the tree. Simply stated, in an *unrooted tree* the ancestral node has not been identified (figure 2.6). Thus, an unrooted tree makes no assumptions about ancestry and provides no "arrow of time." Many phylogenetics packages produce unrooted trees. Rooting the tree simply adds a new node to the tree where the root is hypothesized to occur; thus, a rooted tree will have one more node than its equivalent unrooted tree. The root can be identified along any branch in an unrooted tree. This is often achieved by including an *outgroup* to the monophyletic group of interest (i.e., the *ingroup*), which represents a species or group that is known to lie outside the clade of interest and is thus more likely to possess ancestral characters. The root must (by definition) reside along the branch connecting the outgroup to the ingroup.

Distance-matrix methods. Distance methods calculate a measure of distance between all pairs of species and then generate a tree that is most consistent with this matrix. These methods were originally developed for research in the context of *phenetics* or *numerical taxonomy*, which were very influential (but now largely unused) approaches to classifying organisms according to their overall similarity. Distance-matrix methods are thus often called phenetic methods. The data that can be analyzed by distance-matrix methods are quite varied, ranging from gene sequences to morphological data. The branch lengths reflect amounts of evolutionary change in the characters used to generate the distances, rather than evolutionary time.

An important point to keep in mind is that distance methods do not generate trees based on shared derived character states (*synapomorphies*). As a consequence, such methods can be more easily misled by convergent evolution, or they can group species that share ancestral (rather than derived) character states. Nonetheless, distance-matrix methods often provide reasonable first approximations of phylogeny and can generate these inferences rapidly relative to other methods. Thus, distance-matrix methods, such as neighbor joining (Saitou and Nei 1987), are often used to produce "starter" trees for a variety of other methods that produce more reliable inferences of phylogeny, and a recent article suggests that distance-matrix methods may be more valuable for inferring relationships than typically thought (Roch 2010).

Parsimony. Parsimony attempts to minimize the number of changes over evolutionary time and thus selects the tree that can account for the data with the fewest evolutionary changes (i.e., most "parsimoniously"). Parsimony has played an important role in the development of phylogenetics, as it has long provided an operational criterion for deciding among the many phylogenetic arrangements that are possible for a group of species. This is not a trivial problem, as the number of rooted bifurcating trees is given by the following equation, where n is the number of species:

$$\frac{(2n-3)!}{2^{n-2} \cdot (n-2)!}$$

Thus, for three species, three possible topologies can be constructed, and for ten species, 34,459,425 topologies can be constructed (see Felsenstein 2004). But for a relatively modest dataset of only thirty species, approximately 4.95×10^{38} trees are possible! A parsimony analysis explores this "tree space" either fully or (more commonly) partially, with the goal to find the tree that requires the fewest changes in the characters of interest.

Parsimony is increasingly displaced by other methods, but understanding parsimony—and some of the statistics produced by parsimony—is essential for many comparative methods, such as reconstructing ancestral states and for interpreting past research in evolutionary anthropology that used parsimony. *Evolutionary steps* are the unit of currency for a parsimony analysis. Trees are scored by counting the minimum number of steps needed to explain the distribution of the characters, and a variety of algorithms have been developed to explore the possible universe of trees. The output from a parsimony analysis is a rooted or unrooted tree that requires the fewest steps across the whole tree. In some cases, the program returns many (even thousands) of equally parsimonious trees. For a worked example of a parsimony analysis, see AnthroTree 2.2 and Felsenstein (2004).

The *consistency index* (CI) and *retention index* (RI) are two statistics that are commonly used in parsimony analyses to assess the degree of homoplasy (see Maddison and Maddison 2000). These metrics have become important in evolutionary anthropology, because they are used to study aspects of cultural evolution (Collard et al. 2006a; see also chapter 10). The CI provides a measure of the amount of homoplasy in a tree and can be calculated for an entire set of characters (the ensemble CI) or for a single character (Maddison and Maddison 2000). For a single character, it is calculated as $CI = m \div s$, where m is the minimum number of possible evolutionary steps on a tree and s is the actual number of reconstructed steps. A higher amount of homoplasy results in higher s and thus a lower CI, while a tree without homoplasy has a CI of 1. The RI measures the degree to which potential synapomorphy is exhibited on the tree. For a single character, it is calculated as $RI = (h - s) \div (h - m)$, where h is the maximum number of steps possible for a character and the other variables correspond to those just given for the CI. An RI of 1 indicates that the character is completely consistent with phylogeny (i.e., it shows no homoplasy), while an RI of 0 indicates the maximum amount of homoplasy that is possible. The CI increases as the number species in the analysis increases and thus requires a correction term when comparing values among data sets with different numbers of species (Sanderson and Donoghue 1989; Hauser and Boyajian 1997). *Ensemble indices* are calculated in a similar way, but for the whole set of characters in a data set rather than single characters (Maddison and Maddison 2000). Many programs that reconstruct phylogeny or character evolution calculate the CI, RI and similar measures of homoplasy (AnthroTree 2.3).

When constructing a phylogeny, we would obviously like to know whether clades are supported above some level of chance or, in other words, to have some measure of confidence for the tree (Felsenstein 2004). Several different approaches are available to quantify the support of parsimony trees. Probably the best known of these approaches is the bootstrap (Felsenstein 1985a). Bootstrapping involves inferring phylogeny on datasets that are generated by random resampling (with replacement) of the original dataset. Bootstrap support indices are generated for clades on the phylogeny, represented as the proportion of times that the clade occurs in the bootstrapped data sets. Although some researchers have questioned the appropriate interpretation and statistical validity of the bootstrap (see Li and Zharkikh 1994; Sanderson 1995), it provides a way to assess how heterogeneity in a character matrix affects the outcome of a phylogenetic analysis. Felsenstein (2004) provides further discussion of the bootstrap and other support measures.

An important issue in parsimony analyses involves *long-branch attraction.* This refers to a situation in which two long branches on a phylogeny tend to group together even when the lineages are not actually closely related. Such a problem can arise because longer periods of evolutionary time—reflected in longer branch lengths—create an opportunity for convergence in character states to occur by chance along those branches; thus, such events could make it appear that descendants from two long branches on different parts of the true tree share derived characters. Long-branch attraction tends to be a more common problem in parsimony analyses and in analyses of genetic data, although it can in principle afflict other types of methods or data (e.g., Kolaczkowski and Thornton 2009), and it is worsened when taxon sampling is poor. Bergsten (2005) reviews these issues, along with methods to detect and resolve long-branch attraction problems.

It was noted above that evolutionary steps are the currency in a parsimony analysis. This is relatively straightforward in most cases, with one change in character state equivalent to—or "costing"—one step in the parsimony analysis. Importantly, however, more complex cost structures can be created to reflect that some types of changes are more likely than others. For example, DNA changes can be identified as transitions or transversions. Transitions occur among nucleotides with similar molecular shapes, specifically purines (A and G) and pyrimidines (C and T), and they are more common than transversions, which involve changing a purine to a pyrimidine or vice versa. Thus, it makes sense to weigh these costs differently in a parsimony analysis. Cost frameworks for some additional character types, as well as more general frameworks for incorporating costs flexibly into a parsimony analysis, are relevant to reconstructing ancestral states (discussed in chapter 3).

Parsimony is widely applied. In most areas of phylogenetics, however, researchers have shifted toward using methods that incorporate explicit models of evolution into the analysis. A common misperception is that these "other" methods apply only to genetic data, with the result that morphological data (including characters scored from fossils) can only be analyzed using parsimony or distance-based methods. This is not the case, however, and software is now available that can infer trees from genetic, morphological, behavioral, and cultural data using a variety of explicit models of evolution. We turn next to one of these methods, maximum likelihood.

Maximum likelihood. Maximum likelihood methods make use of the observed trait data, a probability model of character evolution, and a hypothesis of phylogeny with branch lengths. From these, one can obtain the likelihood of the data given the proposed tree and evolutionary model. After conducting such calculations on many tree and parameter value combinations, the tree and model parameters offering the highest likelihood of the data are selected. Support values such as the bootstrap can be used to assess support for particular nodes on the resulting tree. Using likelihood as a criterion for selecting among the many possible trees, maximum likelihood methods aim to find the tree with the highest likelihood, typically using the same techniques as parsimony to search tree "space." AnthroTree 2.4 provides a dataset and guidance on performing a maximum likelihood analysis of genetic data.

The term "maximum likelihood" is thrown around a lot, and it certainly sounds appealing, but what does this term really mean, and why is it important? Maximum likelihood estimation is a statistical method used to fit a mathematical model to data, which is typically used to obtain the likelihood of the data given the model. Maximum likelihood estimates have many desirable statistical properties. In particular, they typically show (1) consistency—that is, they converge to the correct value of an estimate with increasing sample size; (2) efficiency—they have the smallest possible variance around the true parameter; and (3) low bias—they are not consistently over- or underestimated. There is nothing particularly mystical about maximum likelihood estimates. Indeed, many familiar statistical measures can be shown to provide maximum likelihood estimates of the parameter of interest (see also Felsenstein 2004 and Huelsenbeck and Crandall 1997).

An essential part of the maximum likelihood approach to tree inference is the model of character evolution. In the case of molecular sequences, this is a *substitution model* that reflects the relative rates of mutations from one nucleotide base to another. The model might simply reflect that transitions and transversions have different rates, or it might set a rate for all possible changes from one nucleotide type to another. These models can also assign different rates to different sites, for example depending on the codon position of the site (i.e., third-position nucleotides will tend to have higher rates of evolution because changes to this position less often affect the coding of the amino acid than do mutations at other positions). It is important to select a model that is not under- or over-parameterized, and a variety of programs and websites can be used to select substitution models for phylogenetic analyses (AnthroTree 2.5; see also Posada and Crandall 2001b; Felsenstein 2004; Posada and Buckley 2004).

Maximum likelihood methods also can be used with morphological data (Lewis 2001), provided that the traits can be easily classified into discrete codes, such as presence or absence of a particular anatomical feature. Likelihood models for discrete traits must control for *acquisition bias*, because morphological systematists tend to choose characters to measure that exhibit variability across species (Lewis 2001). This differs from molecular studies, where researchers collect data regardless of its variability, and from which rates of evolution can be measured more accurately. While some authors have argued against using morphological characters to construct phylogeny given the advantages of molecular data (Scotland et al. 2003), others have defended the application of morphological phylogenetics on multiple (and very compelling) grounds (Jenner 2004; Wiens 2004). Indeed, morphologically based trees are essential if we are to place fossil taxa within the tree of life in a statistically rigorous way, but maximum likelihood approaches to tree inference have not yet had a major impact on studies of primate fossil taxa, including hominins.

It is important to keep in mind the following about maximum likelihood as it is applied to phylogenetics. First, like parsimony, maximum likelihood is a procedure that aims to identify (in most cases) the single best tree by selecting trees using an optimality criterion. Second, the model of evolution is an integral component of the maximum likelihood approach and should be taken seriously when implementing the method (AnthroTree 2.5). Third, maximum likelihood is computationally expensive. This can hinder effective exploration of tree space. Last, as with all phylogenetic methods, it is important to include measures of statistical support for clades on the tree. The bootstrap, for example, is also applicable to maximum likelihood methods, but it can be very time-consuming (often prohibitively so) to implement.

Bayesian methods. Bayesian methods are the "new kids on the block" in terms of phylogenetic methodology, and they are revolutionizing phylogenetics by providing an efficient approach to searching tree space that is not based on the strict optimization procedures of maximum likelihood or parsimony approaches (Huelsenbeck et al. 2001, 2002; Holder and Lewis 2003). Imagine, for example, that we used maximum likelihood to identify the tree with the highest likelihood. Is this really the correct tree? Many other similar trees might have slightly lower likelihoods. Perhaps stochastic effects led us to choose the tree with the highest likelihood, when in fact this tree is not an accurate representation of the evolutionary history of the organisms. While Bayesian methods help to deal with some of this uncertainty, it is important to keep in mind that errors in the specification of fixed model parameters can still mislead Bayesian analyses of phylogeny, and as the newest method, Bayesian phylogenetics is still undergoing testing and comparison to other methods (e.g., Kolaczkowski and Thornton 2009).

As noted above, the maximum likelihood approach estimates the probability (likelihood) of obtaining the observed DNA sequences (or other data) given the candidate phylogeny; it selects the tree that maximizes the likelihood of the data. Conversely, in Bayesian approaches, one estimates the probability that a candidate phylogeny is correct given the observed data and evolutionary model. In addition to the tree itself, Bayesian methods are particularly well suited for estimating the mean and variances of parameters of the evolutionary model. Thus, Bayesian methods allow the user to obtain a set of trees and model parameters that are sampled in proportion to their *posterior probabilities*, which are based on a combination of the likelihood of the data and the *prior* (i.e., existing knowledge about the phylogenetic relationships or evolutionary parameters). The set of trees obtained reflects the uncertainty in the phylogeny given the substitution model and data, and the user can summarize the set of trees into a *consensus tree*. In fact, the user can create as many trees as he or she wishes—hundreds, even thousands of phylogenies, all fully bifurcating and with branch lengths. As we will see in later chapters, it is possible to run comparative analyses on this sample of trees, and in this way the results of a comparative study are not conditioned on a particular phylogeny or set of branch lengths being correct (Pagel and Lutzoni 2002). AnthroTree 2.6 provides an example file, instructions, and links to programs that run Bayesian phylogenetic analyses. Recently, a Bayesian inference of primate phylogeny has been produced for more than 180 primate species (Arnold et al. 2010). From a dedicated website, users can download a consensus tree, a tree "block" of up to 10,000 trees sampled in proportion to their posterior probabilities, and the raw data files (see Anthro-Tree 2.7 for links to the site).

A method known as Markov chain Monte Carlo, or MCMC, is used to implement Bayesian approaches. Other authors have provided definitive and useful accounts of MCMC and its advantages (Larget and Simon 1999; Huelsenbeck et al. 2001; Pagel and Lutzoni 2002). Briefly, one can summarize this process as follows. Start with a random tree and branch lengths, and obtain the likelihood of the data given this tree and a model of evolution. Next, randomly perturb the tree, for example, by changing the topology, branch lengths, or a parameter value in the evolutionary model, and compute the likelihood for this "new" tree. If the new likelihood is higher, accept this newly proposed tree, and if it is lower, accept the tree probabilistically in proportion to the difference in likelihoods. Otherwise, retain the tree from the previous time step and begin the perturbation process anew. This process is iterated to create a sequence (or chain) of trees, and if run for long enough, the chain reaches a stable distribution in which the likelihood is, on average, not increasing over time. Importantly, the "burn-in" period—that is, the period when the likelihood is climbing to the stable distribution of values—should be discarded (figure 2.7). After burn-in, one can sample the trees along the chain at equal intervals to produce a posterior probability distribution of trees and their branch lengths.

With this posterior probability distribution of trees, it is straightforward to calculate support measures, such as the proportion of the trees in which a particular clade is found (Larget and Simon 1999) and to generate various kinds of consensus phylogenies that effectively summarize the trees. To ensure that the analysis has not settled on a "local" peak, it is important to run the analysis from multiple random starting points. Several statistics can help to ensure that independently initiated runs are settling on the same phylogenetic tree space (e.g., split frequencies in the program MrBayes, Ronquist and Huelsenbeck 2003).

A Bayesian analysis involves much more than is presented here, and I have simplified the situation by not discussing the prior in much detail (which is part of the iterative process just described). Interested users should try the example on AnthroTree 2.6 and consult the relevant sources cited above to learn more. Briefly, what are some of the strengths of Bayesian phylogenet-

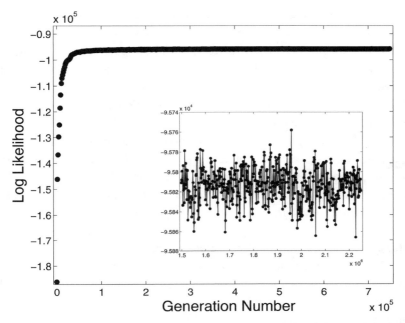

Figure 2.7. Burn-in. This plot shows burn-in for a Bayesian analysis using Markov chain Monte Carlo. Notice that the likelihood climbs rapidly before reaching stationarity, an example of which is shown in the inset panel. The trees (and other parameters of interest) are sampled at regular intervals from this stationary distribution.

ics? First, Bayesian methods often run much faster than the procedures discussed above, particularly maximum likelihood. This is a real advantage, given the tremendous computational effort that is needed for maximum likelihood optimization and calculation of support indices. For large datasets, however, computational constraints are still significant, often making it necessary to run the analyses on computer clusters. Second, the Bayesian analysis returns a set of trees that are sampled in proportion to their posterior probability. This provides an intuitive way to calculate support indices for particular nodes. *Importantly, we will see in later chapters that the set of trees from a Bayesian analysis can be used to investigate comparative variation while controlling for phylogenetic uncertainty* (Huelsenbeck et al. 2000; Lutzoni et al. 2001; Pagel and Lutzoni 2002; Pagel and Meade 2006a; Arnold et al. 2010).

Bayesian methods also have weaknesses that users should keep in mind. For example, Bayesian methods require specification of priors for the parameters in the evolutionary model, and this can affect the results of the analysis (see Huelsenbeck et al. 2002; Yang and Rannala 2005). Similarly, one must ensure that tree space has been adequately sampled by running the analysis from multiple starting points, using appropriate parameters in the "proposal mechanism" for rearranging trees, and allowing the analysis to run long enough to ensure that a stable distribution of likelihoods has been reached (often many millions of generations). In addition, recent work suggests that some characteristics of Bayesian phylogenetic inference can lead to long-branch attraction (Kolaczkowski and Thornton 2009), which, as we saw earlier, can also affect parsimony. It may be necessary to put constraints on the tree topology to prevent cases of long-branch attraction. Last, it is essential to use an appropriate evolutionary model in Bayesian analyses.

In conclusion, Bayesian methods are on the rise and are especially suited for dealing with phylogenetic uncertainty in comparative analyses (Huelsenbeck 2000; Pagel and Lutzoni 2002). However, measures of uncertainty are only as good as the underlying data and evolutionary model, and recent work has identified some weaknesses of Bayesian phylogenetic approaches. Thus, looking into the near future, the other approaches discussed above are likely to continue to play a major role in phylogenetics research in evolutionary anthropology.

Language Trees

It is common when discussing linguistic phylogenies to hear mention of Darwin's observation that the formation of different languages is "curiously parallel" (1871, 90) to the formation of different species (e.g., Atkinson and Gray 2005). And it is all too easy to become habituated to the beauty of this brilliant insight. Few would quibble with the statement that language is one of the most remarkable evolutionary inventions in the history of life on earth, as it has been the stepping-stone that probably played a major role in allowing a hairless and seemingly helpless tropical mammal (us!) to cover the extreme corners of the globe in only fifty thousand years. Among cultural traits, there are few others that have such complexity and deep cognitive roots and that are learned so readily by children. Given that languages can also reveal the history of our species, including the timing of innovations like agriculture (e.g., Campbell and Kaufman 1985; Bellwood 2001; Hill 2001; Brown 2006), who could fail to be fascinated by viewing a portion of that history represented as a tree that is shaped by these innovations? Indeed, as we will see, linguistic evolution has become a major "growth industry" for evolutionary biologists (e.g., Pagel 2000b; Gray and Atkinson 2003; Pagel et al. 2007), which has at times generated significant controversy in the field of linguistics itself (Marris 2008).

The branch of linguistics known as *historical linguistics* or *comparative linguistics* is responsible for constructing language trees. As its name suggests, researchers in this field make use of comparison, and they do so in a way that is both similar to and different from most phylogenetic approaches in biology. In terms of similarities, both fields have recognized the importance of shared derived characters for inferring evolutionary history, and both biology and linguistics have used trees to represent relationships among the taxonomic units of interest (Platnick and Cameron 1977; Hoenigswald 1987; Blust 2000; Atkinson and Gray 2005). Indeed, it could be argued that linguists discovered the value of tree-like historical representations—and the importance of shared derived characters in inferring relationships—before similar intellectual insights among biologists (see Atkinson and Gray 2005).

Major differences also exist between linguistics and biology, however, particularly involving the methods that are used to infer trees. Biologists have typically applied quantitative analytical approaches that implement specific algorithms; molecular biologists further rely on sophisticated laboratory technology to obtain the data used to infer phylogeny. By contrast, historical linguists often have relied on more qualitative assessments, and these are largely based on an individual researcher's knowledge of multiple languages and a system of constructing sound correspondences among those languages (Campbell 2004; Atkinson and Gray 2005). Although research in historical linguistics in the recent past is often less quantitative than similar approaches in biology, it is important to stress that biology and linguistics actually were on similar quantitative tracks until the middle of the twentieth century (e.g., Swadesh 1952; Kruskal et al. 1971). As we will see, these fields are beginning to converge again on similar approaches, with many linguists using quantitative methods to test historical hypotheses (Dyen et al. 1992; Ringe et al. 2002; McMahon and McMahon 2003; Campbell 2004; Croft 2008).

In what follows, I will provide a brief sketch of a typical comparative approach in historical linguistics then review recent attempts to apply biological methods to investigate Indo-European language phylogeny, and I discuss methods from linguistics for dating nodes on language trees.

The comparative method in linguistics. The standard approach used in historical linguistics is to organize languages into related groups based on careful comparison of vocabulary and grammar. A critical component of this *com-*

parative method involves determination of sound correspondences, such as the shift from the sound *p* in Romance languages to *f* in Germanic languages (e.g., Greek *pater* versus English *father*). This process is important for identifying *cognates*, words that have a common ancestry and are thus the linguistic equivalents of homology in biology. From this comparative method (and use of ancient texts, when they exist), it is possible to reconstruct the *protolanguage*, which refers to the characteristics of the sounds, grammar, and vocabulary of an extinct language (see chapter 4). Indeed, when historical linguists can show that words for particular technological "innovations" probably existed in an ancestral language—such as words for wheel, sheep, or maize—this is considered evidence that the ancestral population of humans that spoke the language possessed this skill or behavior.

It would be desirable to automate the process of collating and comparing features across languages, especially given that this brief description overlooks many important details that must be considered when reconstructing relationships among languages (Campbell 2004; Fortson 2004). Compared to molecular biology, however, it is not yet possible to feed dictionaries of different languages into a sequencer, align the data using computer algorithms, and then produce a language tree from the input (but see Brown et al. 2008; Holman et al. 2008). It is only after linguists have painstakingly assembled a set of cognates for a language (e.g., for Indo-European languages; Dyen et al. 1992) that it becomes possible to apply computational methods from biology to reconstruct the history of different languages (Holden 2002; Gray and Atkinson 2003; Rexová et al. 2003; Pagel and Meade 2006b; Gray et al. 2009). Thus, all attempts to reconstruct language history using modern phylogenetic methods rely on the sweat and tears of numerous linguists who have compiled the data and identified cognates across the languages of interest. In this respect, linguistic phylogenetics is more akin to the meticulous data collection undertaken in morphological systematics than to the automated, "high throughput" research that characterizes molecular systematics.

In attempting to build linguistic phylogenies, recent phylogenetic approaches have used a common set of traits across languages. The basic idea underlying this approach is to take a list of words that are part of a "core vocabulary," such as words for body parts, numbers, and common animals. These types of words are important because they are usually found in all languages, and they are more resistant to borrowing by speakers of other languages (Atkinson et al. 2005). Based on work by Morris Swadesh (1952, 1955), these sets of a hundred or two hundred words have become known as the

Swadesh list, and they are commonly used when developing cognate sets, including data used in recent applications of biological methods to language evolution (e.g., Holden 2002; Gray and Atkinson 2003; Pagel et al. 2007). We will return to Swadesh's work in a later section on dating linguistic trees.

Applying biological methods to study language evolution. In response to the need to bring more systematic and quantitative approaches to historical linguistics, a number of linguists have called for greater use of quantitative methods to study language evolution (e.g., Ringe et al. 2002; McMahon and McMahon 2005; Wichmann 2008). Simultaneously, evolutionary biologists and computer scientists have turned their attention to language diversity and evolution (Warnow 1997; Gray and Jordan 2000; Pagel 2000b; Gray and Atkinson 2003; Nakhleh et al. 2005; Pagel et al. 2007; Gray et al. 2009). To give a sense for recent attempts to apply biological methods to linguistic data, I briefly review research that applied these methods to infer Indo-European language history using methods from above. Other examples will be given in later chapters (e.g., Gray and Jordan 2000; Holden 2002; Atkinson et al. 2008), and a worked example is provided in AnthroTree 2.8.

First, one research group used a parsimony approach to reconstruct the relationships among Indo-European languages (figure 2.8), and they published their findings in the biological journal *Cladistics* (Rexová et al. 2003). These authors used a collection of cognates (Dyen et al. 1992) scored in several different ways, and they rooted their tree with Hittite, an extinct language that is preserved in textual form (for an independent investigation using similar methods, see Ringe et al. 2002). In one of Rexová and colleagues' analyses that treated cognate classes as binary presence-absence characters, the analysis of 2,456 characters produced 6 parsimony trees with a length of 4,988 steps and an RI of 0.77 (a respectable value that is comparable to biological data). This tree (figure 2.8c) exhibited some differences relative to trees inferred with the other methods, but still supported some of the major clades. The analyses of Rexová et al. thus demonstrated that specific optimality criteria can be applied to linguistic data.

At around the same time as Rexová and colleagues were conducting their research, Gray and Atkinson (2003) were applying Bayesian methods to infer Indo-European relationships (figure 2.9, see also Pagel and Meade 2006b). Importantly, these authors investigated two hypotheses for Indo-European language origins. Under one hypothesis, Indo-European languages spread with the rise of agriculture; under this scenario, the age of Indo-European should

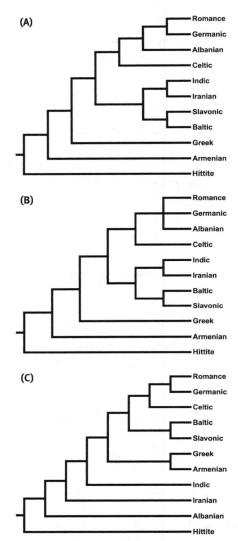

Figure 2.8. Parsimony inference of Indo-European linguistic groups. Trees are based on (a) standard multistate matrix of cognates (Dyen et al. 1992), with character states corresponding to individual cognate classes across two hundred meanings, (b) an altered multistate matrix of cognates designed to reduce subjectivity and excessive splitting of cognate classes, and (c) binary coding of cognates, where each character is a different cognate and thus scored as present or absent. A total of eighty-four languages were used, and the tree was rooted with Hittite. (From Rexová et al. 2003.)

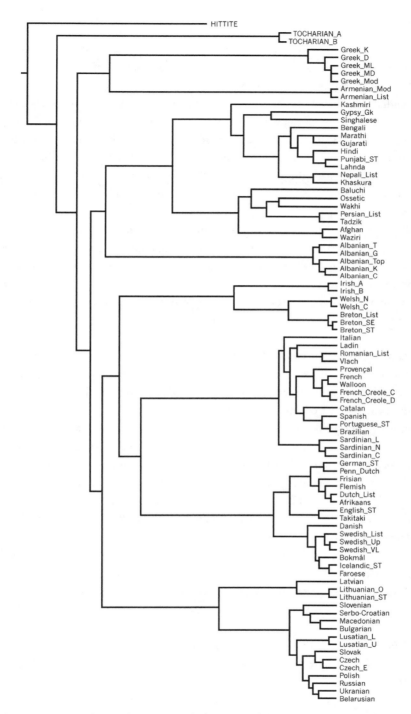

Figure 2.9. Bayesian inference of Indo-European languages. The phylogeny shown here is the majority-rule consensus tree of one thousand inferences of phylogeny from a Bayesian analysis (posterior probability distribution generated through Markov chain Monte Carlo). The phylogeny was rooted with Hittite, and cognates were scored as binary (present-absent) traits. (From Gray and Atkinson 2003.)

match dates for the estimated origins of agriculture in the Fertile Crescent, at approximately 8000 to 9500 years BP (before present; Renfrew 1987). Under the other hypothesis, Indo-European emerged with "Kurgan horsemen" that invaded Europe and Asia around the sixth millennium BP (Gimbutas 1973; Mallory 1989). With rigorous dating of a phylogeny of 87 languages derived from 2,449 lexical items, the authors found support for the agricultural hypothesis. The Indo-European expansion was dated at 7800 to 9800 BP, which is inconsistent with the Kurgan expansion but consistent with the origins of farming. The resulting tree also provided evidence for subsequent diversifications around the time of the Kurgan expansion, suggesting that domestication of horses and their use in warfare might also have contributed to the spread of some Indo-European languages. Whereas standard linguistic approaches cannot produce dated phylogenies, the Bayesian approach used by Gray and Atkinson provides a means to not only date the tree but also to quantify error on the inferred nodal ages. Their results were robust to alternative decisions in terms of data and analysis (see also Atkinson et al. 2005).

Dating a linguistic tree and time depth. For many questions that methods in this book can address, we need a dated phylogeny. Indeed, Gray and Atkinson's analysis (2003; figure 2.9) is only one example of how dates can be useful in the context of linguistics, that is, for testing among hypotheses for the origins of Indo-European languages. Putting dates on a language tree can be as controversial in linguistics as it is in biology, yet the dates are often just as important for interpreting hypotheses for linguistic evolution, the emergence of new cultural behaviors, and human population movements (Nichols 1997; Gray and Atkinson 2003; Gray et al. 2009). A related issue concerns estimating "time depth," and specifically limits to how far back in time we can reconstruct evolutionary relationships among languages. It is widely believed that a "linguistic threshold" exists beyond which resemblances among languages become too obscure to infer the relationships among them (for discussion, see Nichols 1997; Renfrew 2000a; Gray 2005). Thus, when inferring the dates on a tree, we should keep in mind that some language relationships might be undecipherable, which will obviously make dating those relationships impossible.

Putting dates on trees is not a new exercise in linguistics. In fact, linguists have long attempted to develop methods to date language splits. Linguist Morris Swadesh developed quantitative procedures for estimating dates on linguistic trees, with his attempts giving birth to an approach known as *glottochronology* (Swadesh 1952; Lees 1953; Gudschinsky 1956). Using a list of basic meanings for two or more languages, the idea was to identify shared words, or cognates, and plot how many of these exist relative to the time since the two languages were thought to diverge (for a more recent example, see Pagel 2000b). This revealed a negative association between time since divergence and the proportion of shared cognates; in other words, societies that were separated for a longer period of time shared fewer words. The rate at which cognate sharing declines provides, in principle, a way to date language splits. Early analyses of relatively well-studied languages revealed a retention rate of 80.5 percent per one thousand years (Lees 1953). Thus, just as geneticists speak of a molecular clock and archaeologists make use of radiocarbon dating, the glottochronologists proposed that languages exhibit clock-like features, and these can be used to estimate the time since two languages separated.

Glottochronology is a subfield of *lexicostatistics*, which simply refers to the statistical analysis of lexical materials. Whereas glottochronology seeks to estimate absolute dates, however, lexicostatistics refers to inferring the relative relationships among languages (Blust 2000). One of the appealing aspects of these methods is that they are quantitative. In its search for generality, however, glottochronology made strong assumptions of clocklike language change that is constant across languages. Similarly, many lexicostatistical methods failed to distinguish between ancestral and derived characters as a way to infer historical relationships, and thus have weaknesses similar to those for numerical (distance) methods in biological research (Ruvolo 1987; Blust 2000). As evidence accumulated for rate variation across languages and meanings (e.g., Bergsland and Vogt 1962; Blust 2000), the methods fell into disrepute (see Campbell 2004; Atkinson et al. 2005). Thus, glottochronology and its kin were close to extinction in the second half of the twentieth century.

Interestingly, however, lexicostatistical approaches appear to be undergoing a renaissance (Renfrew 2000b), especially as biologists with quantitative phylogenetic toolkits dip their toes into linguistics research—often with impressive outcomes that overcome previous methodological concerns (Pagel 2000a, 2000b; Gray and Atkinson 2003; Pagel et al. 2007; Gray et al. 2009). In general, biologists have more sophisticated methods for inferring evolutionary relationships than do linguists, especially concerning algorithms and informatics approaches to building trees, dating the nodes on trees, dealing with polymorphism, and incorporating heterogeneity in rates of evolution (see, for example, Ruvolo 1987; Warnow 1997; Pagel and Meade 2006b).

To make this more concrete, consider rate heterogeneity for different lineages or for different sites, as applied to data on Indo-European languages.

Several new methods provide a way to deal with the observation that the "glottoclock is not constant" (Atkinson and Gray 2005, 521). For example, rate-smoothing techniques were developed by Sanderson (2002) for biological data, specifically to deal with situations in which rates of molecular evolution vary across branches of a tree. Gray and Atkinson (2003) applied these methods in their study of Indo-European languages. Similarly, substitution models used in evolutionary biology allow for different rates of evolution at sites along the genome, and these rates can be estimated. Pagel and Meade (2006b) and Pagel et al. (2007) applied this basic approach in a Bayesian analysis of the rate at which words change.

Statistical development is also occurring from within linguistics (e.g., Ringe et al. 2002; McMahon and McMahon 2003; Wichmann 2008). For example, linguistics research also has documented slower rates of linguistic evolution for structural features involving the order of parts of speech, verb formation, and other features (Nichols 1992; Dunn et al. 2005, 2007; Wichmann and Holman 2009). Thus, it seems reasonable to expect that the next decade will result in a cascade of new linguistic trees, including relationships among currently irresolvable groups of languages by analyzing more slowly evolving linguistic elements.

Loanwords. Previous research has often tended to shoehorn language history into a bifurcating tree structure, spurning patterns of borrowing as useless "noise" that disrupts the phylogenetic signal in which researchers were ultimately interested. Yet tree-like change might occur only at certain times in a language history, with borrowing being more common at other times. These borrowed words are often called *loanwords*.

For example, Dixon (1997) proposed that languages actually undergo two major types of change. During times of population stability, linguistic features might diffuse among neighboring societies, reducing the ability of the "family tree model" to reconstruct patterns of change. On the other hand, when new innovations occur or when one group gains prestige over neighbors, this can result in the rapid spread of a group of people and their cultural and genetic characteristics through demic expansion. This effect creates a tree-like pattern in the spread of languages. Examples include the Indo-European and Bantu language families following the independent origins of agriculture and the spread of people through Polynesia following development of sea-faring vessels and navigational abilities. By comparison, a greater amount of geographical diffusion may have occurred in Australia. Less resolution (i.e., poly-

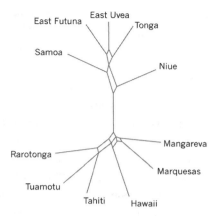

Figure 2.10. "Splits graph" for Polynesian languages. A splits graph shows how pairs of nodes can be linked under alternative evolutionary scenarios that, in the case of languages, are likely to be to the result of borrowing among different language groups. Pairs of parallel lines indicate locations on the network where conflicting signals exist in the data matrix, thus indicating that borrowing has potentially taken place. In this figure, Polynesian languages from the western (Samoa–Niue) and eastern (Rarotonga–Mangareva) clades show evidence of borrowing, with no borrowing between these two clades. (Redrawn from Hurles et al. 2003.)

tomies) closer to the root of a tree may indicate the existence of greater areal diffusion in the early stages of a language expansion.

Given that languages often represent a network of relationships, rather than a strictly bifurcating tree, what methods are appropriate for assessing linguistic history? Some methods from biology can accommodate a reticulating network of relationships, including a method known as split decomposition (figure 2.10; Bandelt and Dress 1992). The method results in a "splits graph," which indicates divisions of the tree into sets of taxa that share different sets of traits (see chapter 10). Another approach to generating splits is taken with a method known as Neighbor Net (Bryant and Moulton 2004), and linguists also have developed network-based approaches (Forster and Toth 2003; Nakhleh et al. 2005). A recent simulation study revealed that Bayesian phylogenetic methods are relatively robust to low levels of borrowing among closely related languages, although split dates tended to be biased toward the present as rates of borrowing increased (Greenhill et al. 2009).

Building Large-Scale Phylogenies

For many questions in comparative biology, the number of species to be investigated is larger than the number of species that are typically included in a phylogenetic analysis. More generally, we might be interested in combining different phylogenies to arrive at a broader view of evolution, and ideally a view that contains all the species in a clade (e.g., Purvis 1995; Bininda-Emonds et al. 2007).

Two basic approaches are available for building large-scale phylogenies, such as would be needed for all primates (which involve between about two hundred and four hundred species, depending on the taxonomic authority that is consulted, e.g., Corbet and Hill 1991; Wilson and Reeder 2005). One of the first and most common approaches is based on building "supertrees" (Purvis 1995; Sanderson et al. 1998; Bininda-Emonds 2004b). The goal of a supertree analysis is to infer relationships among all species in a clade in a systematic way based on source trees created for different subsets of that clade (Sanderson et al. 1998; Bininda-Emonds et al. 2002; Bininda-Emonds 2004a, b). Indeed, one of the first applications of supertree approaches involved reconstructing primate phylogeny (Purvis 1995; Purvis and Webster 1999), and perhaps the best current primate-wide phylogeny is also a supertree (Bininda-Emonds et al. 2007). Supertrees have played a major role in mammalian comparative biology in the past decade (Gittleman et al. 2004).

The basic idea behind a supertree is as follows. If we have trees for different groups of organisms in a larger clade, it is possible to combine these trees in a systematic way to reconstruct the evolution of the group as a whole, provided of course that the individual trees overlap sufficiently that they can be "stitched" together; polytomies will result when portions of the larger tree are not included in the source tree or when source trees conflict. Combining trees can be achieved in several ways (Bininda-Emonds 2004b), but one of the simplest is to represent the "source" trees in matrix form, combine these into one larger data matrix, and run a parsimony analysis on the combined matrix (figure 2.11). Such an analysis is called *matrix representation with parsimony* (Sanderson et al. 1998).

Another approach to building a large phylogeny is called the "supermatrix" approach (Sanderson et al. 1998). In this case, the goal is to combine as much raw data as possible for the species in question. Thus, a supermatrix would consolidate all the gene sequence data from individual studies, rather than pulling together the actual trees and recoding them as matrices. The supermatrix itself can be analyzed using phylogenetic methods described above but like the supertree approach, full resolution of the relationships requires that the clades have at least partially overlapping data. An example of the supermatrix approach for primate phylogeny can be found in the Bayesian analysis

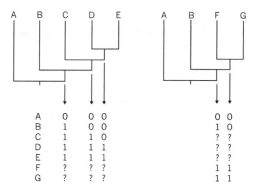

Figure 2.11. Matrix representation with parsimony (MRP). Using multiple-source phylogenies, it is possible to create a character matrix that indicates which species share nodes in the trees. Each node is a character and is coded with a "1" when two species share that node. Thus, for the tree on the left, character 3 represents the node linking species D and E. Species are scored as "0" if they are in the source tree but not a member of the clade represented by a character node. Other species in the data matrix that are not included in the source tree are scored as "?," such as species C, D, and E on the tree on the right. The final matrix combines many such source trees and is examined using parsimony, with unresolved nodes indicated by polytomies. (Figure based on an example given by Sanderson et al. 1998.)

of phylogeny mentioned above (see AnthroTree 2.7; Arnold et al. 2010). One advantage of the supermatrix approach (compared to supertrees) is that it makes use of the original data, rather than summarizing source trees into matrices. From the original data, it is more straightforward to estimate branch lengths and to generate sets of trees that can be used to control for phylogenetic uncertainty (Huelsenbeck et al. 2000; Pagel and Lutzoni 2002).

Summary and Synthesis

Tree thinking pervades attempts to understand organismal and linguistic evolution. Phylogenetic approaches also provide new insights to the origins and distribution of culturally transmitted objects, such as projectile points (Buchanan and Collard 2007), baskets (Jordan and Shennan 2003), and textiles (Tehrani and Collard 2002; see chapter 10). Phylogenetic methods even have been used to investigate manuscript evolution, for example in a phylogenetic analysis of the Canterbury Tales (Barbrook et al. 1998)—just as traits are shared through common descent in biology, errors or modifications of handwritten manuscripts are promulgated through subsequent copying of the manuscript. Phylogenetic methods can thus be applied to investigate this history (Platnick and Cameron 1977; Howe et al. 2001).

What is the future likely to hold for our understanding of primate phylogeny? Knowledge of primate phylogenetic relationships is still in flux, but primate evolutionary relationships are likely to become more certain in the future, at least with regards to topological relationships (Disotell 2008). In particular, genomic insertions provide strong inferences of phylogeny, as the probability of such an insertion at the same place in a genome (i.e., homoplasy) is extremely low (Hillis 1999). Similarly, the accumulation of genetic data make it possible to apply "supermatrix" approaches to infer phylogeny for over 180 species of primates, and to do so in a way that allows users of the tree inference to control for phylogenetic uncertainty in comparative studies (Arnold et al. 2010). While a number of primate clades remain ambiguous, the number of uncertainties has decreased over time and, importantly, new methods provide a way to quantify phylogenetic uncertainty and take this uncertainty into account in comparative research.

Inferring hominin phylogeny imposes a number of serious challenges. One obvious issue involves the limited number of specimens and character states available to study. In addition, the integrated nature of functional morphological traits creates serious statistical problems concerning the independence of the characters (Lieberman 1999; McCollum 1999). Given that many morphological trees of extant primates have been overturned with genetic evidence, it seems foolhardy to expect that we will have better luck inferring hominin phylogeny on the basis of morphology alone. Yet morphology is all that we currently have for hominins, and thus more powerful approaches are needed for the phylogenetic analysis of morphological data. In particular, it is essential to use methods that use explicit evolutionary models and provide an assessment of phylogenetic uncertainty, such as Bayesian or maximum likelihood methods. It is also important to assess the phylogenetic signal in various characters, and to evaluate levels of dependence among characters and to incorporate these correlations into phylogenetic inference methods (e.g., Cheverud 1982; Felsenstein 2002).

Amazing advances have occurred in applying phylogenetic methods to study language evolution. In terms of deeper historical relationships among language groups, however, the future is less certain, as real limits may exist on how far back we can reconstruct relationships among languages. For example, Nichols (1997) makes the sobering observation that approximately three hundred language "stocks" exist that currently cannot be related to one another using standard linguistic approaches. She notes that with diligent application of some new methods, it might be possible to shrink this to two hundred or so stocks, but that "given present knowledge of language change and probability . . . descent and reconstruction will never be traceable beyond approximately 10,000 years" (365). Thus, it is likely to take some time to resolve the massive polytomy that lies near the root of the human language phylogeny, and given the speed with which linguistic characters evolve, it may be impossible to ever create a complete tree of the world's languages. Nonetheless, significant advances are likely to be made in the near future for particular language groups.

Pointers to related topics in future chapters:

- Linguistic trees: chapters 3, 4, 8, and 10
- Reconstructing ancestral languages (protolanguages): chapter 4
- Do cultural traits evolve? chapter 10
- Borrowing of cultural traits (i.e., similar to loanwords): chapter 10
- The consistency and retention indices: chapters 10 and 12
- Species concepts: chapter 11
- Homology in the context of behavior: chapter 11

3 Reconstructing Ancestral States for Discrete Traits

Some of the most fundamental questions in evolutionary anthropology involve reconstructing ancestral states (Ghiglieri 1987; Wrangham 1987; Di Fiore and Rendall 1994; Kappeler 1998; Nunn 1999a; Holden and Mace 2003; Jordan et al. 2009). Consider the following examples: What types of locomotor behavior did the ancestral ape exhibit, and what sort of diet did members of this species have? Was a particular lineage of hominins bipedal? To what extent did the cognitive abilities of the ancestral anthropoid primate change along the lineage leading to great apes? Can we reconstruct the first Indo-European language or the ancestral language of the first Australian? At what point in human evolution did genes allowing the digestion of milk arise, and can we recreate the ancestral proteins that these first genes produced?

These examples reveal the fundamental importance of ancestral state reconstruction for many questions in evolutionary anthropology. Some questions concerning ancestral states can be addressed using fossil material. For example, fossils provide information on limb proportions of extinct species—key information for reconstructing locomotor behavior—and dental features can be used to reconstruct the dietary niche of an extinct organism (Plavcan 2002). Similarly, ancient texts can be used to help reconstruct ancestral forms of the languages in use today, and it is possible to extract DNA from fossil material, such as extinct Neanderthals (Krings et al. 1997) and other hominins (Krause et al. 2010). Inference of characters from fossils often requires knowledge of how traits covary in extant species. To use fossilized tooth morphology to infer the diet of an extinct primate species, for example, we must first understand functional links between diet and dental traits among living primates (Kay 1984; see chapter 9).

While fossil evidence is obviously important for studying evolution, paleontological reconstructions are only possible for characters that leave a trace in the fossil or archeological records. For other cases, we must reconstruct ancestral states using variation among existing species, populations, or languages by "mapping" this variation onto the nodes of a phylogeny. For example, we need phylogenetic reconstructions to infer group size in the ancestral monkey, language features of Polynesian ancestors, cognitive performance of early apes, or patterns of wealth transfer in the early stages of the Bantu expansion.

Pagel (1997) called this approach "statistical paleontology" because it relies on statistical models rather than the acquisition of biological or cultural specimens from the past. As with all statistical models, the effective use of phylogenetic reconstruction requires that we appreciate the underlying assumptions of the methods. While reconstructing ancestral states has long been a focus of evolutionary studies, the past decades have seen a surge of new and more powerful approaches to reconstruct ancestral states. It is possible to use these methods to systematically address fundamental questions about our ancestors—and to do so with a rigor that was previously impossible.

This chapter explores phylogenetic reconstruction methods and how they have been applied to questions in evolutionary anthropology and in biology more generally. I provide examples that use different methods to reconstruct ancestral states, and I review the strengths, weaknesses, and some key assumptions of different approaches (see also Cunningham et al. 1998; Cunningham 1999; Omland 1999). As discussed in later chapters, questions involving trait reconstructions represent the tip of an iceberg of questions that can be addressed phylogenetically; other methods can be used to examine the tempo and mode of evolution (chapter 5), to study correlated evolution of two or more traits (chapters 6 and 7), and to investigate the factors that influence diversification rates (chapter 8).

Background

Reconstructing ancestral states on a phylogeny is essential for understanding evolutionary history, and for many students, these questions represent their first exposure to evolutionary thinking. More generally, reconstructing ancestral states allows us to identify how many times traits evolved, the conditions that favored their evolution, and the lineages in which changes occurred.

Reconstructing an ancestral state is not the same as reconstructing a phylogenetic tree (chapter 2). Reconstructing an ancestral state for some biological trait, such as body mass or diet, usually requires that the tree is already available; the trait is then mapped onto this phylogeny. In nonhuman primates, for example, I used an inference of primate phylogeny (Purvis 1995) to reconstruct the evolution of exaggerated sexual swellings (Nunn 1999a). Using the

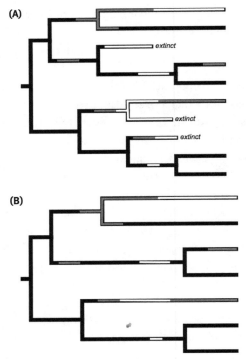

Figure 3.1. Extinct lineages, branches, and evolutionary change. (a) The true evolutionary history of life on earth is full of dead ends, that is, lineages that went extinct. The same is true of languages. (b) Most phylogenies reconstruct evolutionary trees form living (extant) species. Thus, extinct lineages are not represented on the tree. Moreover, evolutionary change can occur along branches, not just at the branching points.

principle of parsimony, I found that exaggerated sexual swellings evolved at least three times, with all cases concentrated in Old World monkeys and apes. Moreover, this trait was lost in two lineages based on my analysis. This immediately informs us that exaggerated sexual swellings have been gained and lost multiple times in Old World monkeys and apes. We will return to this example throughout this chapter to explore different reconstruction methods.

Before proceeding, it is important to be clear about what phylogenetic reconstruction can and cannot do. As an example, consider figure 3.1, which provides hypothetical data on a group of living and extinct organisms. Let us imagine that the data represent coat color, and that the species can have white, gray, or black coat color. The true evolutionary history of coat color and the branching of lineages is shown in figure 3.1a. This tree represents branches in units of absolute time (i.e., it is a dated tree). We see that coat

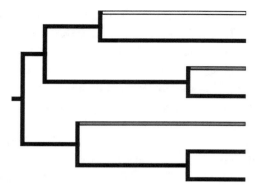

Figure 3.2. Reconstructing ancestral states in the face of imperfect knowledge. Compared to the true evolutionary history depicted in figure 3.1, reconstruction algorithms focus on states at ancestral nodes, indicated on this tree at the point that a lineage splits, rather than where changes occurred along the branches. Thus, our true understanding of ancestral states will always represent only a glimpse of the past. As shown here, these glimpses may be in error (e.g., the ancestor of the top two lineages is reconstructed as black, but it actually was gray in figure 3.1b).

color has sometimes changed multiple times on a single branch, and also that three extinct lineages are included on this tree.

The first important point is that most phylogenetic methods cannot reconstruct the changes that occur along branches—i.e., between nodes on the tree (cf. Garland et al. 1999). Instead, most methods reconstruct the character state in the ancestor of two or more species and say nothing about when the change took place on the branch leading to that node.

A second important point is that phylogenetic analyses can only reconstruct values in extinct lineages when those lineages are included on the phylogeny. In most cases, the phylogeny represents relationships among living species only; branches leading to extinct species or clades are not included. Thus, we are usually faced with the situation in figure 3.1b in which the extinct lineages are "pruned" from the tree. We want to place estimates of values at the branching points (nodes) to indicate whether particular ancestors of *living species* possessed the trait. It is important to realize that when data are missing for extant species, this has the effect of pruning the tree in ways that are similar to the effects of extinct lineages. This pruning of the tree can create sampling biases that affect the outcome of analyses, especially when species with particular character states, such as large body mass or terrestrial habits, are more (or less) likely to be sampled.

A final point is that even with phylogenetic methods, our vision of the past is often imperfect. For example, figure 3.2 shows how one reconstruction algorithm reconstructs the ancestral states on this phylogeny. All nodes are re-constructed as black, whereas one of these nodes was actually gray in the true evolutionary history (see figure 3.1b), and only a fraction of the many changes that actually occurred are reconstructed (i.e., only three changes, all on ter-minal branches). Thus, at the outset we must accept that phylogenetic ap-proaches can capture only a glimpse of the past, and the view from present-day variation often will be misleading.

The challenges of reconstructing ancestral states are further appreciated if you consider the number of possible reconstructions on a tree. With $x = 3$ character states and $n = 6$ internal nodes, there are $x^n = 729$ possible nodal reconstructions on the tree in figure 3.2. Some of these reconstructions are more probable than others, and thus we need to consider the question, "What is the most probable set of reconstructions, and how sensitive are these re-constructions to the phylogenetic hypothesis and our assumptions about the nature of evolutionary change?" While most effort has focused on obtaining a "point" estimate at an interior node, a newer set of methods makes it pos-sible to place confidence limits on the estimate. *Indeed, it is essential to keep in mind that the values shown at ancestral nodes are not themselves data*; they are only estimates generated from the data on different species and subject to a number of assumptions about the evolutionary process. These reconstruc-tions can be extremely sensitive to statistical and evolutionary assumptions, to error in estimating the actual trait values in living species, and to sampling biases regarding the distribution of species that have been studied. Placing confidence limits or probabilities on the estimates can greatly alter our con-fidence in a proposed evolutionary scenario, and is thus an essential compo-nent of ancestral state reconstruction.

Finally, reconstructing ancestral states is often seen as an important step in studies of adaptation (Coddington 1988; Baum and Larson 1991; Brooks and McLennan 1991). The basic argument runs as follows. An adaptation repre-sents a derived trait—that is, a trait that differs from the ancestral form—and the evolution of this trait was presumably the result of natural selection origi-nating on the branch leading to the derived character state. Thus, to identify adaptations, we need a way to identify derived traits, which is accomplished by having access to both a phylogeny and a method for reconstructing ances-tral states. The notion of placing adaptation in a phylogenetic context thus has been extremely influential in evolutionary biology and relies heavily on the ability to reconstruct ancestral states.

Methods for Reconstructing Ancestral States for Discretely Varying Traits

The first step in reconstructing ancestral states is to decide on the method to use. First and foremost, this decision depends on the trait of interest and whether the character states are coded *discretely* or *continuously*. Discrete traits are coded into distinct states, such as the hypothetical coat color example in figure 3.1, and thus are, in effect, integer values (0, 1, 2, etc.). In phylogenetic comparative studies, discretely coded traits are often dichotomous (binary)—coded as 0 or 1—although more than two character states are possible. For example, most primate species can be classified as nocturnal (0) or diurnal (1). Other species are cathemeral, meaning that they are active during both day and night (Tattersall 1987). Cathemerality could thus be added as an intermediate value to give three character states (with adjustment of character states such that nocturnality = 0, cathemerality = 1 and diurnality = 2).

Continuously varying traits are measured quantitatively, often with decimal fractions (e.g., 10.2, 605.97), subject to measurement precision and logical limits to some traits (e.g., length and mass measures cannot be negative). Examples of continuous traits include body mass, group size, and longevity. Sometimes, continuously coded data are recoded into discrete categories. This is especially common when quantitative information is unavailable for all the species of interest, making a complete sample of continuously coded data impossible to obtain. Primates, for example, are often coded as using mainly terrestrial or arboreal substrates, but this actually reflects an underlying variable involving the percentage of time that individuals engage in terrestrial activity. Quantitative data on substrate use are available for only a subset of primate species, which means that researchers must often rely on discrete codes of substrate use to obtain unbiased samples of species and sufficient sample sizes.

In addition to coding the data of interest as discrete or continuous, it is necessary to decide on the phylogeny to use. A number of authors have cautioned against using a tree generated with the traits of interest in the comparative study (e.g., Coddington 1988; Brooks and McLennan 1991), and to avoid controversy in this regard, the best solution is to build the tree with trait data that are independent from the study goals. Articles by de Queiroz (1996) and Swofford and Maddison (1992) provide further perspective. This issue has become more of a historical concern than a current one, given that most biological researchers use molecular data to infer phylogenies, which are then used to examine the evolution of morphological, life history, or behavioral traits. Similarly, in cross-cultural research, the use of genetic or linguistic data to generate phylogenies helps to provide independence between tree reconstruction and trait reconstruction.

In what follows, I consider three methods for inferring ancestral states for discrete variables: parsimony, maximum likelihood, and Bayesian methods. These approaches, which correspond to methods described in chapter 2 for inferring phylogeny, represent a deeper philosophical and methodological consistency, and this chapter will hopefully make these concepts more concrete. Parsimony is included because it is commonly used in evolutionary anthropology and provides background for other methods. Parsimony has weaknesses relative to the other methods, however, and although it is easy to understand and implement, this method is no longer the preferred way to infer ancestral character states.

Reconstruction of continuously varying traits is covered in the next chapter. I decided to separate the two types of data into two chapters, because the methods differ depending on the data and because including analyses for both types of data in one place would have created an exceptionally long chapter.

Parsimony. Parsimony is the most widely used approach in previous research aimed at reconstructing ancestral states. In evolutionary anthropology, parsimony has been used to study primate sociality and mating systems (Sillén-Tullberg and Møller 1993; Mitani et al. 1996b), to reconstruct DNA and amino acid sequences (Krishnan et al. 2004), to study human cultural traits involving family organization (Borgerhoff Mulder et al. 2001), to investigate the evolution of sleeping habits in primates (Kappeler 1998), and in analyses of homoplasy in primate and hominin evolution (Lockwood and Fleagle 1999). Di Fiore and Rendall (1994) used parsimony to investigate the suite of social organization characters that changed during primate evolution. In Old World monkeys, they found that many of these traits hinged on female philopatry (i.e., when females remain in their natal groups rather than dispersing at adulthood). More specifically, they found evidence in this lineage of monkeys for females to group with their kin and exhibit coalitionary behavior, to display high levels of grooming, and to care for other females' infants (allomothering). In another example, Gray and Jordan (2000) mapped a geographic series onto a linguistic phylogeny to assess whether humans colonized Southeast Asia and the Pacific islands more or less in a single wave of migration, compared to other colonization scenarios. They found support for this "express train" hypothesis (Diamond 1988; this example is discussed further in chapter 10).

As discussed in the previous chapter, parsimony refers to procedures for minimizing the number of changes on the tree; it is an optimality approach that prefers evolutionary scenarios that minimize the amount of change. Rather than minimizing the number of changes in a whole matrix of data to construct a tree, however, here the goal is to map a single trait onto a given phylogeny in a way that minimizes evolutionary change in that trait. Figure 3.2, for example, reconstructs only three changes on the tree. If we proposed instead that the ancestor of this clade had white fur, more steps would be needed to account for the distribution of character states at the tips of the tree, and therefore this would be less parsimonious.

Stewart (1993), Cunningham et al. (1998) and Maddison and Maddison (2000) provide nice overviews of the mechanics underlying parsimony reconstructions, and many programs have been developed to reconstruct ancestral states using parsimony (AnthroTree 3.1). The parsimony algorithm operates over two directions on the tree in its most basic implementation: first down the tree (from tips to the root), and then up the tree (from root to tips). At each step in these "downpass" and "uppass" procedures, values from surrounding nodes are used to define sets of similar states, and a final optimization procedure makes use of the stored uppass and downpass sets of the two descendent nodes. The value for a node is therefore a function of the values for all surrounding nodes, with these values determined through two (or more) separate passes through the tree.

Parsimony assumes that branches are equal in length. In other words, it assumes that evolutionary transitions are equally probable on the longest branch and the shortest branch on a phylogeny. It is also important to realize that parsimony cannot always reconstruct a character state unambiguously. In such cases, multiple equally parsimonious reconstructions can result, and equivocal values are indicated with a node having multiple states. This is seen if we use parsimony to reconstruct the evolution of exaggerated sexual swellings in macaques (figure 3.3), where two of the reconstructed ancestral nodes are ambiguous (i.e., the half-white, half-black circles). Sometimes, including an outgroup that is known to possess the ancestral character state can help to resolve ambiguous reconstructions. In the case of sexual swellings, for example, including an outgroup of New World monkeys may help to resolve the ancestral state for Old World monkeys, because New World monkeys lack the exaggerated character state (Nunn 1999a).

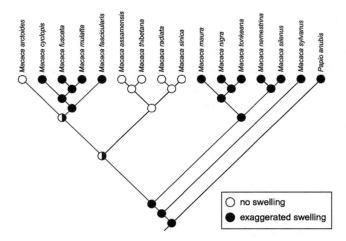

Figure 3.3. Maximum parsimony reconstruction of sexual swellings in macaque monkeys. The presence or absence of sexual swellings was mapped onto primate phylogeny (Arnold et al. 2010) using parsimony, with the cost of gains of swellings equal to the cost of losses. *Papio* was included as the outgroup. Circles that are half white and half black indicate ambiguous reconstructions. The analysis was done in Mesquite, version 2.7.

The previous chapter discussed how cost structures can be created to reflect that some types of changes are more likely than others, for example in the context of transitions and transversions in DNA data. Several other "transformation types" appear in the literature and are worth knowing about (see also AnthroTree 3.2). For example, traits can be *ordered* or *unordered*, and this affects how trait changes on the tree are converted into different numbers of evolutionary steps. In an ordered series of more than two character states, it is assumed that traits pass through intervening states to go from one end of the trait spectrum to the other end. In the analysis shown in figure 3.2, for example, I assumed that changes among all character states are equally likely, and thus count as an equal number of steps in the analysis (table 3.1). This is an unordered character state. However, we might reasonably argue that gray is an intermediate state, which means that a transition from white to black involves a first transition from white to gray, and then a second transition from gray to black (as suggested in the case of cathemeral activity period given above, with cathemerality serving as an intermediate stage between nocturnal and diurnal activity periods). Thus, a transition from white to black would count as two steps in this ordered character (table 3.2).

In addition to ordered versus unordered character states, other transformation types are useful to know (AnthroTree 3.2; Swofford and Maddison

Table 3.1. Step matrix for an unordered character.

		To:		
		Black	Gray	White
	Black	0	1	1
From:	Gray	1	0	1
	White	1	1	0

Table 3.2. Step matrix for an ordered character.

		To:		
		Black	Gray	White
	Black	0	1	2
From:	Gray	1	0	1
	White	2	1	0

1992; Maddison 2000). For example, we might assume that traits are irreversible: once they are gained, they cannot be lost. Or we could assume that it is easier to lose a trait than to gain it, in what is called *Dollo parsimony*. In this case, only one gain is allowed, but multiple losses can occur. Another option is to base the transformation on the ages of strata, or, *stratigraphic parsimony*, which allows paleontologists to assess how well a phylogeny corresponds to stratigraphic age.

Most parsimony analyses assume that the costs of gains are equal to the costs of losses, in other words, that the probability of gains and losses is equal. *Importantly, however, the assumptions embedded in a parsimony step matrix can have very strong impacts on the conclusions that are drawn about trait evolution.* In one example, Omland (1997b) investigated the assumption of equal gains and losses in the evolution of sexual dichromatism in ducks (Anthro-Tree 3.3). He found that by adjusting the probability of gains or losses using a step matrix, the ancestral state of ducks could easily switch from dichromatic—with males exhibiting showy plumage and females dull plumage—to monochromatic, with both sexes exhibiting dull plumage. In fact, to reach the widely held belief that dichromatism is ancestral to all major duck clades, the cost of gains had to be five times as high as losses! In other words, the cost (step) from absent to present (0 to 1) equaled 5, while the cost from present to absent (1 to 0) equaled 1. Thus, as noted by Cunningham et al. (1998), the reconstruction of ancestral states can end up saying more about the reconstruction method itself than about the evolution of the traits of interest.

This previous example shows that it is possible to investigate the effects

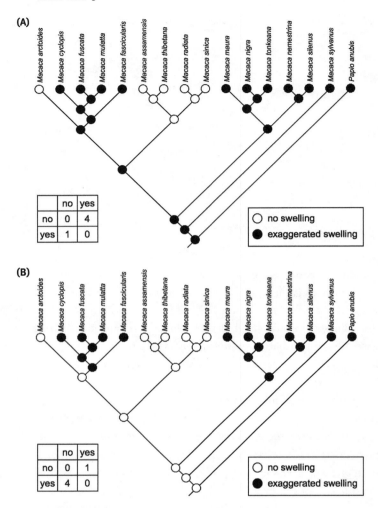

Figure 3.4. Reconstruction of sexual swellings using unequal costs for gains and losses. The presence or absence of sexual swellings was mapped onto primate phylogeny (Arnold et al. 2010) using parsimony, with (a) the cost of gains 4x the cost of losses and (b) the cost of losses 4x the cost of gains, estimated using a step matrix in Mesquite, version 2.7. The step matrix is shown on the lower left of each panel, where "no" indicates no exaggerated swelling, "yes" indicates exaggerated swellings, and the table is read as going from the state on the left column to the state in the top row.

of unequal gains and losses using a range of relative costs (see also Ree and Donoghue 1998; Belshaw and Quicke 2002). This point is illustrated further for the sexual swellings data in figure 3.4. Compared to figure 3.3, we get radically different interpretations of evolutionary history when using different weighting schemes, ranging from exaggerated swellings in the ancestral macaque and all variation due to losses (figure 3.4a; costs of gains is 4× the cost of loss), to a maximum of four gains and no losses and the absence of swellings reconstructed in the ancestral macaque (figure 3.4b; costs of losses is 4× the cost of gains). It seems that any scenario can be conjured up by simply using a different weighting scheme, which is troubling, because how can we justify any particular step matrix?

In the duck example, Omland (1997b) summarized other lines of evidence that support the hypothesis that dichromatism was ancestral in ducks based on biogeography, vestigial characters, and other observations. He was lucky that such information was available, as it provided external validation for deciding on the step matrix to use. In most cases, however, we lack a strong basis to support one hypothesis over another, and so the best we can hope for is that reconstructions are insensitive to changing the costs for gains and losses of traits. While it might seem that the simplest assumption is to set the costs of losses and gains to be equal, it is important to remember that this is also an assumption, specifically, a 1:1 assumption of gains:losses. It is no better justified than weightings of 2:1, 3:1, 10:1, 1:10, or even 2.87:1.2 (the step matrix is not limited to integer values). Ree and Donoghue (1999) provide a framework for investigating how changes in the elements of a step matrix impact reconstructions of trait evolution.

Thus, the person seeking to use parsimony has at hand a wide array of options. This can be both exciting and overwhelming. Maddison and Maddison (2000) discuss assumptions about characters as ordered versus unordered, losses more likely than gains, and the irreversibility of characters. Clearly, there are many aspects of the evolutionary process to consider, and decisions involving parsimony reconstructions require knowledge of the biology of the traits in question and careful scrutiny of the assumptions underlying the reconstruction procedure. More often than not, it is possible only to rule out some kinds of transformation types rather than to identify the one to use. Last, it is important to remember that parsimony analyses fail to take into account branch lengths, despite the very real probability that change is more likely on longer than shorter branches. The next methods deal with this issue explicitly.

Maximum likelihood reconstructions. Maximum likelihood methods use an explicit evolutionary model to reconstruct ancestral states in a way that makes the observed character states in extant species most likely (Schluter et al. 1997; Pagel 1999b). This class of methods makes use of a Markov model, such that the probability of change on a branch is independent of changes on other branches and depends only on the state at the beginning of a branch. The key parameters in this model involve estimates of the rate at which a character state is gained and lost. It is possible to estimate these parameters given a phylogenetic tree and species data, and then to use the estimates to obtain the likelihood of different ancestral states on internal nodes. Schluter et al. (1997), Pagel (1994a, 1997, 1999b), and Mooers and Schluter (1999) provide overviews on maximum likelihood reconstructions of ancestral states. An example is provided in AnthroTree 3.4.

Maximum likelihood methods offer many advantages over reconstructions based on parsimony. For example, maximum likelihood methods use information on branch lengths; this is important, because a longer branch offers more opportunities for change to occur than a shorter branch. Another important advantage of maximum likelihood is that it provides a way to assess statistical support for one reconstructed ancestral state over one or more other reconstructed states. Maximum likelihood thus moves away from treating these nodal estimates as actual data, and instead explicitly shows that they represent output from a statistical model (Pagel 1994a).

Schluter et al. (1997) compared parsimony and maximum likelihood reconstructions of discrete traits in real-world biological systems. Figure 3.5 shows their reconstruction of diet in Darwin's finches (Grant 1986). Although parsimony and maximum likelihood reconstructions were similar, some important differences emerged. For example, the common ancestor of the ground finches (*Geospiza*) and tree finches (*Camarhynchus* and *Playspiza*) was reconstructed as originating from an insectivorous ancestor under parsimony, with transitions to granivory and folivory found within this clade. Using maximum likelihood reconstructions, however, the reconstructed evolutionary history changed. In this case, the ancestor of these two major groups of finches was reconstructed as granivorous, with shifts from this state to insectivory along the tree finch lineage.

Details on the statistics underlying maximum likelihood reconstructions are provided elsewhere (Pagel 1994a, 1999b; Schluter et al. 1997). Briefly, the method involves construction of a transition matrix for the rate at which traits change from one state to another; this is generated from a dated tree and current species values and assumes that rates of change are constant throughout the tree (but independent across branches). The transitions represent instantaneous transition rates and are estimated using maximum likelihood. If we have a trait with two states, we end up with a transition matrix like that shown

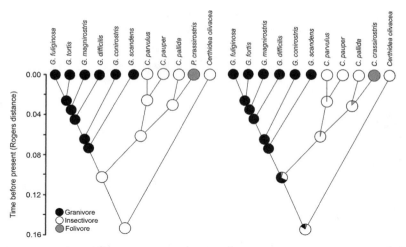

Figure 3.5. Reconstructing diet in Galapagos finches using maximum likelihood. This shows reconstructed dietary categories among eleven species of Galapagos finches, for genera involving *Geospiza*, *Camarhynchus*, *Platyspiza*, and *Certhidea*. On the left, ancestral states were reconstructed using parsimony with equal gains and losses. Branch lengths are shown but not taken into account in the analysis. On the right, maximum likelihood estimates are provided based on equal transition rates ($q_{01} = q_{10}$) and branch lengths were used. (Redrawn from Schluter et al. 1997.)

in table 3.3. In this table, the transition rate from 0 to 1 is given by q_{01}, and the transition rate from 1 to 0 is given by q_{10}. The transitions q_{00} and q_{11} represent rates of no change, with $q_{00} = 1 - q_{01}$, and $q_{10} = 1 - q_{11}$. If the trait has more than two states, larger transition matrices are constructed to reflect all the possible transitions (table 3.4). From these instantaneous rates we can estimate probabilities of trait change on branches of different length.

With estimates of the transition rates, it also becomes possible to obtain the likelihood of a particular state at an interior node given the data and phylogeny. In effect, the user is setting the value of the node to take a particular state, the likelihood is calculated, and this is repeated with the other possible states at that node (and across other nodes on the tree).

Branch lengths are important in the reconstruction of maximum likelihood estimates (Pagel 1999b). The example given earlier of Darwin's finches demonstrates the important influence of branch lengths on reconstructed values: one reason for the greater contribution of granivory to the ancestor of ground and tree finches in figure 3.5 involves the shorter branch leading to ground finches (*Geospiza*), compared to tree finches, from their common ancestor. The change in diet is more likely to occur on the longer branch; thus, the longer branch leading to tree finches is assigned the change (with gra-

Table 3.3. Transition matrix for a two-rate (asymmetric) evolutionary model.

		To:	
		0	1
From:	0	q_{00}	q_{01}
	1	q_{10}	q_{11}

Table 3.4. Transition matrix for a three-state model.

	0	1	2
0	q_{00}	q_{01}	q_{02}
1	q_{10}	q_{11}	q_{12}
2	q_{20}	q_{21}	q_{22}

nivory thus identified as being more probable in the ancestor). It is important to keep in mind that branch lengths, which are typically estimated from genetic data, are assumed to reflect the opportunity for change in the characters that are being mapped, with more changes expected over longer evolutionary branches. Although this assumption depends on the model of evolution (see chapter 5), it is more reasonable than assuming that branches are equal, as in a parsimony analysis.

Based on their analyses, Schluter et al. (1997) drew the following conclusions about maximum likelihood reconstructions (see also Zhang and Nei 1997). First, maximum likelihood reconstructions are similar to parsimony reconstructions when the rates of change are low. Second, some differences between maximum likelihood and parsimony arise due to the incorporation of branch lengths into the maximum likelihood analysis. This is an advantage rather than a disadvantage, as we expect that more changes will occur when more time is available (i.e., longer branches). Third, reconstructing nodes farther from the tips—especially at the root—are more uncertain than reconstructions closer to the tips of the tree. Finally, the examples in Schluter et al. (1997) revealed that maximum likelihood reconstructions are often uncertain even when parsimony reconstructions appear to be clear-cut (i.e., unambiguously reconstructed as 0 or 1 under parsimony; see figure 3.5).

Compared to other fields, maximum likelihood methods for trait reconstruction have been used less commonly in evolutionary anthropology. In one study, Holden and Mace (2003) used maximum likelihood reconstructions to study matrilineal descent and cattle-keeping in Bantu-speaking people (see also chapter 10). They found that it was not possible to distinguish whether matrilineal or patrilineal descent characterized the root of the tree, with neither state being statistically supported over the other in a likelihood model. In addition, reconstructions of cattle use revealed that this cultural trait was absent in the deepest branches on the tree (i.e., toward the root).

By estimating transition rates, maximum likelihood approaches also provide a means to assess whether rates of gains and losses of a trait are equal—an interesting question in its own right, as it deals explicitly with the underlying evolution of traits and whether some traits are more likely to be gained or lost. More specifically, the rates of change from 0 to 1 and 1 to 0 can be set to be equal in a *symmetric model* ($q_{01} = q_{10}$), or they can be estimated separately in an *asymmetrical model*, with independent estimates for q_{01} and q_{10} (see Schluter et al. 1997; Mooers and Schluter 1999; Pagel 1999b). In figure 3.5, these transitions were set to be equal. Sanderson (1993) and Ree and Donoghue (1999) describe how fitting two separate rates can be used to test for directional evolution, thus specifically considering whether gains are more likely than losses. Odd results can sometimes emerge in maximum likelihood analyses of transition rates, however, especially when rates are biased heavily in favor of gains or losses (see Cunningham 1999; Pagel 1999b). In such cases, it often takes some detective work to determine exactly what the rates mean in the context of a trait, including by testing whether a two-parameter model offers a significantly better fit than a one-parameter model.

An example by Lutzoni et al. (2001) shows how transition rates can be used to address evolutionary questions, and further illustrates how one can control for uncertainty in phylogenetic relationships (see also Pagel and Lutzoni 2002). These authors examined the evolutionary origins of lichens among fungi using maximum likelihood, and remarkably, they did so across nearly twenty thousand (!) phylogenetic trees that were generated using the Bayesian methods described in the previous chapter (Huelsenbeck et al. 2000). Hence, their results were not conditioned on a single phylogeny. Lutzoni et al. (2001) showed that in over 90 percent of the trees, losses of lichenization exceeded gains (figure 3.6). Take a moment to reflect on what is shown in this figure. While at first it might be more appealing to end an analysis with a single phylogeny and sharply contrasting black and white nodes—as might be produced by a parsimony analysis with a single tree—phylogeny itself is uncertain, and different phylogenetic hypotheses can produce different estimates of transition rates and ancestral states. This figure demonstrates the importance of

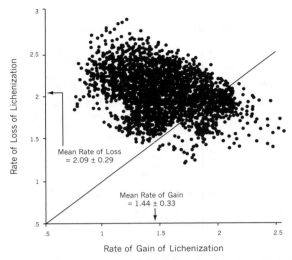

Figure 3.6. Unequal rates of gains and losses of lichenization. Plot shows the estimated rates of losses and gains using maximum likelihood across 19,900 trees generated using Bayesian methods. The line shows equal rates of gains and losses, revealing that most estimates (90.6 percent) fall above this line. (Redrawn from Lutzoni et al. 2001.)

the phylogeny used, and further shows that when higher rates of gains occur, losses are necessarily reconstructed as occurring at a lower rate.

This discussion of transition rates raises an important question for maximum likelihood reconstructions more generally. When reconstructing ancestral states, should one use a single rate, with $q_{01} = q_{10}$, or an asymmetric model with the two rates estimated separately? In one of the original articles describing the reconstruction of ancestral states using maximum likelihood, Schluter et al. (1997) suggested that sufficient data are rarely available to justify estimating two parameters; hence, they tended to prefer a one parameter model, which has benefits in terms of computation and interpretation (Pagel 1999b). In another study, Mooers and Schluter (1999) tested whether using a two-rate asymmetric model improves the statistical fit of the model, based on data obtained from twenty-eight published articles. In only two of these studies were asymmetric transition models supported significantly over one-parameter models. They concluded that fitting two parameters was most likely to improve the fit on larger trees with many transitions and when the majority of tips or internal nodes are reconstructed to be in the derived state. In an application to a real-world dataset on flower morphology, however, Ree and Donoghue (1999) found that a two-parameter model of evolution was generally supported. Thus, it is worthwhile to investigate whether a two-parameter model offers a significant improvement in the likelihood (AnthroTree 3.5).

To put maximum likelihood reconstructions into practice, let us return to the example of reconstructing exaggerated sexual swellings in macaques (see AnthroTree 3.4 and figure 3.3). For the two-state sexual swellings character, I used maximum likelihood to reconstruct ancestral states throughout the tree with rates of gains and losses set to be equal (a symmetric model where $q_{01} = q_{10}$), rates of gains and loss estimated separately (an asymmetric model), and for an asymmetric model when branches were set to be equal (as an example of how branch length information can affect the results). In the applications of this approach presented here, I report proportional likelihoods—which reflect relative support for different character states—and take significance levels from analyses in Mesquite, version 2.7.

Results for the analysis of exaggerated sexual swellings are shown in figure 3.7. The maximum likelihood estimates are similar to those from the parsimony analysis with equal gains and losses, but they enable quantification of uncertainty and allow for statistical tests. Take as examples the two nodes that were ambiguous under parsimony in figure 3.3 (labeled "1" and "2" in figure 3.7) and another node that was unambiguously reconstructed as lacking swellings under parsimony (labeled "3" in figure 3.7). A model with symmetric gains and losses gives a higher probability for nodes 1 and 2 to have possessed exaggerated sexual swellings (figure 3.7a), but the support was not statistically significant. Node 3 shows more uncertainty than in the parsimony reconstruction, but the maximum likelihood model significantly favors the absence of swellings at this node. At the base of the macaque lineage, we have clear statistical support for reconstructing the ancestor as possessing exaggerated sexual swellings.

We can further investigate how changing the symmetry of the evolutionary model and use of equal branch lengths affects the results. Thus, with asymmetric rates, statistical support becomes more ambiguous throughout key nodes on the tree (Figure 3.7b), but it still supports the absence of exaggerated sexual swellings at node 3. What should we conclude in terms of whether to use a model with symmetric versus asymmetric transition rates? The one-parameter model gives an overall likelihood of -7.60, while the two-parameter model gives an overall likelihood that is only slightly higher, -7.19 (i.e., it is less negative). This difference is not significant ($p = .52$) in a likelihood ratio test (LRT; see AnthroTree 3.5). Thus, we would favor using the

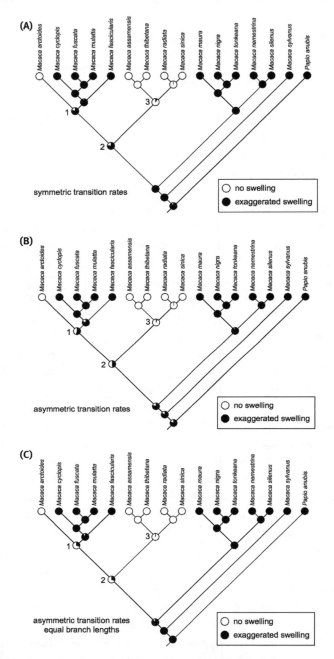

Figure 3.7. Reconstruction of sexual swellings using maximum likelihood. The presence or absence of sexual swellings was mapped onto primate phylogeny (Arnold et al. 2010) using (a) a single transition rate, that is, a symmetric model with $q_{01} = q_{10}$, which was estimated as 0.032; (b) a two-parameter asymmetric model, that is, $q_{01} \neq q_{10}$, with estimates of $q_{01} = 0.091$ and $q_{10} = 0.029$; and (c) a two-parameter asymmetric model assuming equal branch lengths (estimates of $q_{01} = 0.152$ and $q_{10} = 0.060$). Analyses were conducted in Mesquite, version 2.7. Pie charts at internal nodes reflect the proportional likelihoods of the two states. For the ancestral macaque, the proportional likelihood of having swellings is 96.9 percent, 82.8 percent, and 97.6 percent for each of panels a, b, and c, respectively. Numbered nodes are discussed in the text.

one-parameter model (Figure 3.7a), and we would conclude that exaggerated swellings have been lost either once or twice in macaque evolution.

If we assume equal branch lengths (Figure 3.7c), we start to obtain somewhat greater support for an absence of swellings at nodes 1 and 2 (although neither is statistically significant). Given that we have actual estimates of branch lengths, the use of equal branch lengths is not recommended in this case; it is for illustrative purposes only, specifically to emphasize the importance of using branch lengths when they are available.

In summary, maximum likelihood approaches provide a valuable alternative to parsimony reconstructions, with the possibility to both reconstruct ancestral states and give statistical confidence in these estimates. It takes "statistical paleontology" to a new level (Pagel 1997; see also Oakley and Cunningham 2002). An important conceptual difference between parsimony and likelihood approaches is that likelihood reconstructions are not penalized by the number of gains (see Schluter et al. 1997). Thus, maximum likelihood can produce results inconsistent with a strict philosophical stance on minimizing evolution. On the other hand, when changes are common, maximum likelihood will fail, just as we saw with parsimony. We should not be overly concerned about this, because, as noted by Schluter et al. (1997), it underscores an intuitively simple tenet in reconstructing evolution, namely that high rates of evolutionary change should make any reconstruction method less certain.

Bayesian reconstructions of discrete traits. An exciting development in recent years is the application of Bayesian methods to phylogenetics, including for reconstructing ancestral states (Schultz and Churchill 1999; Huelsenbeck and Bollback 2001; Pagel et al. 2004; Ronquist 2004). In this application of Bayesian methods, the user estimates the probability that an ancestral node has a particular character state based on data involving the character states in extant species and how they are related (a dated phylogeny), and also takes into account *mapping uncertainty* by quantifying uncertainty in the transition rate parameters that are used to reconstruct trait evolution (Ronquist 2004). As mentioned in chapter 2, another advantage of Bayesian methods is that they can take into account error in the phylogenetic topology and branch lengths (Huelsenbeck and Bollback 2001; Pagel et al. 2004), or what Ronquist (2004) calls *phylogenetic uncertainty*. For this, the user needs a set of trees (e.g., Arnold et al. 2010).

Pagel and Meade (2006a) used a Bayesian approach described by Pagel et al. (2004) to map the evolution of exaggerated sexual swellings while also accounting for uncertainty in the phylogeny of Old World anthropoid primates. For the root node of the tree shown in figure 3.7, the probability of exaggerated sexual swellings was estimated to be 0.98 ± 0.02. The essential addition that a Bayesian analysis adds to this effort is the ± sign, which represents a credible interval based on uncertainty in the underlying evolutionary model (e.g., rates of transitions) and uncertainty in the phylogeny. A worked example is provided in AnthroTree 3.6.

The theoretical principles that underlie Bayesian ancestral state reconstruction are identical to those described for estimating phylogeny in chapter 2. To briefly review, we can think of a Bayesian analysis as aiming to obtain the probability distribution of some parameter, whether it is an inference of phylogenetic relationships and branch lengths, an ancestral node, or the rate parameters of the evolutionary model. This is called a *posterior probability distribution*, because in Bayesian statistics it is the conditional probability of the parameter after taking the data and the prior probability of the hypothesis into account. It is difficult (if not impossible) to solve this problem analytically, but a solution can be obtained by using a sampling procedure known as Markov chain Monte Carlo (MCMC). At each step in the chain, a new set of parameters is proposed, and an algorithm from Bayesian inference is used to accept or reject the new parameters. In essence, the chain "roams" through parameter space, covering this space in proportion to the posterior probability—and many very cool things can then be done with the information that the chain collects while it samples parameters. Importantly, this sampling must be done only after the model has reached stationarity, or, "burn-in" (see figure 2.7).

How is all of this achieved in the context of reconstructing ancestral states? In the approach developed by Pagel et al. (2004), the authors sought the posterior probability distribution of the transition rates. Thus, rather than estimating the reconstruction based only on optimization of the transition rates, they examined the distribution of reconstructed states given potential error in estimating these rates. Importantly, one can also sample different phylogenies using MCMC, and this effectively deals with the problem that different trees can produce different reconstructions of evolutionary history (Huelsenbeck and Bollback 2001; Lutzoni et al. 2001; Pagel et al. 2004). Pagel et al. (2004) provide a "most recent common ancestor" method to handle situations in which the ancestral node of interest is unstable across trees, that is, because different sets of species derive from the node in different phyloge-

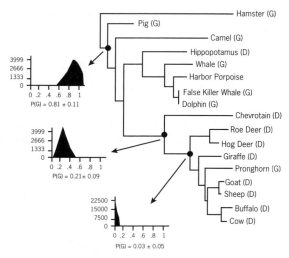

Figure 3.8. Ribonuclease evolution in mammals. Bayesian analysis has been used to reconstruct genetic evolution in the ribonuclease gene. Plots show the probability distribution of a particular nucleotide (G) based on a Bayesian analysis that also incorporated phylogenetic uncertainty (shown here is the consensus tree). (Redrawn from Pagel et al. 2004.)

nies. Huelsenbeck and Bollback (2001) took a slightly different approach by restricting the reconstruction to only those trees that possessed the node of interest.

Pagel et al. (2004) applied the method to study genetic evolution in artiodactyls (figure 3.8). Their analysis had two main parts. First, they ran the Bayesian analysis to reconstruct artiodactyl phylogeny. For this, the authors downloaded sequence data, and before inferring the phylogeny they excluded the genetic data that they wished to reconstruct. From their Bayesian phylogenetic analysis, they sampled five hundred trees after the model reached burn-in to incorporate phylogenetic uncertainty (Ronquist 2004). Second, they constructed an MCMC model of trait evolution to estimate the transition rate parameters that were necessary to reconstruct ancestral states. At each step, a new combination of rate parameters (q_{01} and q_{10}) was selected along with one of the five hundred trees. The likelihood of this new combination was then computed and accepted or rejected accordingly. The authors sampled the coefficients from the resulting chain every twenty generations and used these parameters to reconstruct ancestral states for the corresponding phylogeny. This second step therefore addressed issues concerning mapping uncertainty (Ronquist 2004), and by applying this approach across trees selected from the Bayesian analysis, the authors simultaneously dealt with phylogenetic uncertainty.

This seemingly complicated set of analyses yielded several important findings. First, Pagel et al. (2004) discovered that different trees produced different transition parameters and even larger differences in log-likelihood scores. This demonstrates the importance of taking into account phylogenetic uncertainty; such error is impossible to detect if only a single tree is used in the analysis. Second, they discovered marked heterogeneity in the reconstructed ancestral states. In some cases, the probability of a reconstruction approached zero; in other cases, however, the probabilities were more intermediate. Last, they demonstrated that their "most recent common ancestor" method helps to address conflicts among the topologies from the MCMC analysis.

As with maximum likelihood, Bayesian approaches can be used to assess whether a symmetric model of equal gains and losses is adequate (i.e., $q_{01} = q_{10}$). In their study of artiodactyls, for example, Pagel et al. (2004) examined the correlation between the rate coefficients for gains and losses at a particular genetic locus. They found little correlation between the estimated rates, suggesting that the model of trait evolution is not drawn towards a situation in which $q_{01} = q_{10}$.

Bayesian approaches also have been applied to study human cultural traits. Fortunato et al. (2006) investigated the ancestral states for patterns of bride-price and dowry in Indo-European cultural groups. This is an important question, because significant amounts of wealth have been transferred at marriages throughout human history, with resources typically transferred from either the bride's or groom's family (dowry and bride-price, respectively). Among the Kipsigis of Kenya, for example, the bride-price payment averages one-third to one-half of the groom's father's livestock (Borgerhoff Mulder 1988). Linguistic data involving 2,449 characters were used for the MCMC inference of phylogeny, and data on 52 societies exhibiting either bride-price or dowry were mapped onto 1,000 of the trees using Bayesian reconstruction methods. Fortunato et al. (2006) found that in most analyses, dowry was the ancestral state for this group of societies, and this was generally true when different data coding procedures were used (figure 3.9). Only when the outgroup was coded as exhibiting bride-price did this conclusion change, with the ancestral state becoming ambiguous. In another application of Bayesian reconstruction methods to cross-cultural data, Jordan et al. (2009) investigated residence patterns in Austronesian societies.

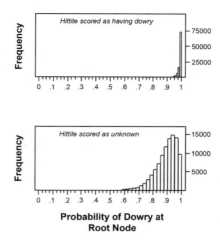

Figure 3.9. **Bayesian reconstruction of dowry in Indo-European societies.** Analyses indicated that dowry was the most likely ancestral state at the root node. Analyses were run with different ways of scoring the outgroup (Hittite, an extinct society) and polymorphism. Only when the Hittites were scored as having bride-price did the results change substantially. (Redrawn from Fortunato et al. 2006.)

Summary and Synthesis

We have just covered three methods for reconstructing ancestral states for discretely varying characters—and probably some of you have become more skeptical of previous attempts to reconstruct ancestral states! If so, I consider this chapter to be a success. A reasonable level of skepticism is a good thing, especially given the central importance of ancestral states in glimpsing the remote past and inferring adaptation (Coddington 1988, 1994; Brooks and McLennan 1991).

However, a critical perspective on trait reconstruction does not require that we "throw out the baby with the bathwater." Reconstructing ancestral states is not an invalid approach; it just needs to be done carefully relative to the assumptions that are made, as is true of any statistical method. Some reconstruction methods are better suited for some questions than others, and the challenge is to identify the approach that is best suited for the task at hand. Moreover, we need to think about ancestral state reconstruction in more explicitly statistical terms. Or, as noted by Pagel, "The correct representation of a character at each of the internal nodes of a phylogeny is a probability distribution of character states. The probability distribution represents our implicit belief that all states of the character are possible at each node, even if some are more likely or more parsimonious than others" (1994a, 37).

What method should be used? On the whole, maximum likelihood and Bayesian approaches have distinct advantages over parsimony—they take into account branch lengths, they offer a more solid statistical framework for quantifying uncertainty, and it is possible to explore alternative evolutionary models with these methods. Given that branch lengths vary remarkably across primate phylogeny (Purvis 1995; Purvis and Webster 1999; Bininda-Emonds et al. 2007; Arnold et al. 2010), methods that ignore the possibility of more change occurring on longer branches—such as parsimony—are likely to produce estimates that are less accurate than methods that make use of branch lengths. In general, parsimony will provide a compelling set of reconstructions only when rates of change are low, and even then, it requires the user to make assumptions about the costs of different transitions types. As we saw, the step matrix can have a major impact on the reconstructed evolutionary history in a parsimony analysis (Omland 1997b), and it is difficult to decide what this step matrix should be for a given data set.

If we rule out parsimony, the major choice for reconstructing ancestral states involves a decision between maximum likelihood and Bayesian methods. The advantage of Bayesian methods is that they provide a way to assess statistical error across model parameters and phylogenies; in contrast, maximum likelihood reconstructions provide a point estimate based on a particular set of transition probabilities and (usually) a single tree (cf. Lutzoni et al. 2001; Pagel and Lutzoni 2002). While these transition rates might often be valid for a tree, the study by Pagel et al. (2004) of artiodactyl genes shows that different trees can produce different reconstructions of evolutionary history (see also Lutzoni et al. 2001). In the case of primate comparative studies, a recent Bayesian inference of primate phylogeny makes applying Bayesian reconstruction methods relatively easy (see AnthroTree 2.7 and Arnold et al. 2010).

With any of the methods, reconstructions at the root of the tree will tend to be most uncertain (e.g., Schluter et al. 1997). One possibility is to choose a set of taxa so that the node of interest is not at the root (i.e., by broadening the taxonomic scale to include more distantly related species), and to obtain data on as many species in this larger clade as possible (Salisbury and Kim 2001). It might also be useful to choose some additional traits for the clade of interest, with some traits recognized to be evolving slowly and others thought to be evolving more quickly. One could then compare the ability to reconstruct these traits with the trait of interest. If the focal character has probabilities more similar to the slowly evolving characters, this would tend to give more credence to the reconstruction.

Some researchers have used computer simulation to assess how well reconstruction methods estimate nodal values. The basic approach is to simulate data under a known evolutionary model and phylogenetic tree, record the actual values at the internal nodes, and then test how well a method successfully reconstructs the ancestral character states. In one study, for example, the authors compared several reconstruction methods in the context of amino acid sequences (Zhang and Nei 1997). They found that maximum likelihood outperformed parsimony when rates of evolution were high, especially when the phylogeny contained long branches (see Chapter 2). In another study, Salisbury and Kim (2001) examined the effects of taxon sampling and showed that increased sampling improves ancestral state reconstructions (see also Swofford and Maddison 1992).

A number of issues remain outstanding. One issue concerns reconstruction of a trait that also influences diversification rate of the species studied. For example, if a particular character state tends to result in increased speciation, as proposed for some sexually selected traits (Lande 1981; Barraclough et al. 1995; Turelli et al. 2001), estimates of transition rates among character states could be biased (Maddison 2006). These biased estimates could then impact attempts to reconstruct ancestral states using model-based approaches, such as maximum likelihood. A new method makes it possible to investigate both the evolution of the trait and its effect on phylogenetic diversification in a single analysis (Maddison et al. 2007; see also chapter 8). A related aspect involves sampling biases. If species with a particular character state are more often studied and included in the analysis, this can bias interpretations of evolutionary history.

Another issue concerns intraspecific variation. All of the methods just discussed typically assume that the trait of interest is fixed for all individuals of a species. This is often not the case, with some individuals of a species expressing a trait and others failing to do so (Strier 2003). Some parsimony approaches allow for polymorphic characters (e.g., polymorphism parsimony; Felsenstein 1979), and other methods also allow for characters that are coded polymorphically among extant species (BayesTraits; see AnthroTree 1.1). At present, however, we lack a general understanding of how intra-specific variation impacts the reconstruction of ancestral states.

Finally, it is worth remembering that only a subset of methods have been discussed here. Several methods were not discussed because they are no longer widely used, such as distance methods (Zhang and Nei 1997). Other new methods are coming on the scene, such as stochastic character mapping

(Huelsenbeck et al. 2003; Bollback 2006). Overall, this is a rich and important area of evolutionary research. Be sure to check AnthroTree 3.0 for updates to the material in this chapter.

Pointers to related topics in future chapters:

- Integrating reconstruction and functional associations: chapter 9
- Reconstructing ancestral states in human evolution: chapter 12

4 Reconstructing Ancestral States for Quantitative Traits

The previous chapter considered the kinds of questions that can be addressed concerning origins and losses of traits, including questions about how many times traits evolved, the conditions that favored their evolution, and the lineages in which origins and losses occurred. These are all questions involving discretely coded characters for which the presence or absence of a trait in living species can be assigned.

What about traits that vary continuously? These traits, such as body mass, are always present—an animal will always have some mass, and the same is true for many other quantitative traits, such as metabolic rate and home range size. Simply focusing on gains and losses is less appropriate for continuously varying traits. Instead, we might wish to reconstruct the ancestral value for primate body mass or to identify the lineages on which larger amounts of evolutionary change in body mass occurred, compared to other branches on the tree.

A variety of interesting questions can be addressed using reconstructions of continuously varying traits. One study, for example, used reconstruction methods to synthesize the vocalizations of extinct frogs (Ryan and Rand 1995). The authors then played these calls to living species to test hypotheses related to species recognition. Another classic system for ancestral reconstructions involves swordtail fish and their close relatives in the genus *Xiphophorus*, where researchers have investigated female choice and preexisting biases for male "swords" (a colored extension of the caudal fin). By reconstructing the ancestral state in this clade of fish as lacking the swords and then showing that females of species without swords prefer them in males, Basolo (1990) suggested that the ancestor of swordtails had a preexisting bias for swords. This preference may have driven the evolution of the male sword in some lineages.

When one first glances through the comparative biology literature, it appears that several sets of methods are available for reconstructing the values of continuously varying traits. These include generalized least squares (GLS; Martins and Lamont 1998), maximum likelihood (Schluter et al. 1997), squared-change parsimony (Maddison 1991), and independent contrasts (Garland et al. 1999; Garland and Ives 2000). This is somewhat confusing, however, because the methods actually represent different ways of finding the maximum likelihood estimate of a trait at an interior node under a particular model of evolution. In other words, the differences relate mainly to implementation, and as we will see, the methods all give remarkably similar results. Another method, known as linear parsimony, is fundamentally different from the others in that it minimizes the sum of the absolute changes on a tree. In comparison to the other methods discussed, it will tend to concentrate most changes on fewer branches of the tree (Maddison and Maddison 2000). This method is not used widely, however, and so will not be discussed further.

Reconstructing the evolution of continuously varying traits requires methods that differ from those used to study the evolution of discrete traits, hence their separation into this chapter. In addition, this chapter serves as an introduction to using specific models for investigating the evolution of continuously varying traits—a topic that is the focus of the next chapter and integral to many of the chapters that follow. We will find that one particular evolutionary model, known as *Brownian motion*, is the basis for most reconstructions of continuously varying traits. The next chapter expands this framework to consider other models that are available for investigating the evolution of continuously varying traits.

In what follows, I give an overview of methods to reconstruct ancestral states. Throughout, I use a single example from primates to show that the different approaches all give similar results. I then consider another possible application of reconstructing ancestral states, namely, to obtain a mean value for a group of organisms, which is commonly of interest when describing how traits differ in different groups. Averages are often calculated incorrectly as a simple mean for the species in a clade; as we will see, this can produce erroneous results when a particularly species-rich group within the clade exhibits trait values toward one of the extremes in the range of trait values for a group of organisms. This chapter also reviews how evolutionary trends impact the reconstruction of ancestral states. New phylogenetic methods provide a way to investigate trends even in the absence of a fossil record. Last, I provide a brief overview of reconstruction methods in linguistics.

Three brief comments are worth keeping in mind before we proceed. First, the issues of sampling bias raised in the previous chapter are also relevant in this chapter. For example, if large-bodied species are more likely to be studied than small-bodied species, then reconstructions of body mass at the root of the tree may be biased upward. Second, rather than simply producing point estimates, it is more valuable to quantify uncertainty in the estimates by using confidence intervals and similar measures. Examples that follow will reveal why this is crucial for drawing conclusions from the reconstruction analysis. Third, when conducting actual research, it is worthwhile to select clades of particular interest to investigate a priori rather than blindly estimating the trait at every node in the tree. The issue is how can one estimate $n - 1$ nodal reconstructions and $n - 1$ standard errors from only n observed species? This situation suggestions that pseudoreplication exists. In many examples here, I provide reconstructions for all nodes as a way to illustrate the methods, but it is often worth carefully considering which nodes are most important to construct and limiting reconstructions to that subset of nodes.

Maximum Likelihood Ancestral State Reconstructions

Schluter et al. (1997) used maximum likelihood methods to reconstruct the evolution of continuous characters under a Brownian motion model of evolution. Brownian motion is a common model that underlies many comparative methods for continuously varying traits, and this is a good point to introduce this important concept. Brownian motion assumes that evolutionary changes in a trait are randomly distributed around a mean of zero, independent of previous changes and changes on other branches, and that larger changes are more likely to occur on longer branches (more specifically, that variance accumulates proportional to time; see also chapter 5). Succinctly, Brownian motion emulates a random walk of the character along the different branches of the phylogeny. With this evolutionary model and a dated phylogeny, one can obtain values at interior nodes that make the tip data most likely.

Let us apply this approach to a real-world example from primates involving a well-studied morphological measure, the intermembral index (IMI). The IMI is calculated as the ratio of forelimbs to hindlimbs, multiplied by 100; it is important in studies of primate morphology, because it correlates with locomotor behavior (Napier and Walker 1967; Napier 1970; Martin 1990). More specifically, the IMI approximates 70 in species that exhibit vertical clinging and leaping, 70–100 in quadrupedal primates, and 100–150 in primates that exhibit suspensory locomotion (Martin 1990). Given this strong association with locomotor mode, the IMI has been used to reconstruct locomotor behavior in the fossil record (e.g., Napier and Walker 1967; Jungers 1978; Martin 1990). Apes exhibit a wide variety of locomotor styles, which range from sus-

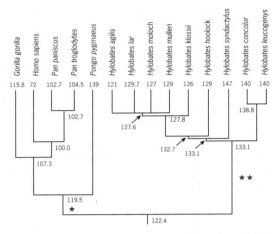

Figure 4.1. Reconstructing the intermembral index in apes using maximum likelihood. Analysis was run using a recent phylogeny (Bininda-Emonds et al. 2007) and the computer program ANCML (Schluter et al. 1997). Values at nodes are reconstructions, while values at the tips are the data for extant species. The analysis takes into account branch lengths. The stars indicate branches relevant to the discussion of branch lengths in the text.

pensory specialists in the small-bodied brachiating gibbons (*Hylobates*, also known as lesser apes) to obligate terrestrial bipeds in larger-bodied humans. Hence, the apes represent an interesting clade for investigating the evolution of anatomical features related to the IMI, and that is what we will do here.

Maximum likelihood estimates for the IMI are shown in figure 4.1, and data and instructions for running the analysis are provided in AnthroTree 4.1. Ancestral state reconstructions are equal to or greater than 100 on all internal nodes, with the root node reconstructed as having an IMI of 122.4. Based on this point estimate, we would conclude that the ancestral primate was a brachiator. Some branches on figure 4.1 also reveal how reconstructions can provide insights to evolutionary rates (see chapter 5). For example, there appears to be a high rate of evolution on the lineage leading to humans, where the evolution of bipedalism has clearly impacted the relative lengths of forelimbs and hindlimbs (see also chapter 12). In addition, the difference between the root node and the ancestor of the great apes (indicated by a single star on figure 4.1) is smaller than the difference between the root node and the ancestor of lesser apes (indicated by two stars). This is due to the shorter branch connecting the great apes and the common ancestor of all apes at the root of this tree.

These estimates give some sense of the evolution of IMI among the apes,

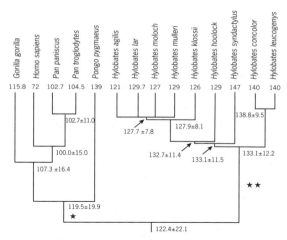

Figure 4.2. Reconstructing the intermembral index in apes using maximum likelihood. Reconstruction of the intermembral index and 95% confidence intervals based on Schluter et al. (1997), with analyses conducted using ANCML (see AnthroTree 4.1 and 4.2).

but how much confidence can we place in the reconstructions? To interpret the results, it is essential to place confidence intervals on the reconstructed values. For example, at the root we estimated a value of 122, which would seem to suggest that the ancestral ape was a brachiator. But our conclusion would be weakened considerably if the confidence interval was ± 100. In such a case, we could not confidently place the ancestor into categories of brachiation, quadrupedalism, or even vertical clinging and leaping, rendering the reconstruction effectively uninformative (but not useless—quantifying uncertainty is always useful in comparative studies).

Schluter et al. (1997) provided a way to place confidence limits on maximum likelihood ancestral state reconstructions. When they applied the method to investigate lizard morphology and bird wing lengths, they found that the method produces extremely wide confidence intervals. In many cases, the confidence interval encompassed all of the variation in the extant species for that clade! A similar effect was demonstrated by Garland et al. (1999), who provided an approach to obtaining confidence intervals that is based on the method of independent contrasts (see chapter 7).

To explore this issue further, I used the method of Schluter et al. (1997) to place 95% confidence intervals on nodal values of the IMI in primates (figure 4.2; AnthroTree 4.2). This produced a confidence interval of 100.3 to 144.5 for the root node. Two important conclusions about the root node can be drawn from this analysis. First, the root node falls within the brachiating category (i.e., even the lower bound is greater than 100, although only by a hair). Second, the value for humans lies below the confidence interval for all ancestral nodes, including the root. Thus, the method is able to detect the large amount of morphological change that occurred on the evolutionary lineage leading to humans. Interestingly, *Hylobates syndactylus*, the largest bodied gibbon with an IMI of 147, lies just outside the upper confidence limit.

How well do these methods work? Emília Martins (1999) addressed this question using a computer simulation approach. Her analyses revealed that methods for placing standard errors on reconstructed values are often narrower than they should be, and should therefore be viewed as a minimum estimate of the error in a reconstructed state (see also Rohlf 2001, who provided corrected formulae). More encouragingly, however, the analysis revealed that ancestral states were often reconstructed accurately (at least in the cases she simulated, e.g., without evolutionary trends in the trait values). This study is discussed again below, as it compared different methods of reconstructing continuously varying traits.

Squared-change parsimony. Another approach to finding the maximum likelihood estimate at a node is to use the method of squared-change parsimony (Maddison 1991; Schluter et al. 1997; Maddison and Maddison 2000). Squared-change parsimony minimizes the sum of squared changes on a tree (Huey and Bennett 1987; Losos 1990; Maddison 1991), as illustrated in figure 4.3. Maddison (1991) provided a recursive algorithm to obtain the reconstructed values by moving through the tree, similar to the parsimony algorithm described in the previous chapter for reconstructing discrete traits. Branch lengths are taken into account by minimizing the sum of squared changes divided by the associated branch lengths; in effect, more change will be reconstructed on longer branches using this algorithm.

Squared-change parsimony has played a prominent role in comparative studies of physiology and ecology. In a classic study, Losos (1990) used squared-change parsimony to examine character displacement in lizards (see also Butler and Losos 1997). In another early and influential comparative study, Huey and Bennett (1987) used squared-change parsimony to reconstruct ancestral states in thermal preferences and sprinting speed in lizards, and then used these reconstructions to examine whether these traits have changed in a correlated fashion. The authors found evidence for correlated evolution. A study of the concentration of substances in the blood (plasma

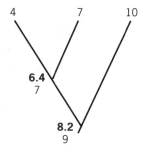

Figure 4.3. Squared change parsimony. Squared-change parsimony reconstructs ancestral states by mini-mizing the sum of the squared changes along the branches of the tree. At the two internal nodes on this tree, the upper numbers reflect values that minimize this sum, giving a parsimony score of 12.6. If instead we used the lower numbers (7 and 9) given for the two nodes, the amount of squared change across the tree would be higher (14); it therefore would be a less parsimonious reconstruction.

osmolarity) serves as an excellent introduction to squared-change parsimony more generally and to methodological issues involving phylogenetic uncer-tainty and derivation of confidence intervals on reconstructed states (Garland et al. 1997).

When the reconstructions are weighted by branch length, the squared-change parsimony estimates are equivalent to a maximum likelihood estimate under Brownian motion (Maddison 1991; Schluter et al. 1997; Maddison and Maddison 2000). Thus, it is not surprising that squared-change parsimony produces estimates of ancestral states that are identical to those found using maximum likelihood (AnthroTree 4.3).

In addition, branch lengths can have a major impact on reconstructed values. Indeed, when branch lengths are assumed to be equal throughout the tree, an equal amount of evolutionary change would be reconstructed in the two starred branches in figure 4.1. This affects the reconstructed root node and surrounding values: with equal branches, the reconstructed ancestral value for the IMI in great apes is 128.2, for lesser apes is 136.6, and for the root node is 132.4 (giving change of 4.2 in each of the basal branches). Setting branches to be equal has been a common practice, especially when branch length estimates are unavailable or deemed unreliable. This example shows that such assumptions should be made carefully because they can impact the reconstructions (see chapter 5 for ways to investigate this assumption using the branch length scaling parameter κ).

Another important question concerns the roles of fossils. If we include fossils in the reconstructions, does this change ancestral state reconstruc-tions and help to narrow the confidence intervals? To investigate this possibility in the context of IMI evolution in apes, I added *Dryopithecus laietanus* along the branch leading to *Pongo pygmaeus* (at 9.5 million years ago, the midpoint of the branch) and set its IMI to 114 (Moyà-Solà and Köhler 1996). Using squared-change parsimony, the root node estimate became some-what smaller, to 117.1. While the confidence intervals did not narrow mark-edly, they began to encompass values under 100 (95.3 to 138.9), which would slightly weaken support for a brachiating ancestor of the apes. If instead this fossil was placed on the midpoint of the branch from the root to the great apes (i.e., at the single star in figure 4.1), the confidence intervals narrow con-siderably (estimate, 116.1; 95% confidence interval, 105.2–127.1), and the varia-tion is maintained firmly in the range of a brachiating ancestor to the great apes.

It is important to keep in mind that this analysis treats the IMI for the fossil taxon as if is measured without error. The original paper reveals that this as-sumption is unlikely to be true, given that the complete skeleton is unavailable (Moyà-Solà and Köhler 1996). Incorporating error, either in the fossil or in any of the other species on the phylogeny, would tend to widen the confidence intervals on reconstructed states. Thus, it is not always the case that adding a fossil to the analysis will make ancestral state reconstructions more certain, especially when character states in the fossil are also uncertain.

Generalized least squares. A number of authors have placed ancestral state reconstruction in a more flexible statistical framework based on generalized least squares (Martins and Hansen 1997; Pagel 1999a; Garland and Ives 2000; Rohlf 2001). A major advantage of GLS is that it becomes possible to investi-gate the effects of intraspecific variation, including measurement error, and it is possible to implement a wider range of evolutionary models (see Martins and Lamont 1998).

Martins and Lamont (1998) applied the method to study the evolution of displays in the iguana genus *Cyclura*. They showed how to put confidence in-tervals on reconstructed values, and they argued that incorporation of mea-surement error or intraspecific variation is important for reconstruction methods. They also used standard errors to identify branches where signifi-cant amounts of change occurred. In evolutionary anthropology, Pagel (2002) used GLS to reconstruct the evolution of cranial capacity in hominins. We will return to this example in a later section of this chapter that discusses how to reconstruct trait values when evolutionary trends have occurred. Return-ing to the example used above, I applied GLS to estimate the root node for the

IMI in apes using the program BayesTraits, version 1.0 (see also Pagel et al. 2004; AnthroTree 4.4). This produced an estimate of 122.4, which is consistent with the other maximum likelihood estimates given above.

Another approach is based on a method known as "independent contrasts" (Garland et al. 1999), which is mathematically equivalent to GLS (see chapter 7). This method reconstructs ancestral states using only descendent lineages from a node for the estimate, that is, those pointing towards the "fork tines" of the phylogeny (Felsenstein 1985b). Therefore, it does not make use of all the information on the phylogeny above and below each node and is more accurately termed a "local estimate" (Garland et al. 1999; Rohlf 2001). At the root node, however, the estimate is based on all descendent lineages and thus is "global."

I applied Garland et al.'s approach (1999) using Mesquite, version 2.5 and the PDAP module in Mesquite, version 1.08(2). This produced an ancestral value for the IMI in the ancestral ape of 122.4, with standard error of 10.23. The confidence intervals again ranged from 100.3 to 144.5.

Bayesian methods. A Bayesian implementation of GLS for continuous traits is available, based on similar approaches used for discrete traits (Pagel et al. 2004). In this implementation, the parameters used in the GLS model are sampled by Markov chain Monte Carlo, just as they are in Bayesian analyses of phylogeny (see chapter 2) or reconstruction of discrete traits (chapter 3). Hence, the Bayesian approach accounts for uncertainty in the model parameters (i.e., mapping uncertainty; Ronquist 2004) and provides a way to put credible intervals on the reconstructions and model parameters. In addition, one can sample a posterior probability of phylogenetic trees to control for phylogenetic uncertainty.

Using this method for the IMI data in apes, the ancestral state was estimated to be 122.9 (figure 4.4; AnthroTree 4.5). After a "burn-in" of fifty thousand generations (e.g., see figure 2.7), I sampled the Markov chain every hundred generations for more than one million generations, resulting in ten thousand samples of the IMI at the root node. These samples of the posterior probability distribution ranged from 73.6 to 181.7, although most values clustered around the mean. I used this distribution to place a 95% credible interval on the mean, which ranged from 100.4 to 144.8. As with the other analyses in this chapter, humans and *Hylobates syndactylus* fell outside these limits. This variation reflects mapping uncertainty rather than phylogenetic uncertainty, as only a single phylogeny was used.

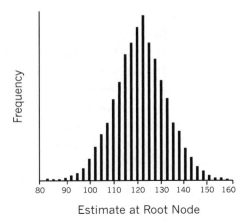

Estimate at Root Node

Figure 4.4. Bayesian reconstructions of the intermembral index in apes. Estimates were obtained from the computer program BayesTraits (Pagel and Meade 2004; 2006a). The rate deviation was set to ensure that the acceptance rate fell within the range of 0.2 to 0.4 (mean = 0.287). A flat prior distribution was used and the first 50,000 sampled estimates were discarded as burn-in. The standard deviation of the estimates was 11.3, and 95 percent of the estimates fell within the range 100.4 to 144.8—almost identical to the confidence limits found by the other methods described in the chapter.

Obtaining a Phylogenetic Mean

In addition to viewing the reconstructions as character states of extinct ancestors, we can also treat a root node reconstruction as a phylogenetically correct estimate of the mean for a group of organisms (Garland et al. 1999). Such a measure might be useful, for example, when describing the average value of a trait in a clade, or to document variation among subclades on a phylogeny.

The standard way to calculate a mean value of a group of organisms is to simply average the values of the species in the clade of interest—in the example given here, one would calculate an average IMI across all the apes. This is, of course, how people have traditionally calculated averages for a group of organisms, with little regard for the phylogenetic relationships of species within the sample. This approach might be flawed, however, if a particularly large subclade has trait values at either the upper or lower ends of the range for that group; these many data points could tilt the mean toward a particularly speciose clade. In the apes shown in figure 4.1, for example, there are many species of gibbons, and they are smaller in body mass than the great apes. Using a simple mean calculated across species, the average body mass of all apes is close to 22 kilograms (AnthroTree 4.6). By using the estimated value at the root node weighted by branch lengths as a phylogenetically correct esti-

mate of the mean, the *phylogenetic mean* is 33 kilograms, which is 50 percent higher! Thus, the phylogenetic approach avoids the over-representation
of species in larger subclades, such as the smaller-bodied gibbons among the
apes. However, the 95% confidence interval on the phylogenetic mean is
fairly wide, ranging from 11.1 to 54.8 kilograms and encompassing the simple
average across species.

Evolutionary Trends and Reconstructing Ancestral States

Directional evolutionary trends pose a challenge when reconstructing continuous characters using phylogenetic methods. If there has been a trend in
evolution, it will be difficult—if not impossible—to accurately reconstruct
the ancestral states using only statistical methods (Garland et al. 1999). This,
of course, makes good sense. When evolutionary trends occur, the values at
the tips of the tree should be representative for today's species, but not for
those in the past. More importantly, we know good examples of evolutionary trends, such as Cope's rule that lineages tend to evolve toward larger body
mass (Stanley 1973; Peters 1983).

The challenges posed by evolutionary trends were demonstrated in a study
of a bacteriophage phylogeny and traits of the bacteriophage reconstructed
on the tree (bacteriophages are viruses that infect bacteria). In this study,
Oakley and Cunningham (2000) used an experimentally generated phylogeny to investigate the ability of several methods, including squared-change
parsimony, to reconstruct ancestral states. Virus speciation was directed by humans in the laboratory, with populations split experimentally, eventually leading to eight descendent populations from an ancestral viral stock (figure 4.5).
The speciation events were imposed at equal intervals, resulting in equal
branch lengths. For the traits, the authors used frozen stocks of the bacteriophage to measure viral characteristics on plates of host bacteria; this was done
in the eight descendent populations and, to obtain the "actual" values at each
internal node, for the frozen stocks at the split points on the tree. The viral
lineages underwent strong directional selection in several attributes. After
reconstructing ancestral states, analyses revealed that these estimates were
wildly incorrect (figure 4.5). A computer simulation by these authors further
demonstrated that as the strength of a directional trend increases, the ability
to reconstruct ancestral states declines precipitously.

Another study also found that strong directional selection reduces the ability of reconstruction methods to produce accurate estimates, although the

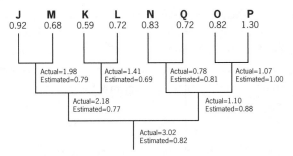

Figure 4.5. Empirical test of reconstruction methods using a known phylogeny. The tree shows the actual evolutionary history and ancestral states for plaque diameter among eight lineages of bacteriophages, plus the reconstructed ancestral states based on squared change parsimony. (Redrawn from Oakley and Cunningham 2000.)

results were not as disheartening as in the previous example. Webster and Purvis (2002b) turned to the fossil record in their evaluation of reconstruction methods by using the well-preserved remains of Foraminifera from the Pleistocene fossil record. In their dataset, all but one of the characters showed increases in size in the fossil record. They compared six different methods for reconstructing ancestral states, including three methods that gave confidence intervals. When comparing the reconstructions to the actual data from the fossil record, they found that all methods produced inaccurate estimates of ancestral states. Importantly, however, confidence intervals bracketed the actual value in 92 percent of the cases, possibly because the intervals were very wide. Another study that compared reconstructions to fossil values—in this case involving molar teeth dimensions in carnivores—found that confidence intervals bracketed the observed value on four nodes (figure 4.6).

The evolution of life has likely been characterized by trends in many traits (Alroy 1998; McShea 1998), and as we just saw, a trend can reduce the performance of phylogenetic methods to reconstruct ancestral states. Clearly, it would be useful to identify trends and take them into account when reconstructing character states (see also Garland et al. 1993). Two main solutions are available to identify trends and incorporate them into ancestral reconstructions. One solution is to use fossil evidence. Finarelli and Flynn (2006) showed how incorporation of fossil data can improve character reconstructions—this is true not only in terms of the point estimate for a particular node, but also for the confidence intervals that can be placed on these estimates (see also Oakley and Cunningham 2000; Organ et al. 2007). The ex-

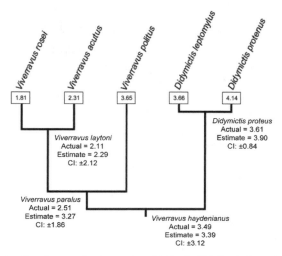

Figure 4.6. Reconstruction of ancestral states compared to fossils. This figure shows data on the area of the first lower molar from Polly (2001). All species represented are fossil Carnivora. Reconstructed values were estimated using squared change parsimony. Confidence intervals (CIs) are 95% and based on Martins and Hansen (1997).

ample from above using IMI data estimated from *Dryopithecus* provides an example of one way to incorporate fossil evidence. Whereas incorporating fossil evidence had little effects on the estimate of IMI in apes, it should have a bigger effect when fossils have traits that are very different from those of living species.

Second, it is possible to investigate evolutionary trends using a dated phylogeny and trait values, and even to incorporate a trend into the ancestral state reconstruction (Pagel 1997, 1999a, 2002). This "directional" method requires branch length information, but cannot be used with an ultrametric tree in which all the tips are lined up at the end of the tree; in other words, the distance from the root to each of the tips should vary across lineages, as would usually be the case with a phylogeny where branch lengths reflect the amount of genetic (or other) change on those branches (or on a tree that includes fossil taxa). With such a tree, one then tests whether tips that are closer to the root—that is, have a shorter path length from root to a living species—have consistently smaller or larger values, compared to longer paths on the tree (figure 4.7; see also Webster and Purvis 2002b). The underlying logic is that if a trend exists, then larger values of the trait should be found on longer paths of the tree. Thus, the directional method assumes that rates of genetic change

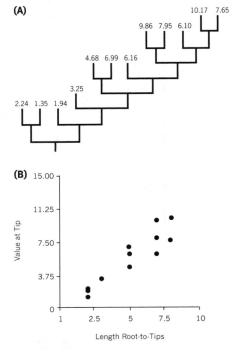

Figure 4.7. Detection of evolutionary trends using generalized least squares (GLS). (a) Phylogeny and data are a fictitious example to illustrate the principle behind Pagel's directional test (1997; 1999a). Branch lengths are set to 1, and values at the tips are a function of root-to-tip distance plus a random value. (b) In testing a directional model using GLS, the main issue concerns the distance from the root to the tips and how this relates to tip values. If a trend exists, then longer paths represent more time for evolutionary change to occur, and should therefore be associated with larger values at the tips. Thus, we expect an association between root-to-tip distance and trait values, as found for this dataset ($r = 0.93$; $n = 12$; $p < .0001$). The actual analysis in GLS suggests that a directional model has a significantly higher log-likelihood than a random walk model (−16.06 vs. −24.85; $p = .00003$ in a likelihood ratio test; for data and analysis see AnthroTree 4.7).

used to generate the tree correlate with rates of change in the traits of interest. The directional model can be compared to a random walk model using a likelihood ratio test (LRT), which compares twice the difference in likelihoods to a χ^2 distribution (see AnthroTree 4.7). If the directional model offers a significant improvement, the ancestral states can be estimated with the directional model (Pagel 1999a).

It would be valuable to apply the directional method to questions in primate evolution where trends are known to have occurred based on fossil evidence, such as increases in body mass over primate evolution (Gebo 2004). In fact, Pagel (2002) did exactly this on an even better example—the evolution

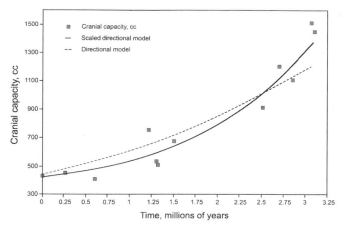

Figure 4.8. Brain size evolution in hominins. Plot shows observed cranial capacity among a sample of hominins spanning three million years, and the fit of two models of directional evolutionary change. In the standard directional model (dashed line), the cranial capacity in the most recent hominins (*Homo*) is underestimated. The fit is improved with a scaled directional model (solid line), which estimates a branch-length scaling factor ($\delta = 1.55$; see chapter 5). The directional models offered a significantly better fit than a random model, which is not shown here. (Redrawn from Pagel 2002.)

of hominin cranial capacity. He found that the directional model fit the data much better than a random walk model (figure 4.8), thus confirming that cranial capacity has increased over time (or, more accurately, confirming that the method can detect increases that were already known to exist from the fossil record). For another approach to fitting a directional model to fossil taxa, see Hunt (2007).

Other researchers have applied the directional method, with more mixed success. In the example given above involving fossil Foraminifera, Webster and Purvis (2002b) applied the method to detect trends in the evolution of size and shape variables. In this case, the directional model failed to detect the trend observed in the fossil record (see also Moen 2006). In a recent simulation study, Laurin (2010) found that the directional method shows fairly good statistical properties, compared to several other approaches to detecting evolutionary trends.

Linguistic Reconstructions

For comparison to quantitative approaches derived from biology, it is worthwhile to briefly touch on how languages are reconstructed. Reconstruction of "ancestral states" plays a central role in historical linguistics. In studies of language relationships, it is widely viewed as essential to reconstruct the ancestral language, or *protolanguage*. The protolanguage can itself be used to make inferences about the behavior and ecology of its speakers.

Protolanguages are reconstructed using the linguistic equivalent of the comparative method (Fortson 2004). Although many different approaches have been used in historical linguistics, the procedures can be described roughly as follows. One starts by compiling words with the same meaning in different languages, with the goal to identify *cognates*—words with a common origin—in two or more languages. Cognates are usually recognized by comparing the phonetic structure of basic vocabulary terms, such as numbers and body parts. Loanwords, which are borrowed from another language, are avoided. Cognate identification therefore involves homology assessment, as would be done with base pair alignment in molecular systematics.

Next, regular sound changes are established among the cognates. This process is often based on detailed knowledge of the languages and how specific sounds correspond across languages, and effort is generally placed on minimizing the number of sound changes. This step gives insights to how sounds have changed since the languages diverged, and from this procedure one can begin to make inferences about the ancestral word for a particular meaning in the group of languages being compared.

Repeating the procedure across many meanings and languages, the historical linguist hopes to see consistency in the grouping of languages and in the sound changes that have occurred. This consistency provides greater support for the proposed sound changes and the linguistic reconstructions. As an ancestral state, protolanguages are usually reconstructed as simplified versions of the descendent languages. Of course, it is important to keep in mind that the language spoken by the ancestors is likely to be as rich in complexity as the languages upon which the reconstruction is based.

The reconstruction of a protolanguage not only gives a sense of how people communicated; it also informs us what they were communicating about (Fortson 2004). Specifically, it can provide insights to the cultural behaviors, the geographic "homeland" of the language, the environment in which the speakers lived, and even the factors that might have led that particular language to expand and leave numerous daughter languages, while other languages went extinct. In the case of proto-Indo-European (PIE), for example, historical linguists have used reconstructions to document evidence for a strongly hierarchical society, a well-developed economic exchange system, sophisticated agricultural practices, and wheeled vehicles pulled by animals (Fortson 2004; Anthony 2008). The question of the PIE homeland has prompted much interest but less resolution.

The process of reconstructing protolanguages is a complex task, and it requires extensive training and familiarity with the languages involved. While the methods may often produce ancestral languages that are approximately correct, the methodology is less quantitative than one might prefer, given existing computing power and the desire for repeatability in science. Moreover, many of the reconstructions fail to make use of the actual branching structure of the descendent languages, especially in terms of the amounts of time that have separated different languages and their ancestors (i.e., branch lengths). Recent methods based on phylogenetic approaches in biology are making headway in linguistic research and promising to revolutionize how language history is reconstructed (Gray and Atkinson 2003; Pagel et al. 2007; Gray et al. 2009). Importantly, phylogenetic methods appear to be robust to borrowing of words among linguistic groups (Greenhill et al. 2009), although inferences may be stronger when "core vocabulary" and other terms that are resistant to borrowing are used.

Summary and Synthesis

One remarkable aspect of methods for reconstructing continuous traits is that they produce very similar results, at least among the methods discussed here. This is easily seen in the reconstructions of anatomical characteristics in apes, shown in figures 4.1, 4.2, and 4.4, with most of the values clustered around 122. Another point worth amplifying involves the wide confidence intervals, as shown in figures 4.2 and 4.4. That humans (and one gibbon species) fall outside this limit is unusual and interesting, as previous applications of these methods found that confidence intervals encompass the values for all living species in the clade (Schluter et al. 1997; Garland et al. 1999). Clearly, the selection pressures on human bipedality have produced morphological characters that differ greatly from our last common ancestor with quadrupedal chimpanzees and bonobos (see also chapter 12). As one would hope, the major morphological changes associated with bipedality are captured in the reconstructions of trait evolution. In addition, reconstructions are influenced by estimates of branch lengths, thus emphasizing the importance of generating trees with branch lengths for comparative studies (and highlighting the risks of blindly assuming branch lengths to be equal without testing this assumption explicitly; see chapter 5).

As with discrete characters, methods for reconstructing continuously varying traits will depend on whether all relevant species are included in the sample. This sampling bias problem is potentially serious, because it is usually impossible to obtain data on all relevant species. Moreover, as noted in the case of discrete traits, it will be important to integrate models of evolutionary change, diversification in relation to particular character states, and ancestral state reconstruction (Maddison 2006; Maddison et al. 2007).

Another important issue concerns the quantification of confidence on ancestral reconstructions of continuous characters. The issue here concerns the ability of existing methods to accurately estimate both the reconstructed values and error on these estimates. As noted above, Martins (1999) developed a simulation approach to investigate the ability of different methods to accurately reconstruct ancestral states. Specifically, she compared the statistical performance of squared-change parsimony, maximum likelihood, and GLS under two different models of evolution. She also assessed the accuracy of standard errors on estimated values for maximum likelihood and GLS. The simulations revealed that the methods performed reasonably well when data were simulated using a Brownian motion model of change, especially for more recent ancestors. Under a situation of stabilizing selection, however, all methods fared poorly—including the implementation of GLS that was designed for this model of trait evolution!

What is the future likely to hold for reconstructing continuously varying characters? Two areas in particular are likely to bear fruit in the near future. First, as we will see in the next chapter, a number of statistical models have been developed to describe the evolution of continuous characters, including effects of directional and stabilizing selection (Hansen and Martins 1996). At present, however, most of the widely used approaches to reconstructing continuously varying characters assume that traits evolve by one model of evolution, namely Brownian motion. As with attempts to integrate trends and ancestral state reconstruction, incorporating appropriate models of evolution into trait reconstructions is likely to be important in future work, especially in terms of placing confidence limits on the results. Second, computer simulations are needed to more fully evaluate approaches for reconstructing ancestral states and the confidence intervals placed on these estimates.

Pointers to related topics in future chapters:

- Brownian motion and evolutionary models: chapter 5
- Rates of evolution: chapter 5
- Independent contrasts and GLS: chapter 7
- Integrating ancestral state reconstruction and functional relationships: chapter 9
- Evolutionary change in the IMI in humans: chapter 12
- Directional model of evolution: chapter 12
- Reconstructing ancestral states in human evolution: chapter 12

5 Modeling Evolutionary Change

The previous chapters considered methods to reconstruct the evolutionary history of traits in a group of closely related organisms. This process involves estimating the states of characters in extinct species, as represented by internal nodes on the phylogeny. Reconstructions are essential for describing a history that, for many types of characters, would otherwise be forever lost to the sands of time. These narratives are rooted in explicit evolutionary models. As such, they can be evaluated statistically, and thus lend credence (or doubt!) to particular evolutionary scenarios.

While these details of history are important, it can be equally important to understand the tempo and mode of evolutionary change. We can think of tempo as generally referring to the rate of evolutionary change, while mode refers to the mechanisms underlying change. Thus, given a dated phylogeny and data on a continuously varying trait, we might ask, "Did evolutionary change in the trait occur early in the radiation of a clade, late in its evolution, or evenly throughout the clade?" Another important and related question is, "Do the traits show an association with phylogeny?" Some traits show strong phylogenetic association, meaning that species that shared an ancestor more recently have more similar trait values (figure 5.1). Measuring this *phylogenetic signal* is an important step in applying methods that take phylogeny into account.

More generally, we can ask, "How can we characterize the evolutionary process, and what new insights does this provide for a comparative study?" Indeed, evolutionary models are essential to all that is discussed in this book, including the previous chapters on reconstructing ancestral states. Thus, for reconstructing discrete traits in chapter 3, different "transformation types" can be thought of as representing different models of trait evolution in parsimony analyses, while maximum likelihood approaches provide an explicit framework to investigate whether rates of gains and losses are equal. Similarly, the reconstruction methods discussed in chapter 4 were based on a Brownian motion model of evolution.

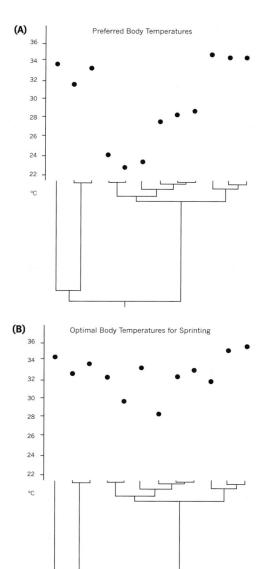

Figure 5.1. Example of phylogenetic signal. Plots show two examples of phylogenetic signal using Australian skinks as an example. In (a), preferred body temperatures show strong phylogenetic signal; more closely related species have more similar trait values. In (b), optimal body temperatures for sprinting show weaker phylogenetic signal. Based on a randomization test, results in (a) are statistically significant, while results in (b) are not statistically significant. (Redrawn from Blomberg et al. 2003.)

Table 5.1. Three major models of evolutionary change.

Model name	Description
Brownian motion	A model of randomly fluctuating selective regimes or neutral evolution. Trait change is drawn from a normal distribution with variance proportional to branch length. Speciational change can be implemented by setting branch lengths to be equal.
Ornstein-Uhlenbeck (OU)	A model of stabilizing selection. Trait change is random and can be either positive or negative but is more likely to be drawn toward a central value.
Niche-filling	A model of adaptive radiation. New niches arise and species closest in niche space diversify to fill empty niches, which results in greater change early in the radiation of a group of organisms.

A number of evolutionary models have been proposed, where "evolutionary model" refers to an explicit framework for considering how traits change over time. After covering some preliminaries that are essential to all that follows, this chapter summarizes three widely discussed general evolutionary models (table 5.1). I then review some quantitative measures of the evolutionary process and phylogenetic signal, including additional, flexible ways to model evolutionary change for continuously varying characters. The chapter concludes with a discussion of evolutionary rates.

Preliminaries

Before digging into the models and how to investigate them, two important issues deserve attention. First, the concept of branch lengths will arise multiple times in this chapter, because the methods typically require a plausible set of initial branch lengths. Many of the parameters discussed can actually be used to transform these "starter" branches to better represent amounts of evolutionary change in the traits of interest; indeed, many methods require that the branches reflect expected amounts of evolutionary change (e.g., Garland et al. 1992). It is useful to recall that branches can be represented in two ways: as units of absolute time since species shared an ancestor (which produces a dated phylogeny or ultrametric tree, with all extant lineages lined up at the "top"; see figure 5.1), or as amounts of evolution in the characters used to build the tree, which typically results in variable distances from the root to the tips (a phylogram; see figure 3.8). Most comparative methods can be used to examine trait evolution on either type of tree. Using a phylogram might be important if we expect variable rates of trait evolution throughout the tree.

For example, lineages of organisms with shorter generation times might experience higher rates of evolution in both the traits used to construct the phylogeny and in the phenotypic traits studied with the phylogeny. The critical assumption is that rates of morphological and molecular change are correlated, which can be tested statistically (Omland 1997a; Bromham et al. 2002).

Second, a major challenge in evolutionary biology is to identify the *process* of evolution from the *pattern* that we observe (Simpson 1944; Stanley 1979; Eldredge and Cracraft 1980). We will see in the next section that the models are formulated in terms of specific processes of evolution. In general, however, the parameters used to estimate these models in real data are often unable to distinguish among the possible evolutionary processes that generated the patterns; in other words, we cannot usually go from pattern to process with comparative data and a phylogeny. For perspectives on this important issue of pattern and process in comparative biology, see Felsenstein (1988), Leroi et al. (1994), Hansen and Martins (1996), Blomberg and Garland (2002), Hohenlohe and Arnold (2008), and Revell et al. (2008).

Three Evolutionary Models for Continuous Characters

Brownian motion. Brownian motion is a good starting point for discussing evolutionary models. The name comes from physics, where Brownian motion describes the random movement of particles suspended in a gas or liquid. The Brownian motion model states that trait change along any given branch is independent of previous changes and changes on other branches, and that larger changes (both positive and negative) are more likely to occur on longer branches, with the expected variance of trait change proportional to branch length (see AnthroTree 5.1). This model further assumes that rates of evolution are constant over time, with neither increased nor decreased rates of change as one approaches the extant species at the tips of the tree. Methods can test for different rates of character evolution across the tree (e.g., Garland 1992; O'Meara et al. 2006).

One common point of confusion is that Brownian motion is viewed as a random or neutral model of evolution, and this seems to conflict with our aim of studying adaptive function—you might be thinking, "surely an adaptive trait cannot be neutral, and so Brownian motion models must be incorrect!" As just noted in the discussion of pattern versus process, however, Brownian motion should be seen as simply describing the distribution of observed trait changes rather than the mechanism that produced these changes, and in fact it is consistent with many models of adaptive evolution (Hansen and Martins 1996; O'Meara et al. 2006). The driver of these supposedly "random" changes can be natural selection, with *changes in the selective regime* assumed to occur independently of previous changes and to be more common on longer branches. Thus, Brownian motion should not necessarily be viewed only as a neutral or random evolutionary model but rather as a simplifying assumption that, in many cases, is a perfectly adequate description of adaptive evolutionary change. Calling the model a "constant variance process" (e.g., Freckleton et al. 2002; Pagel 2002) might avoid confusion concerning the "randomness" conveyed by the term Brownian motion. Because the term "Brownian motion" is used so widely, however, I stick with the established terminology in this book.

The Brownian motion model of evolution is important to comparative biology because it forms the basis for a number of widely used statistical methods. We saw this when reconstructing continuously varying traits in chapter 4, and we will see it again when we learn about phylogenetically independent contrasts in chapter 7. Thus, in addition to possibly being an accurate description of the pattern of trait evolution, the Brownian motion model provides a foundation to statistically test evolutionary hypotheses. Comparative data and phylogenetic branch lengths often can be transformed to meet the assumptions of Brownian motion (or more accurately, transformed so that the assumptions of Brownian motion are not violated statistically, see Garland et al. 1992).

An important variant on the Brownian motion model is the *speciational model* of evolution (Garland et al. 1993). Under this model, all changes are assumed to occur at speciation points on the tree and to be drawn from the same underlying distribution (i.e., equal variances), and recent studies provide evidence for speciational evolution (Atkinson et al. 2008; Venditti and Pagel 2010). To illustrate the idea that we can adjust branch lengths to reflect different evolutionary models, speciational change can be represented as a tree with equal branch lengths (e.g., see figure 4.7 and AnthroTree 5.1). Understanding these sorts of transformations is crucial to many modeling approaches in comparative methods, the crucial point being that the branches can be transformed to reflect the expected amount of phenotypic change under different evolutionary scenarios (see Mooers et al. 1999).

The name "speciational" is somewhat misleading unless extinction rates are zero and all extant species in the clade of interest are sampled (an unlikely scenario in almost all biological systems; see chapter 8). This is because many speciational events will be "hidden" along branches of a tree representing extant species, that is, when lineages descending from those events are lost through extinction (see figure 3.1). In addition, it is important to distinguish between speciational and punctuational evolution (Rohlf et al. 1990). In speciational evolution, it is assumed that change occurs on both branches that emanate from a node; by comparison, punctuational evolution usually refers to change on only one of the two branches (see also Garland et al. 1993). Not all authors follow this distinction, and so the terminology used here might differ slightly in the writings of different comparative biologists.

Ornstein-Uhlenbeck and stabilizing selection. A second influential model focuses on stabilizing selection and is known as the Ornstein-Uhlenbeck (OU) model (Felsenstein 1988; Garland et al. 1993; Hansen and Martins 1996; Hansen 1997; see also the historical overview in Lavin et al. 2008). It arises from the reasonable belief that once an adaptive complex of traits has evolved, selection will act to stabilize trait values—provided, of course, that the ecological context and selective regime remains similar.

The OU model envisions that a trait has an optimum value that can itself "wriggle about" over macroevolutionary timescales due to unmeasured variables or stochastic effects. Specifically, the trait wanders around an optimum following a Brownian motion process, but with a restraining force (α_{OU}) that tends to pull the trait back to the optimum (figure 5.2 and AnthroTree 5.1). This force is linearly proportional to the distance of the trait from the optimum (hence it is sometimes called a "rubber band" process); indeed, when $\alpha_{OU} = 0$, the model is equivalent to Brownian motion (see figure 5.2a). This model is appropriate for a trait that wanders around an adaptive peak under genetic drift (Felsenstein 1988), and it is possible to examine the fit of models that include different peaks on different branches of the phylogeny (Hansen 1997; Butler and King 2004).

In contrast to Brownian motion, the OU model can act to limit evolutionary variation. This is important, because many of the traits of interest to biologists and anthropologists are likely to have constraints on their maximum or minimum size (Garland et al. 1993; Freckleton et al. 2003). Another important aspect of the OU model is that trait history tends to be "forgotten" or lost as a trait wanders around the adaptive peak. Specifically, the character will begin to retrace its steps around the peak, and the same will be true of characters in other closely related lineages that are evolving on the same adaptive landscape. Thus, trait values may be equally similar in closely and distantly

(A)

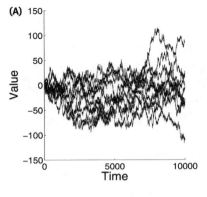

(C)

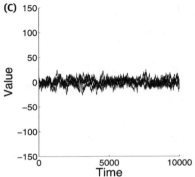

(B)

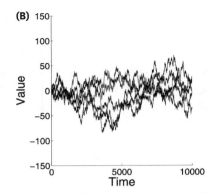

Figure 5.2. **Trait evolution under stabilizing selection.** Trait evolution was simulated under the
OU model, that is, $t_{i+1} = t_i + r_n - t_i * \alpha_{OU}$, where t
is the trait value at time i, r_n is a random normal
variate with standard deviation of 1 and mean
of 0, and α_{OU} is the restraining factor in the OU
model. Each panel shows five realizations of
ten thousand steps of the model using different
values of α_{OU}. (a) Trait evolution with no restraining force ($\alpha_{OU} = 0$), which is equivalent to Brownian motion evolution. The other plots show in-
creasing values of α_{OU}, with (b) $\alpha_{OU} = 0.001$ and
(c) $\alpha_{OU} = 0.01$. Thus, increasing values of α_{OU} reflect increasing stabilizing selection, and this
can have a dramatic effect on the evolutionary
change in a trait over time (see AnthroTree 5.1).

related species throughout a clade that is experiencing the same selective regime, which is relevant in discussing "phylogenetic signal" below. Stabilizing
selection in general has had some appeal among those interested in cross-
cultural studies of humans (e.g., Borgerhoff Mulder 2001; Borgerhoff Mulder
et al. 2001), but it has not yet been applied to systematically study the evolution of human cultural traits.

Ecological niche-filling model. A third influential model of trait evolution involves adaptive radiation and has been called the "adaptive radiation model"
(Price 1997), the "niche model" (Harvey and Rambaut 2000), or (as will be
used in this book) the "ecological niche-filling model" (Freckleton et al. 2003;
Freckleton and Harvey 2006). It builds on ideas of phylogenetic niche conservatism by Harvey and Pagel (1991), which is a process describing how similarity among species can arise through the radiation of an existing species into
niches that are similar to their ancestor's niche. In contrast to the unbounded

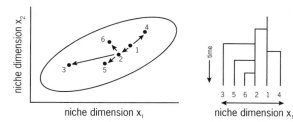

Figure 5.3. The ecological niche-filling model. The ecological niche-filling model starts by assuming that two or more traits (x_1 and x_2) are correlated in "niche space" (oval), and that new niches are filled by species closest to that niche in this niche space. The niche is originally filled by one species, labeled "1," which is assumed to arrive on this virgin landscape with no competitors. As subsequent niches open at random places in niche space (which are here numbered consecutively), the species closest in this bivariate space experiences a speciation event to fill the new niche. Thus, niche 2 is filled by a diversification involving species 1, niche 3 by a diversification involving species 2 (since species 5 and 6 have not yet arisen), and so on. From this process, it is possible to derive a phylogeny as speciation events take place. For each node, change occurs only on the branch leading to the new niche, while the "trunk" lineage remains unchanged. Thus, lineages maintain trait values that form the material for new lineages in adjacent niches later in time (see also Harvey and Rambaut 2000; Nunn and Barton 2001). (Redrawn from Nunn and Barton 2001.)

variation that characterizes the Brownian motion model, it is possible to put limits on the range of values that traits can take in the niche-filling model (Freckleton et al. 2003).

Under a simple version of the niche-filling model, it is assumed that a single species arrives on a landscape with at most one available niche and that new niches appear on this landscape over time (figure 5.3; see also Nunn and Barton 2001). The model assumes that this landscape represents the functional linkage among environments and traits— "niche space." It further assumes that the species that is closest to a newly arisen niche is the one that is best positioned to fill it. Thus, the origin of a new niche results in a diversification event, with the most proximal species in the niche space splitting into two lineages—one for the old niche and one for the new niche. From this process, it is possible to construct a phylogenetic tree that reflects the order of niche openings and the species that filled the new niches (see figure 5.3). Note that for each node, change occurs only on the branch leading to the new niche, while the "trunk" lineage remains unchanged (represented as vertical lines on the figure). Hence, the model has similarities to the punctuational model discussed above.

Because there are fewer species at the beginning of a radiation than later, earlier stages of the radiation are accompanied by larger shifts in trait values when new niches arise (Freckleton and Harvey 2006). Conversely, as the niche space fills up over time, newly arising niches will tend to lie in closer proximity to already filled niches; this should result in less change later in the radiation of a group of organisms than at earlier stages of the adaptive radiation. In other words, the rate of evolution should decline through time, which could be represented on the tree by transforming branches so that they are relatively longer near the root and shorter near the tips. Understanding and diagnosing this model is important for studying correlated evolution using methods that rely on Brownian motion (Price 1997; Harvey and Rambaut 2000; Nunn and Barton 2001; Blomberg et al. 2003; Freckleton and Harvey 2006).

Freckleton et al. (2003) review the ecological niche filling model and variants on the basic model described here, while adaptive radiations are discussed more generally by Schluter (2000). The niche-filling model provides a quantitative framework that could be applied in evolutionary anthropology, ranging from the structure of primate communities (Fleagle and Reed 1999; Cooper et al. 2008) to the adaptive radiation of hominins (Ackermann and Smith 2007).

Summary. To recap, three different models have been discussed most commonly in comparative biology: a Brownian motion model in which change occurs independently of previous changes and variance accumulates along a lineage proportional to time, producing unbounded variation; the OU process of stabilizing selection, with a trait pulled back to an adaptive peak by a restraining force α; and an ecological niche-filling model in which speciation events are closely tied to the opening of new niches (see table 5.1). All of these models are theoretically plausible, different models may apply to different types of traits or systems, and variants on these models can be envisioned (e.g., Hansen and Martins 1996; Mooers et al. 1999). By using an explicit statistical framework, the models help us articulate more clearly the pattern of evolution and discriminate among the alternatives while also giving some sense of the processes that might have generated that pattern (although as noted above, it is often impossible to unambiguously identify the process from pattern alone).

Several parameters and diagnostic tests have been developed that can be used to diagnose the model of evolutionary change for a particular trait in a biological or cultural system. Many of these diagnostics relate to phylogenetic signal, which is a crucial first step in many comparative tests; indeed, proper assessment of phylogenetic signal can be essential for interpreting statistical results of a comparative study (e.g., Capellini et al. 2010).

Phylogenetic Signal: Are Trait Values Linked to the Tree?

It is common to read statements that invoke "evolutionary constraint," "phylogenetic inertia," and similar concepts. In studies of primates, for example, Cheverud et al. (1985) investigated sexual dimorphism in primates using a method to distinguish "phylogenetic effects"—which are inherited from a common ancestor—from "specific effects" that represent independent evolution. They found evidence for phylogenetic effects, with approximately 50 percent of the variation in sexual dimorphism due to inheritance from a common ancestor.

What do phrases invoking constraints and inertia really mean, and why should we care? Despite widespread use and intuitive appeal, these concepts are remarkably fuzzy. Some researchers have used the constraints and inertia terminology to indicate the degree to which evolutionary patterns can change when new selective pressures arise. Others have focused on the mechanisms that drive evolutionary change and the conditions under which these mechanisms might restrict change, such as developmental or genetic constraints. And yet others have invoked the term "inertia" when phylogeny-based comparative tests produce non-significant results that are significant when phylogeny is not taken into account (i.e., a "nonphylogenetic" test). The implication is that some factor has prevented evolutionary change.

In an insightful and probing article, Blomberg and Garland (2002) dissected the use of terms related to phylogenetic inertia and constraint (see also Antonovics and Van Tienderen 1991; Leroi et al. 1994; Burt 2001). They concluded that more rigor is needed in how terminology is used, and they proposed using the term "phylogenetic signal" to reflect the "tendency for related species to resemble each other more than they resemble species drawn at random from the tree" (Blomberg and Garland 2002, 905). By using the term "phylogenetic signal," the authors emphasized that the best we can do is diagnose the degree to which traits correlate with phylogeny. They argued that it is impossible to interpret phylogenetic signal as an alternative to natural selection or as otherwise indicating limits to the evolutionary process, because the underlying mechanisms cannot easily be reconstructed from comparative data. This important point was illustrated by Revell et al. (2008), who simulated data under a wide range of evolutionary models. They found that phylogenetic signal did not show a tight, predictive relationship with the evolutionary model used to simulate the data.

I follow Blomberg and Garland (2002) and avoid using terms that invoke

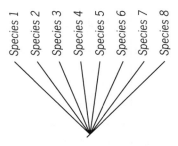

Figure 5.4. A "star" phylogeny. The assumption underlying a star phylogeny is that all the species radiated simultaneously from a single ancestor (Felsenstein 1985b). This produces a set of species that are equally dependent on the traits of other species, and hence, this is equivalent to a complete lack of phylogenetic signal. Although we rarely expect that a star phylogeny represents the history of a group of species, it can be a useful device for representing an absence of phylogenetic signal and the assumption that underlies a nonphylogenetic analysis.

constraints or inertia in this book. Instead, I use the term phylogenetic signal, and thus focus on the pattern rather than the (unknown) process by which closely related species tend to resemble one another. While most of the work on estimating phylogenetic signal has focused on cross-species variation, it is important to note that biological populations may also show evidence for phylogenetic signal (Edwards and Kot 1995; Ashton 2004), and more closely related languages are also more similar to one another (often enabling speakers of one language to understand speakers of the other without formal training). Thus, studies of phylogenetic signal are also useful for research on variation among human populations (Moylan et al. 2006).

More generally, it is important to assess the linkage between trait variation and a phylogeny because phylogeny-based methods assume such a correlation. If traits fail to show a significant association with phylogeny, then many of the phylogeny-based methods in this book will be of little value in understanding trait variation.

As noted above, an important point of this chapter is that one can represent different evolutionary models by adjusting the branch lengths that are used. For example, the concept of phylogenetic signal is closely related to the idea of a "star" phylogeny (Felsenstein 1985b). Specifically, when phylogenetic signal is very low, species are effectively independent from one another, which can be represented as a star phylogeny in which all lineages emanate from a single node (figure 5.4). In such a case, phylogeny-based methods are likely to provide few insights into trait evolution, and it may be appropriate to examine variation among traits without controlling for phylogeny (e.g., Abouheif 1999). As we will see, the concept of a star phylogeny is important because it allows us to move between nonphylogenetic and phylogenetic tests by simply scaling the internal branch lengths from zero to non-zero values (i.e., branches between internal nodes rather than from a node to a tip). In this way, the degree of phylogenetic signal can be incorporated into the branch lengths based on specific scaling algorithms that are described later in this chapter, and this allows comparison of different evolutionary models, including a star phylogeny model in which species values are independent of one another (Spoor et al. 2007; Lavin et al. 2008; Gartner et al. 2010).

With this background, let us now recap the reasons for estimating phylogenetic signal. When phylogenetic signal is high, it is often possible to make predictions for values of species for which data are lacking (e.g., Garland and Ives 2000; Organ et al. 2007) or to reconstruct the evolutionary history of a trait. If a trait does not exhibit phylogenetic signal, applying phylogenetic methods to reconstruct ancestral values is not likely to be very useful (e.g., Laurin 2004). In addition, evidence for phylogenetic signal can be used to determine whether phylogeny-based methods are needed to investigate the correlated evolution of traits (Gittleman and Luh 1992; Abouheif 1999; Revell 2010). Last, many recent measures of phylogenetic signal represent branch length adjustments that can be incorporated into statistical models of correlated evolutionary change, allowing for more flexible and rigorous assessment of how two or more traits covary through time.

General examples. In a study of parasite richness in primates, my colleagues and I tested whether more closely related species exhibited more similar levels of parasite richness (Nunn et al. 2003). Such an analysis is important, because parasite richness is not an evolved characteristic of an organism in the same way as brain size, diet, or life history traits, all of which probably have a significant genetic basis (see also Garland et al. 1992). Parasitism could show phylogenetic signal, however, if most of the parasites are transmitted vertically, if communities of parasites survive relatively intact across generations of hosts, or if parasite richness is strongly correlated with host traits that are shared through common descent (see Nunn and Altizer 2006). We found that for anthropoid primates, measures of parasite richness among different primate species were in fact correlated with phylogeny, and we therefore used phylogeny-based methods to investigate hypotheses for variation in parasite richness (Nunn et al. 2003). In other analyses, my colleagues and I documented phylogenetic signal in malaria prevalence (Nunn and Heymann 2005) and primate white blood cell counts (Nunn 2002a), indicating that diverse measures of parasitism and host defenses covary with phylogeny in primates. In regression models, a recent simulation study showed that phylogenetic signal should be evaluated in the residual variation when deciding whether to use a phylogenetic comparative method (Revell 2010).

Phylogenetic signal also has been estimated for human cultural traits. For example, Moylan et al. (2006) used two different methods (Maddison and Slatkin 1991; Abouheif 1999) to investigate phylogenetic signal using a data set on cultural traits for thirty-five African societies (Borgerhoff Mulder et al. 2001). In this case, we used a linguistic tree to represent the relatedness among the societies, and thus measured linguistic signal rather than phylogenetic signal. The analyses revealed that only eighteen of fifty-five traits (33 percent) showed statistically significant linguistic signal. Although no single type of trait showed signal across all tests, traits involving ritual, politics and ecology tended to show somewhat higher correlations with linguistic history (see also chapter 10).

Quantifying phylogenetic signal. A wide variety of methods has been developed to quantify phylogenetic signal, including methods designed to examine how variation accumulates at different taxonomic levels using nested analysis of variance (ANOVA; see Clutton-Brock and Harvey 1977; Smith 1994) and autocorrelation approaches, such as those used by Cheverud et al. (1985) and others (Gittleman and Kot 1990; Gittleman et al. 1996). Another researcher developed randomization tests to investigate phylogenetic signal using variation along the tips of the tree (Abouheif 1998). And yet other approaches have been developed to investigate phylogenetic effects in continuous characters (e.g., Lynch 1991; Diniz-Filho et al. 1998; Desdevises et al. 2003) or have addressed phylogenetic signal in the context of discrete traits (Maddison and Slatkin 1991; Abouheif 1999).

Here, for continuous traits, I focus on two newer methods for continuously varying traits developed by Blomberg et al. (2003) and Freckleton et al. (2002), as both offer some clear advantages over previous approaches and are relatively easy to implement. I also consider one method for assessing phylogenetic signal in discrete traits (Abouheif 1999). Before proceeding, it is important to note that simply comparing the results of phylogenetic and nonphylogenetic statistics, such as correlations between characters, is insufficient for demonstrating an association between the data and a phylogeny (see Blomberg and Garland 2002). The tests described below are more principled and powerful approaches to quantifying phylogenetic signal.

Blomberg et al. (2003) provided several useful statistical tests to investigate phylogenetic signal and other aspects of the evolutionary process. Here, I focus on a statistical test of phylogenetic signal based on the similarity of trait values in close relatives, such as sister lineages, relative to a null model of variation. This method is derived from the method of independent contrasts, which will be discussed in chapter 7. Briefly, when phylogenetic signal is high, the differences among close relatives will tend to be small because they share more similar trait values; subtraction of one value from another value that is very similar results in a small difference. By comparison, if phylogenetic signal is low, sister lineages on the tree will be relatively more different, and the differences will tend to be larger in magnitude, approaching what would be found when traits are distributed randomly on the tree.

Based on this logic, we can compute a set of differences for a trait across the tree, obtain the variance of these differences, and compare the variance to a null model with no phylogenetic signal. If the observed variance is low relative to the null model, this constitutes evidence for phylogenetic signal. How should this null model of no phylogenetic signal be constructed? Recall that Blomberg and Garland (2002) defined phylogenetic signal as the degree to which closely related species resemble one another more than they resemble species drawn at random from the tree. Thus, an appropriate null model is to compare the observed distribution of contrasts to a null distribution in which the traits are randomly permuted (shuffled) on the tree. If the variance of the observed contrasts is significantly less than the variance of the contrasts from the permuted data, this indicates significant phylogenetic signal (calculations also can be accomplished using mean square error from GLS, see Blomberg et al. 2003).

This permutation-based test is useful for examining phylogenetic signal in a single clade, but it would also be useful to have a standardized value to compare across different groups of organisms and for different traits. Blomberg et al. (2003) provided another statistic, K, for comparing levels of phylogenetic signal (AnthroTree 5.2). K is calculated using GLS, specifically involving calculation of mean square error and its expectation under a Brownian motion model of evolution. K is independent of tree size and shape. A K value less than one indicates departures from Brownian motion evolution, such that species are less similar than one might expect based on their phylogenetic relationships; a K value greater than one indicates greater similarity than expected based on the Brownian motion model of evolution. The authors also developed two statistics that are relevant to measuring phylogenetic signal

Table 5.2. Phylogenetic signal for four primate traits.

Trait	Number of species	p	K	d	g
Log female body mass	105	<.001	1.549	1.001	3.271
Log (male/female mass)	105	<.001	0.315	0.961	0.001
Log body mass	29	<.001	1.52	1.002	2.125
Log relative testes mass	29	<.001	0.38	0.955	0.001

Source: Bloomberg et al. 2003.
Note: The p value reflects statistical significance based on the randomization test for phylogenetic signal. K is a measure of phylogenetic signal that is comparable across traits and trees, while d and g reflect the OU model and acceleration-deceleration (ACDC) transformation parameters, respectively. Relative testes mass refers to size-corrected testes mass.

but are also useful for understanding the OU model and the ecological niche-filing model. These statistics—known as d and g—are described below.

Blomberg et al. (2003) used these two approaches—the permutation test and K—to compare phylogenetic signal in morphological, life history, physiological, and behavioral traits across a wide array of taxa, with adult body mass included as a separate type of character (rather than grouping it with morphological traits). The authors found that 92 percent of the traits with sample sizes of at least 20 species showed significant phylogenetic signal ($n = 53$ traits in total). Based on 121 traits and 35 phylogenies, they found that behavioral traits generally exhibited the lowest signal, and that physiological traits have lower signal than body mass. The lower values for behavioral traits could reflect greater measurement error for these traits (see chapter 9), as this would also reduce the correspondence between traits and the tree (see Blomberg et al. 2003; Ives et al. 2007).

At least four of the data sets analyzed by Blomberg et al. (2003) involved primate morphological traits, and all four of the primate traits exhibited statistically significant phylogenetic signal (table 5.2). In another study, O'Neill and Dobson (2008) investigated phylogenetic signal in body mass and morphological traits in primates using K. Most of the traits examined showed phylogenetic signal, with weaker effects in some morphological traits after controlling for body mass. Thierry et al. (2008) used this method to investigate phylogenetic signal in macaque monkey behavioral traits involving social interactions among individuals within groups, such as reconciliation. They also found statistically significant phylogenetic signal for three of four traits, depending on the data set and phylogeny used (table 5.3 summarizes results from one dataset and phylogeny). As a final example, Ross et al. (2004) applied the method to study cranial characters and also found evidence for phylogenetic signal.

Table 5.3. Phylogenetic signal in four social traits in macaques.

Trait (sample size)	K	MSE data	Mean MSE permutation	p (signal)
Conciliatory tendency between non-kin (9)	1.30	270.3	680.1	.002
Counter-aggression (8)	1.14	403.5	878.9	.017
Explicit contact (9)	1.87	222.2	673.7	.004
Kin bias (9)	0.82	108.4	169.9	.130

Source: Results are from Thierry et al. (2008), focusing on analyses that averaged traits for each species.
Note: A wider range for K was obtained when examining variation among populations (captive groups) and different phylogenies. The p value does not reflect significance level of K; rather, it indicates whether the mean square error (MSE) for the dataset is significantly lower than the mean MSE on permuted datasets, as such a result would indicate significant phylogenetic signal (Blomberg et al. 2003).

The second set of "new" methods were developed and applied by Freckleton et al. (2002) around the same time as Blomberg et al. (2003). Freckleton and colleagues took a slightly different approach to the problem by drawing on previous work by Pagel (1999a), although it should be noted that Blomberg et al. (2003) also discuss and apply branch length scaling parameters.

Specifically, Freckleton et al. (2002) focused on λ (lambda). This parameter is used to scale the variance-covariance matrix, which is simply a way to represent the tree structure in terms of expected variances and covariances in trait change (figure 5.5; see also Cunningham et al. 1998; Freckleton et al. 2002). Along the diagonal of this matrix is the length from the root to each of the tips (the expected *variance* in a Brownian motion model); on the off-diagonal is the shared portion of this evolutionary history for pairs of species, measured from the root to the last common ancestor of the two species (i.e., their expected evolutionary *covariance*). Importantly, then, the off-diagonal portion of the matrix represents the internal branches of the tree (compared to other branches that lead to species values at the tips of the tree, that is, the terminal tips).

It is possible to generate a model of trait evolution that uses this variance-covariance matrix, and within this model, to scale the off-diagonal elements by λ. The maximum likelihood estimate of λ can be used to transform the branch lengths for other analyses, including studies of correlated trait evolution, rather than assuming that the "given" branch lengths are an accurate representation of evolutionary change (Pagel 1999a). More specifically, λ is defined as a phylogenetic transformation that maximizes the likelihood of the data given a Brownian motion model (Freckleton et al. 2002). When $\lambda = 0$, all off-diagonal values are zero, indicating that no covariance is expected and effectively forcing all internal branches to have zero length; this is equivalent

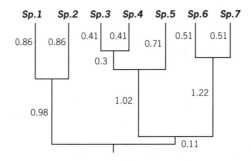

Figure 5.5. A variance-covariance matrix. A dated phylogeny is shown, along with the variance-covariance matrix that describes the tree. The matrix gives variances along the diagonal (shaded numbers) and co-variances on the off-diagonal (unshaded). The variance reflects the evolutionary time from root to the tip (1.84 units in this ultrametric tree), while the covariance reflects the shared evolutionary history for a pair of species.

to a "star phylogeny" (see figure 5.6). When $\lambda = 1$, the traits are consistent with Brownian motion evolution based on branch lengths represented by the variance-covariance matrix. Values between zero and one indicate less phylogenetic signal than expected under a pure Brownian motion model, while values greater than one indicate more signal than expected (although λ is not always defined for values greater than one; Freckleton et al. 2002). Estimating the maximum likelihood of λ is not sufficient to decide whether significant phylogenetic signal exists; one should also compare this likelihood to the likelihood of models when $\lambda = 1$ and $\lambda = 0$ using a likelihood ratio test (LRT). AnthroTree 5.3 provides an example data set and instructions for estimating λ and interpreting the results.

Freckleton et al. (2002) used this approach to investigate phylogenetic signal in ecological traits, which have often been proposed to have low phylogenetic signal (see Harvey 1996). They acquired 26 data sets comprising data on 106 characters, ranging from prevalence of parasitic infections ($\lambda = 0.885$) to mandible length in tiger beetles ($\lambda = 0.0$). The overall data set included some studies of mammals, but none that were limited to only primates. For 60 percent of the traits, the maximum likelihood estimate of λ was significantly dif-

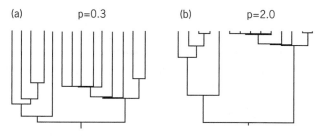

Figure 5.6. Grafen's ρ applied to ape phylogeny. Two examples of ρ transformations for the phylogeny from figure 4.1. On the left, ρ < 1, which tends to stretch the branches leading to the tips. On the right, ρ > 1, which tends to lengthen the branches closer to the root. When ρ = 0, this produces a star phylogeny (as in figure 5.4). As we will see in chapter 7, the branch lengths are an essential part of comparative studies focused on correlated evolution; hence, transformations based on ρ, λ, or other scaling parameters can have large effects on analyses of correlated evolution.

ferent from 0 (indicating phylogenetic signal), and in only 24 percent of the traits was λ significantly different from 1 (indicating departures from Brownian motion expectations). In the remaining tests (16 percent), maximum likelihood estimates of λ were not significantly different from 0 or 1—a real possibility for data sets, and thus something to be aware of when applying the method. In these cases, however, the sample sizes were generally small, suggesting that low statistical power can be a concern when sample sizes are less than about 30 (see Freckleton et al. 2002 for details on the statistical power of this test). It is also important to appreciate that maximum likelihood estimates of λ can differ significantly from both 0 and 1 (e.g., Capellini et al. 2010; AnthroTree 5.3).

A measure related to λ deserves some comment. Grafen (1989) developed a statistic that he called ρ (rho), which also can be used to scale the phylogeny to provide a better fit between the model of evolution and the data. As one of the pioneering papers in comparative biology, Grafen's (1989) paper is well known, and so is ρ. Grafen developed ρ in the context of estimating branch lengths for trees with inaccurate or no branch length information. The parameter is a power transformation that alters the heights of the nodes by compressing the tree at the top or bottom. When ρ is between 0 and 1, branches near the tips are expanded while branches near the root are compressed, and when ρ > 1, the branches near the tips are compressed while those near the root are expanded; this latter situation is similar to what was suggested above as a branch length transformation to represent the ecological niche-filling model. Figure 5.6 provides an example for ρ < 1 and ρ > 1 using the ape phylogeny from the previous chapter (see figure 4.1 for the original branch lengths).

Because ρ allows the user to move from a more star-like tree to a less star-like tree, it can be thought of as a measure of phylogenetic signal, and it is used as a branch length transformation in some comparative methods packages (AnthroTree 5.4). Freckleton et al. (2002) used a simulation approach to compare the performance of λ and ρ under known evolutionary models (note that ρ was applied to the shared path lengths rather than to node height; cf. Grafen 1989). They found that ρ failed to detect phylogenetic signal more often than it should have. In addition, ρ tended to decline when there were more species in the simulation, indicating bias and lack of comparability across data sets. In contrast, the simulations revealed that λ is unbiased and has acceptable statistical properties. Freckleton et al. (2002) therefore advocated use of λ over ρ when quantifying phylogenetic signal. However, it is reasonable to estimate both λ and ρ when attempting to find the best fit between phylogeny and data; in some cases, ρ can provide a better measure of association between the traits and the tree (e.g., Spoor et al. 2007).

Abouheif's method for discrete traits. One of the reasons for quantifying phylogenetic signal in continuous traits is to check whether phylogeny-based methods are appropriate for analyzing character evolution. For this reason, it also can be useful to investigate phylogenetic signal in discrete traits.

Abouheif (1999) developed a statistical approach to measure the degree of nonrandomness in a series of discrete values, such as binary traits along the tips of a phylogeny. The method makes use of the fact that the series of zeros and ones in the species should occur nonrandomly if they exhibit phylogenetic signal. Hence, he suggested that the runs test (Sokal and Rohlf 1995) could be used to investigate whether fewer or more sequences of ones or zeros occur than expected by chance. Because lineages can be rotated around any node without changing the phylogenetic relationships involved (see chapter 2), the results of the test are sensitive to how the species are represented topologically. Abouheif therefore provided an easy-to-use simulation approach that considers different arrangements and can be used to generate a null distribution of statistics. In a one-tailed test, a trait is judged as exhibiting significant phylogenetic signal when fewer than 5 percent of the simulated datasets are below the observed value (AnthroTree 5.5). For another approach to estimating phylogenetic signal in discrete traits, see Maddison and Slatkin (1991), and for application of these tests to discretely coded cross-cultural data, see Moylan et al. (2006).

Summary of phylogenetic signal. Characterizing phylogenetic signal is an extremely valuable first step in any comparative analysis, as it can guide further data analysis decisions. For example, many of the statistics discussed here can be used to scale the branches to better meet the assumptions of other methods, such as investigating correlated trait evolution. If the methods reveal that the traits are clearly independent of phylogeny, this also can be useful. In such cases, for example, it makes little sense to include phylogeny in the analyses of trait evolution, and ancestral state reconstructions are likely to be very uncertain. An absence of significant signal also could indicate that sample sizes are too small, that stabilizing selection is strong, or that high variation within species or populations is obscuring useful variation among them (Blomberg et al. 2003; Ives et al. 2007). The diagnostics in the next section can provide further insights to trait variation.

Diagnostics for Characterizing Evolutionary Patterns

I organized the following diagnostics according to the different evolutionary models discussed above. Keeping in mind the discussion of pattern and process from above, however, they should not be viewed strictly as applicable to only a single model, and are often useful for a wide variety of questions.

Brownian motion. Estimating branch length scaling parameters, such as λ, provides one way to assess whether a trait meets the assumptions of Brownian motion evolution (Grafen 1989; Freckleton et al. 2002; Blomberg et al. 2003). For example, λ is expected to equal one under a Brownian motion model of evolution. As demonstrated in AnthroTree 5.3, it is possible to compare the likelihood of a model in which λ is estimated to the likelihood of a model in which λ is forced to equal one. If the likelihood of the fitted model is not significantly higher than the likelihood when λ is forced to equal one, this is consistent with Brownian motion evolution.

To illustrate, I applied this approach to the data that I simulated under a Brownian motion model (and thus λ should equal one; see AnthroTree 5.1 for the tree and sample data). In this case, the maximum likelihood estimate of λ was 0.958, giving a log likelihood score of 28.65. Forcing λ to equal one produced a log likelihood score only slightly lower, of 28.61. The LRT compares twice the difference in log likelihoods to a χ^2 distribution with one degree of freedom, which is not significant in this case (LRT statistic = 0.086, $p = .77$).

AnthroTree 5.3 provides an example in which λ is significantly different from zero and one. Additional parameters described below (κ, δ, d, and g) can be used to investigate the fit of data to a Brownian motion model of evolution (see also Blomberg et al. 2003; Revell et al. 2008).

Another approach for investigating the Brownian motion model is to examine whether trait change accumulates proportional to time (Garland et al. 1992). For example, we might expect that differences in trait values for pairs of species will increase with increasing time since the species last shared a common ancestor. This is a commonly used and important diagnostic test for Brownian motion in the context of correlated evolution, and is therefore discussed again in chapter 7.

Another descriptor of Brownian motion is to quantify the variance of trait change along each branch. This parameter, which is typically called σ^2, captures a measure of evolutionary change under Brownian motion (AnthroTree 5.6). It is also possible to simulate evolution with defined values of σ^2, which simply reflects the variance of the distribution from which trait changes are drawn (see AnthroTree 5.1).

Finally, it is possible to fit different models of evolution to the data and statistically compare the likelihood scores of these models, including Brownian motion on the hypothesized phylogenetic relationships with branch lengths (AnthroTree 5.7). For example, Mooers et al. (1999) formulated five different hypotheses for trait evolution in a group of birds and then tested which of these models best accounted for trait variation among extant species. The models were represented by having different branch lengths, which thus reflect different amounts of expected evolutionary change; three of these models are illustrated in Figure 5.7.

Mooers et al. (1999) found that different evolutionary models better characterized different traits. Variation in plumage coloration, for example, was best explained with a model that used the phylogenetic topology but fitted branch lengths to the amount of change (a "free" model), while variation in other traits was best accounted for by Brownian motion change on fixed branch lengths (see also Mooers and Schluter 1998, who applied this approach to study vertebrate body size). Similarly, Butler and King (2004) compared a Brownian motion model to several more complicated models, including several versions of an OU model. Although they found that one of the OU models outperformed Brownian motion, the Brownian model was not the least supported of the models (see also Lavin et al. 2008; Gartner et al. 2010).

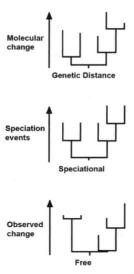

Figure 5.7. Fitting evolutionary models to data. Mooers et al. (1999) examined the fit of five different evolutionary models to data on cranes using maximum likelihood. Three of the models are shown here. The model at top is a gradual-genetic model, in which branch lengths reflect the amount of genetic change in the genes used to construct the phylogeny. In the middle, a speciational model is shown, with all branches equal and thus change proportional to the number of nodes on the phylogeny. The bottom shows a "free model," in which the branches are estimated using the comparative data of interest; it reflects varying trait evolution along branches of the tree and requires more parameters than the other models. (Redrawn from Mooers et al. 1999.)

Speciational trait evolution. As noted above, it is possible to represent speciational evolution by setting all branch lengths to be equal (usually branch lengths are set to unity). This implies that an equal amount of evolutionary change has occurred along each branch and is thus a speciational model (e.g., see AnthroTree 5.1). Pagel (1999a) developed κ (kappa) to model the degree to which a continuously varying trait fits a speciational model. κ is a power function that raises each of the branches on the tree to the power κ (see also Pagel 2002). Under a speciation model, κ should approach zero, which results in branch lengths of one (i.e., a value raised to the power 0 equals 1). As with the other parameters developed by Pagel (1999a), it is possible to assess the significance of κ by estimating the likelihood of the model using the fitted value, and then compare it to the likelihood of models in which κ = 0 or κ = 1. This general approach is described for a different parameter, λ, in AnthroTree 5.3, and can be explored with the data set that was simulated under a speciational model in AnthroTree 5.1.

As an example of the use of κ in evolutionary anthropology, Holden and Mace (2003) studied matrilineal descent and cattle keeping among Bantu-speaking people. They found that κ = 0.45 when taking into account dependencies (i.e., correlation) between use of cattle and matrilineal descent. This indicates that these cultural traits are to some extent a function of speciation events and of branch length, where branch lengths were quantified as the number of linguistic innovations in a parsimony analysis.

In another example, Pagel (2002) examined the evolution of hominin cranial capacity. In addition to testing a directional model (see figures 4.7 and 4.8), he estimated other scaling parameters under the directional model, including κ. He found that the maximum likelihood estimate of κ was 1.18, which was significantly different from 0 but not significantly different from 1. This finding suggests that cranial capacity has evolved gradually on the tree of known hominins, with change accumulating proportional to time rather than in evolutionary bursts at speciational events.

Stabilizing selection. The OU model provides a means to model the effects of stabilizing selection (table 5.1) and was explored by Hansen (1997). It is possible to estimate parameters associated with the OU model, and to compare the fit of different evolutionary models, including OU, to comparative data (Butler et al. 2000; Butler and King 2004; Lavin et al. 2008; Gartner et al. 2010). An example is provided in AnthroTree 5.7.

Blomberg et al. (2003) provide a parameter d that can detect stabilizing selection. As with λ and κ, d operates by transforming the branch lengths. In a simple version of the OU model in which traits move randomly but with a tendency towards a mean, a value of $d = 1$ corresponds to Brownian motion (cf. figure 5.2a, where Brownian motion corresponded to $\alpha_{OU} = 0$). In contrast, values of $d < 1$ are consistent with stabilizing selection. As noted above, an important aspect of very strong stabilizing selection is that it tends to obliterate the history of a trait, due to the fact that the traits are pulled towards a central mean (Felsenstein 1988; Hansen and Martins 1996; Blomberg et al. 2003). In extreme cases, this is equivalent to having no phylogenetic signal, that is, a star phylogeny (see figure 5.4). Thus, d also can provide insights to phylogenetic signal, with values of d approaching 0 being consistent with less phylogenetic signal in the data. All four primate traits examined by Blomberg et al. (2003) and summarized in table 5.2 have estimates of d close to 1. Similarly, O'Neill and Dobson (2008) found that most of the morphological traits in their study had values of d that were not significantly different from 1, with a handful of exceptions for humeral measurements after correcting for body mass.

The ecological niche-filling model. Remember that one of the characteristics of the ecological niche-filling model is that more trait change should occur earlier in evolutionary history than later (see above and Price 1997; Harvey and Rambaut 2000; Freckleton and Harvey 2006). Here I focus on several methods that can be used to detect this effect. Example data and analysis are provided in AnthroTree 5.7.

Blomberg et al. (2003) developed a parameter, g, that can be used to detect temporal variation in rates of trait evolution on a tree. The basis for this parameter lies in a model they call ACDC, for accelerating (AC) or decelerating (DC) evolution in a clade over time. Although not explicitly linked to the niche-filling model, it has similarities and could provide information on patterns that are consistent with it. More specifically, $g > 1$ indicates a deceleration in evolutionary rate over time for a particular trait, and thus is consistent with predictions from the niche-filling model. As with the parameter d described in the context of OU evolution, g can provide insights to phylogenetic signal in trait data, with phylogenetic signal declining as g approaches 0. Among the four primate traits analyzed by Blomberg et al. (2003), g varied widely (table 5.2). Body mass exhibited $g > 1$ in both tests, while estimates of g for body mass dimorphism and relative testis mass were both 0.001. A recent study found evidence for a deceleration in rates of body mass evolution across all mammals, and also in primates (Cooper and Purvis 2010).

Pagel (1999a, 2002) described a branch length transformation that can be used to assess whether the data fit an adaptive radiation model. This statistic, called δ (delta), can detect whether the rate of trait evolution has increased or decreased over time. A value of δ significantly less than 1 indicates that traits changed rapidly early in the origin of group of species and then became more stable closer to the tips of the tree, consistent with an adaptive radiation. Conversely, $\delta > 1$ suggests that most trait change in the tree has occurred closer to the tips than to the root (see also Pagel 2002; Freckleton et al. 2003).

In his analysis of hominin cranial capacity, Pagel (2002) found that $\delta = 1.55$ (see figure 4.8). This indicates that the rate of brain size evolution has increased over time, being higher in the longer branches leading to humans and Neanderthals. As another example, I applied this model to investigate patterns of evolution in geographic range size in 104 primate species and found that most change in these traits occurred late in primate evolution ($\delta = 2.83$), probably reflecting that closely related species can have very different geographic range sizes (and thus also low phylogenetic signal, with $\lambda = 0.015$; but see Jones et al. 2005 and Purvis et al. 2005a for analyses that found stronger phylogenetic signal in primate geographic range size).

Another approach to studying adaptive radiation was taken by Harmon et al. (2003), who investigated the partitioning of morphological trait variation within and between subclades in four independent lineages of lizards, and they compared this to the rate at which lineages arose on the tree using methods that will be described in chapter 8 (lineages through time plots). Harmon and colleagues found that when lineages arose more rapidly early in evolution, greater morphological differences were found among than within subclades. This suggests that more rapidly radiating groups of organisms fill the niche space more rapidly, providing fewer opportunities for trait disparity to accumulate within subclades.

Harvey and Rambaut (2000) examined how a standard Brownian motion assumption check fares in a simulation test of the niche-filling model. Specifically, they checked whether data simulated by the niche-filling model violate the assumption that the absolute value of differences in values in sister lineages is independent of the branch lengths in those lineages (as discussed above, see Garland et al. 1992). Harvey and Rambaut originally claimed that this assumption was violated in nearly 100 percent of the analyses, but this result was later found to be a result of a programming error (see Freckleton and Harvey 2006).

In a later paper, Freckleton and Harvey (2006) provided two diagnostic tests designed to detect the niche-filling model, specifically by testing whether rates of evolution decline as niche space becomes filled by more species. First, they investigated whether contrasts between lineages deeper in the tree (closer to the root) are larger than shallower contrasts involving, for example, pairs of sister species on the tips of the tree. Second, they used a randomization test to determine whether the observed contrasts are consistent with a Brownian motion model. They found that both diagnostics had higher power to reject Brownian motion evolution than the simple diagnostics used by Harvey and Rambaut (2000), although the effectiveness of the test statistics depended on the model of speciation and the correlation between the traits.

More research is needed to investigate diagnostics for the niche-filling model. Consider, for example, the parameter ρ. As illustrated previously in figure 5.6, values of $\rho > 1$ stretch deeper branches, which could be used to represent greater phenotypic change at earlier stages of an adaptive radiation. Thus, ρ would seem to offer a good solution to adjusting the branch lengths to represent an ecological niche-filling model (see also Harvey and Rambaut 2000).

Quantifying Rates of Evolution

Another important question concerns the rate of evolution. Do some traits evolve more rapidly than others, and can we detect these differences using phylogenetic information? Quantifying rates of evolution has a long history. Paleontologists and experimental biologists have used the term "darwin" as a measure of evolutionary rate (Haldane 1949; Gingerich 1983; Reznick et al. 1997). A darwin is defined as change by a factor of e per million years, where e is the base for natural logarithms; a unit of one darwin is thus equivalent to one logarithmic unit of change per million years, and thereby captures proportional change rather than absolute change. Gingerich (1983) compiled 521 estimates of evolutionary rates in darwins across diverse groups of organisms and different time intervals (1.5 years to 350 million years). He found that estimating evolutionary rates over longer intervals of time is associated with slower rates of change, probably reflecting artifacts that include averaging over time, with periods of stasis tending to reduce the overall amount of change that is estimated over a longer time span (see also Gittleman et al. 1996).

A variety of methods have been developed to examine evolutionary rates with dated phylogenies. Differences among sister lineages standardized for time can be viewed as a measure of evolutionary rate (e.g., Garland 1992; Webster and Purvis 2002a). These are independent contrasts, which will be discussed in chapter 7. Martins (1994) also developed an approach to investigate evolutionary rates based on principles derived from independent contrasts.

More recently, O'Meara et al. (2006) provided a new, more powerful approach to investigate rates of evolution. Their maximum likelihood approach estimates rates of evolution, and then tests whether rates in different clades are different (AnthroTree 5.8). Similarly, Revell and Harmon (2008) and Revell and Collar (2009) provide a maximum likelihood framework to investigate rates and correlations. Model-based methods have been applied to study evolutionary rates across a broad range of organisms and suggest that a process that includes a fitness optimum can account for phenotypic divergence (Estes and Arnold 2007). A neutral model provided a poor fit to the data on diversification used by these authors.

In an empirical study of evolutionary rates, Gittleman et al. (1996) investigated whether rates of evolution are higher in behavioral traits, compared to traits related to body mass and life history. Their data covered major groups of mammals, including primates, and behavioral traits included group size and home range size. Consistent with the hypothesis that behavioral traits are more labile, they found that rates of behavioral evolution were higher than for other traits (see also Blomberg et al. 2003). Importantly, however, most measures of phylogenetic signal should not be considered valid estimates of evolutionary rates (Revell et al. 2008).

Summary and Synthesis

Investigating the underlying model of trait evolution is an essential component of a comparative study. In this chapter, I reviewed common evolutionary models that can be used to quantify evolutionary patterns. The parameter κ, for example, assesses the fit of a speciational model, with trait change being a function of the number of divergence events on the tree used rather than branch lengths (e.g., Venditti and Pagel 2010). This chapter and the associated entries on the AnthroTree website also provided basic details on using LRTs, which provide a way to compare different models of evolution (Mooers et al. 1999; Pagel 1999a). As phylogenetic methods are implemented in *R* (*R* Development Core Team 2009), model evaluation, including evaluation based on information theoretic approaches (Burnham and Anderson 1998), is likely to play a greater role in comparative studies.

The approaches described in this chapter have rarely been applied to questions in evolutionary anthropology, yet great potential exists for their application to anthropological questions. For example, does body mass tend to evolve in "speciational" bursts, or in a more gradual way? This question could be addressed by fitting parameters described above, particularly κ (see also Atkinson et al. 2008). Similarly, in the origins of a technological advance, such as new forms of pottery or farming implements, do we see rapid evolution in the early stages of cultural development, and a slower rate of change in later periods? This would be equivalent to an ecological niche-filling model in biology, and the methods for understanding this process in biology could be fruitfully applied to archeological data. Last, to what degree do human cultural traits exhibit phylogenetic signal, and what factors influence variation in phylogenetic signal among different types of human cultural traits? While a few studies have investigated this question using phylogenetic methods (Moylan et al. 2006) or have measured "phylogenetic" signal in human traits across populations (Mace et al. 2003), much more research is needed.

Pointers to related topics in future chapters:

- Independent contrasts: chapter 7
- Incorporating intraspecific variation and measurement error: chapter 9
- Modeling evolutionary change in human evolution: chapter 12

6 Correlated Evolution and Testing Adaptive Hypotheses

Probably no subject in biology has attracted as much attention—or as much controversy—as the concept of adaptation (Williams 1966; Gould and Lewontin 1979; Reeve and Sherman 1993; Rose and Lauder 1996). This attention and controversy is well deserved: adaptation is an essential component of biological investigation, yet it can be remarkably difficult to define operationally and is only rarely possible to test experimentally. Points of (sometimes heated) debate include the following questions. Is it necessary to show that a putatively adaptive trait has a genetic basis? Must one demonstrate the fitness effects of *having* versus *not having* the trait, and if so, how is this achieved if all individuals of a species exhibit the trait? How can we rule out alternative explanations for the evolutionary basis of a trait, and how should we interpret traits that probably originated through selective pressures other than those that maintain them today? Is it necessary to identify the ancestral state of a putative adaptation, and at which deeper node should this ancestral state (and its function) be evaluated?

Comparative methods are commonly used to study adaptation (Ridley 1983; Harvey and Pagel 1991; Garland and Adolph 1994; Mace and Pagel 1994; Martins 2000); indeed, they are often the only way to assess how environmental variation affects traits that are inflexible within species (as this makes it impossible to assess the fitness of different character states among individuals of the same species). As stated by Harvey and Pagel, "Stripped to the bone ... the evidence for adaptive evolution revealed by comparative studies is correlated evolution among characters or between characters and environments.... Our goal is to identify the variable or variables responsible for variation in some other variable" (1991, 11). In other words, we can investigate adaptation by studying the environmental features and biological traits that covary with the trait of interest.

At the outset, it is important to keep in mind that the comparative method is only one of several ways to assess the adaptive basis of a trait. It is also important to appreciate the limitations of the comparative approach. One important limitation is that comparative biology is inherently nonexperimental; thus, it could be that some unmeasured variable influenced the evolution of the trait, rather than the variable of interest to the investigator. Moreover, for biological traits, genetic data are rarely available on the traits of interest, making it difficult to assess whether the variation is inherited genetically and thus subject to natural selection. We will return to some of these limitations of the comparative method later in this chapter.

And what about phylogeny? Some researchers have proposed that adaptation requires an explicit phylogenetic context and details on the historical function or performance of traits (Gould and Vrba 1982; Coddington 1988; Baum and Larson 1991). This obviously gives phylogenetic comparative methods a central role in studies of adaptation. However, regardless of your opinion about the necessity of incorporating evolutionary history in inferring adaptation (e.g., Reeve and Sherman 1993), we will see that *phylogeny has a crucial role to play in comparative analyses by providing the scaffolding to identify evolutionary origins of traits.* These evolutionary origins are crucial, because they represent statistically independent tests of whether traits show correlated evolution (Pagel 1994b; Doughty 1996). Independent origins are also used to make inferences of the evolutionary process, such as rates of gains and losses of traits, which is required by some comparative methods (Pagel 1994a).

Examples. To begin, let us consider a few snapshots of comparative findings that have contributed to our understanding of adaptation in evolutionary anthropology. Many other examples are found throughout this book, including those given in chapter 1.

The comparative method has been the major approach for understanding the adaptive basis for primate behaviors, morphological traits and cognitive abilities (see chapters 9 and 11; see also Martin 1990; Nunn and Barton 2001). In studies of primate behavior and morphology, for example, the comparative method has been used to infer that living in a larger group selects for larger relative neocortex size (figure 6.1; Dunbar 1992, 1998), that primate territoriality favors longer day ranges (relative to home range area; figure 6.2, Mitani and Rodman 1979), and that despotic social relationships among females are selectively favored by contest competition over food within the social group (Van Schaik 1989; Sterck et al. 1997). Indeed, some of the fundamental breakthroughs in our understanding of primate behavior and ecology emerged from inter-specific comparisons—both quantitative and qualitative—long

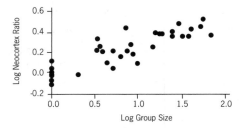

Figure 6.1. Neocortex ratio and group size in primates. The neocortex ratio is calculated as the neocortex volume divided by the difference in neocortex volume and total brain volume (i.e., neocortex / (whole brain − neocortex). Values were averaged for species within genera and plotted. Note that many species in this dataset were recorded as having a group size of 1, which after $\log_{10}$ transformation results in a cluster of points around 0. (Figure produced using data from Dunbar 1992.)

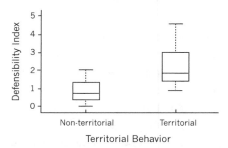

Figure 6.2. Territoriality and the defensibility index. Mitani and Rodman (1979) found that a measure of "home range defensibility" correlates significantly with territorial behavior—species in which the day range relative to home range is large are more likely to be territorial. Boxes show the lower quartile, median, and upper quartile; and whiskers show 1.5 times the interquartile range. (Figure produced using data in Mitani and Rodman 1979.)

before phylogeny-based methods became available (Crook and Gartlan 1966; Crook 1972; Goss-Custard et al. 1972; Clutton-Brock and Harvey 1977).

The comparative approach also has played a decisive role in understanding human behavioral and ecological adaptations (see chapter 10 and Mace et al. 2003). For example, Gaulin and Boster (1990) examined the distribution of dowry, which is money or property offered by a bride's family to the husband at marriage. They focused in particular on the hypothesis that dowry represents a form of female-female competition. Such competition should be most common in monogamous societies characterized by variation in male wealth (i.e., stratified societies). The authors found that this prediction was supported using a large dataset on 1,267 societies from G. P. Murdock's *Ethnographic Atlas*. In another example, cross-cultural comparison demonstrated that the diversity of infectious diseases increases close to the equator in humans (Guernier et al. 2004), similar to a latitudinal gradient in protozoan richness in primates (Nunn et al. 2005). This result therefore suggests that, all else being equal, parasite pressure will be a stronger influence on behavior and immune function of humans living closer to the equator (see also Low 1990).

Recently, comparative methods have been used to uncover links between genetic, phenotypic, and ecological traits in primates, including studies of molecular evolution of the gene *GRIN2A* in relation to an ecological proxy for spatial memory (Ali and Meier 2009), an association between MHC evolution and parasite richness (Garamszegi and Nunn 2011), evolution of the gene *enamelin* in relation to dietary evolution (Kelley and Swanson 2008), and research on sequence evolution in the *ASPM* gene and its association with evolutionary change in the cerebral cortex (Ali and Meier 2008). Similarly, a recent study (Wlasiuk and Nachman 2010) took a more focused genetic perspective on previous comparative work that suggested an association between increased promiscuity and measures of immunity in primates (Nunn et al. 2000). Wlasiuk and Nachman (2010) found a positive association between evolutionary changes in mating promiscuity and rates of protein evolution among immune system genes that interact closely with pathogens. We are likely to see many more studies of this kind as molecular approaches increase our knowledge of primate genes.

Last, comparative approaches have provided an essential foundation to make inferences about hominin behavioral evolution (e.g., Foley and Lee 1989, 1996), including reconstructing the behavior of extinct species through detailed study of trait correlations in nonhuman primates (see chapters 9 and 12). For example, Plavcan and Van Schaik (1997a) used patterns of dimorphism and competition in living primates to investigate sexual selection in hominin evolution and found that measures of dimorphism often provide conflicting inferences of sexual selection in extinct hominins. In *Australopithecus afarensis*, for example, estimated body mass dimorphism predicted high levels of male intrasexual competition, while several measures of canine dimorphism predicted lower levels of male intrasexual competition (figure 6.3).

The examples just given involved studying how two or more traits covary, and they reveal how evidence of correlated trait evolution has provided insights to fundamental questions in evolutionary anthropology—including addressing key questions in primate behavior, making sense of human cultural diversity, understanding molecular evolution, and reconstructing hominin behavior in the fossil record. These examples represent only a tiny sliver of the interesting comparative investigations that have been completed, and the future likely holds many studies that will elaborate on the studies just cited.

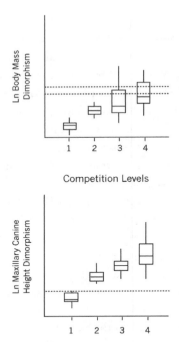

Figure 6.3. Inferring male intrasexual competition in extinct hominins. The top graph shows body mass dimorphism, the bottom shows maxillary canine crown height dimorphism. Competition levels are given on the x axes. On top, the dashed lines represent two estimates of dimorphism in *Australopithecus afarensis*. On the bottom, the dashed line shows canine height dimorphism from *A. afarensis*. (Redrawn from Plavcan and Van Schaik 1997a.)

This chapter lays out the foundations of comparative tests of adaptive hypotheses. I focus in particular on two issues related to phylogeny: the critical role of evolutionary convergence (i.e., independent origins) to make inferences concerning adaptation (Pagel 1994b), and an important assumption involving the independence of species data points that is often violated in comparative studies across populations and species. More specifically, the correlative statistical tests underlying comparative tests of adaptive hypotheses usually assume that the data points are statistically independent of one another, yet trait values for different species or populations are often clustered according to phylogeny and/or geography. This chapter links closely to material in the previous chapter and the next chapter. Thus, we will draw on some diagnostic tools in chapter 5, especially methods for estimating phylogenetic signal (Freckleton et al. 2002; Blomberg et al. 2003), because these diagnostics can be used to assess the degree of statistical nonindependence in a comparative dataset. Looking forward, chapter 7 reviews three methods for controlling for the nonindependence of species data points in comparative studies of continuous and discrete characters.

Using Evolutionary Convergence to Study Adaptation

Pagel (1994b) provided an introduction to what he called the "convergence" approach to studying adaptation with comparative methods. He argued that "the recurrent fit between a trait and some environmental feature provides some of the best evidence available that the trait is an adaptation to its current function" (29). In this view, convergence approaches focus on repeated evolutionary origins, or homoplasy, to make inferences about adaptation. Pagel contrasted this view with the cladistic or homology approach, which aims to study single evolutionary events along lineages. This alternative approach uses detailed data on the ecological context to assess the likely ancestral condition, thus focusing on the factors associated with the origin of a trait (Coddington 1988, 1994; Carpenter 1989).

Whereas convergence approaches make statistical assessments of whether traits covary, the cladistic approach develops detailed, often plausible explanations, but usually shies away from formal statistical tests of the hypotheses. This book generally follows the convergence approach because it provides a statistical framework to test adaptive hypotheses. Indeed, most biologists and anthropologists have gravitated towards the convergence approach, thus "voting with their feet" for a method that offers clear guidance for quantitative hypothesis testing.

How does phylogeny mesh with the central role of convergence in studying adaptation? Phylogeny provides a way to assess whether traits covary evolutionarily, and thus treats the units of analysis as evolutionary change rather than static representations of species values at the tips of the tree. Phylogeny is thus integral to the convergence approach, and it helps to deal with the fact that species values are not statistically independent data points in studies of correlated trait change (more on that soon).

To illustrate these concepts, let us return to an example used earlier involving the factors that favor a larger brain. Primates exhibit remarkable variation across species in brain size, ranging from approximately 1,680 mm³ in mouse lemurs to about 1,250,000 mm³ in humans (Dunbar 1992)—a range that covers three orders of magnitude. A number of scientists have investigated the ecological and social correlates of brain size in broad phylogenetic context. One particularly well-known set of results involves the association between sociality and the size of the neocortex—a part of the brain that is involved in many higher-level cognitive processes such as spatial reasoning and motor commands. Specifically, Dunbar (1992, 1998) and others showed that relative to the size of the rest of the brain, the neocortex is larger in species that live in larger social groups (see figure 6.1; see also Sawaguchi and Kudo 1990; Dunbar 1995; Barton 1996; Deaner et al. 2000; Lindenfors 2005; Dunbar and Shultz 2007b). The association between brain size and sociality has been interpreted in the context of Machiavellian intelligence: a larger social group selects for more sophisticated strategies involving alliances, mating competition, and access to limited resources (Barton and Dunbar 1997). To support this hypothesized function, comparative studies would benefit from further research on the links between brain size and cognitive performance (Healy and Rowe 2007).

Under the convergence approach, we can bring a phylogenetic perspective to the data shown in figure 6.1. Specifically, a phylogeny provides a means to directly examine whether *evolutionary changes* in neocortex size covary with *evolutionary changes* in sociality, rather than implicitly assuming that such correlated changes have occurred based on the distribution of character states among living species. Under a scenario of correlated evolution, we expect to see along branches of the phylogeny that *evolutionary increases* in sociality are associated with *evolutionary increases* in neocortex size on some branches, while *evolutionary decreases* are found in both traits along other branches of the tree. Such an analysis of *correlated evolution* can be conducted with a versatile and widely used method called phylogenetically independent contrasts (Felsenstein 1985b; Garland et al. 1992), which is presented in more detail in chapter 7. Correlated evolution analyses support the predicted link between the evolution of neocortex size and the evolution of group size across primate genera (Spearman $r_s = 0.71$, $n = 12$, $p = .01$ [Dunbar 1995]), as shown, for example with independent contrasts calculated from species level data, in figure 6.4.

The rationale behind using evolutionary changes might seem like a neat statistical trick (or an annoyance, depending on your perspective). But using a phylogeny to assess evolutionary change is far from a minor issue; indeed, phylogenetic methods often produce results that differ radically (and sometimes unpredictably) from analyses of species data. In a review of 566 statistical results in the biological literature, for example, Carvalho et al. (2006) found that it was possible to obtain significant results when examining evolutionary change but not when looking at the variation across species (see also Ricklefs and Starck 1996; Price 1997; Rohlf 2006). In 27 percent of the

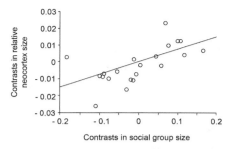

Figure 6.4. **Phylogenetically independent contrasts in group size and relative neocortex size.** Analyses of species-level data revealed that evolutionary changes in group size covary with evolutionary changes in relative size of the neocortex ($r = 0.59$, d.f. $= 20$, $p = .003$). Relative neocortex size was calculated as residuals from the regression of contrasts in neocortex size on contrasts in size of the rest of the brain. (Redrawn from Barton 1996.)

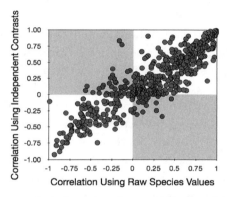

Figure 6.5. **Relationship between results from independent contrasts and nonphylogenetic studies.** Analyses based on independent contrasts and species data tend to produce similar results, but this cannot always be assumed. Values in shaded quadrants indicate situations in which coefficients from contrasts-based and nonphylogenetic tests differ in sign. (Redrawn from Carvalho et al. 2006.)

cases where species data produced a significant result, use of a phylogenetic method failed to confirm the nonphylogenetic finding; conversely, in 16 percent of cases when phylogenetically based tests produced statistically significant results, analyses of species data produced nonsignificant outcomes. In addition, slope estimates obtained from species values and phylogenetic analyses often differed in sign (i.e., those in shaded quadrants of figure 6.5). As we will see, simulation studies have demonstrated convincingly the importance of controlling for phylogeny (e.g., Martins and Garland 1991; Purvis et al. 1994; Harvey and Rambaut 1998). Methods and phylogenies needed to run phylogenetically informed comparative analyses are now so readily available that researchers have little justification for not using phylogenetic methods.

Limitations of the Comparative Method

A central point of this chapter and the next is that the convergence approach in comparative biology is one of the most powerful ways to assess the adaptive significance of a trait or set of traits. To make the most powerful use of this approach, however, it is important to know its limits. Here I focus on several key issues when applying the comparative method (see also Garland et al. 2005).

Causality. One major issue concerns *correlation* versus *causation*—a point I have seen in more reviews of comparative research than I can count. Hypotheses typically assume that one factor or character selects for, that is, *causes*, variation in another character. In most comparative studies, however, we end up with a statistical test that provides evidence for a correlation among traits rather than direct evidence for a causal relationship.

In some comparative studies, researchers grappled with assessing causal mechanisms. Consider, for example, Mitani and Rodman's demonstration that territoriality correlates with more intensive use of the home range (see figure 6.2). From this correlation, they reasonably wondered if selection for home range defense produced more intensive use of the home range, or if longer day ranges of some species arose through other advantages that subsequently facilitated territorial defense. In other words, does territoriality favor more intensive ranging, or does more intensive ranging arise due to other factors and then incidentally lead to territorial behavior? For the second scenario, they reasoned that if day range simply enables territorial defense, then the day range of territorial species should be larger than the day range of nonterritorial species. After controlling for foraging group size and percentage of leaves in the diet, they found no significant difference in day ranges among territorial and nonterritorial species. Their multiple regression models further indicated that day range declines with increasing foliage in the diet. Thus, Mitani and Rodman argued that foraging ecology is the primary determinant of ranging patterns, and as a consequence, of territoriality, that is, "primates will defend a range only if their foraging regimes allow them to do so" (1979, 248).

Another approach to address the correlation-causation issue is to identify which of two traits evolved first. If one trait is found to evolve before the other on a phylogenetic tree, this would be consistent with a causal mechanism, especially when the order of events occurs repeatedly across many evolutionary origins of the character states on the tree (Maddison 1990; Harvey and Pagel 1991). As discussed in the next chapter, some phylogenetic methods provide a way for the researcher to detect the order of events for discrete characters (Pagel 1994a) or to detect evolutionary lag in continuous characters, that is, when one trait evolves more quickly than another trait (Burt 1989; Deaner and Nunn 1999).

Another way to help alleviate the correlation versus causation problem is to choose an independent variable that reflects the selective regime and cannot be influenced by the dependent trait of interest. It can be especially helpful in this regard to choose environmental variables. As noted above, for example, disease risk is thought to covary with latitude in humans (Guernier et al. 2004) and nonhuman primates (Nunn et al. 2005). It seems most likely that environmental factors associated with latitude influence protozoan diversity, rather than increased protozoan diversity acting as a selective pressure for some lineages of primates to move toward the equator! Of course we have to accept in this case that latitude stands in for several other potential driving variables, such as rainfall and temperature, which are likely to be the ultimate drivers of the latitudinal gradient. Hence, this example—like most others—does not fully address questions of causality.

Last, it is important to keep in mind that field or lab experiments can be essential in the quest to resolve causality when testing adaptive hypotheses (Garland and Rose 2009). Comparative perspectives often can help researchers identify species to study experimentally (see chapter 13), although experimental approaches obviously require different methods than are considered in this book.

To summarize, adaptive hypotheses usually propose causal links among traits, but results from comparative studies are typically correlational. When we lack independent evidence for causality, we should be cautious in the terminology used to describe trait associations. Specifically, it is more appropriate to call these associations "evolutionary correlations," rather than stating "x causes y."

Confounding variables and alternative explanations. Above, it was mentioned that several other factors correlated with latitude might account for the association between disease risk and proximity to the equator. This more generally raises questions about confounding variables and alternative explanations. Two traits might be correlated because both are causally related to a third variable that is unmeasured. A classic (and possibly fictitious) example is the positive association between murder rates and ice cream sales. If faced with results suggesting a link between these two measures over time in a society, we might reasonably ask if some other factor independently influences both murder rates and ice cream sales. It might be, for example, that on summer nights more people are out on the streets, with some purchasing ice cream and others becoming involved in disagreements that lead to murders. Thus, ice cream sales and murder are not linked causally, but are instead influenced by another factor, namely, the tendency for people to be outside when the weather is nice.

Alternative explanations for a pattern often emerge based on biological grounds when multiple plausible explanations are possible. In chapter 1, for example, I reviewed comparative research on testes mass, which showed that testes are larger (relative to body mass) in species characterized by female mating promiscuity (see figure 1.1). It is also possible, however, that the association simply reflects variation in breeding seasonality (Harcourt et al. 1981). Such a confound could occur because in seasonal breeders, males would be expected to mate with more females in a shorter period of time, and this might favor males that have larger sperm reserves or greater ability to produce sperm. Moreover, seasonality could reduce the ability of a single male to defend access to females (resulting in females mating with multiple males; Ridley 1986; Nunn 1999b). Thus, in a follow-up study, Harcourt et al. (1995) investigated both mating promiscuity and seasonality. They found support for the effects of mating patterns, but not for seasonality.

In general, the better examples of comparative research acknowledge alternative hypotheses for a pattern when such alternatives exist, and the best examples of comparative research explicitly test alternative explanations.

Limitations of comparative data. Another critical issue for the comparative approach concerns *data quality* and *completeness*. This is worth emphasizing, because most recent efforts have tended to focus on methodological advances in incorporating phylogeny rather than more basic concerns about the underlying data that are used (Freckleton 2009). Thus, researchers might be very quick to use an existing compilation of data that they take as "given" by previous authors—which is easier to do as more informatics data are compiled and downloadable from the Internet—and then invest most (perhaps even 99 percent) of their time in applying the phylogenetic or other methods to these data. It is therefore worthwhile to consider some crucial issues involving comparative data, which I hope will convince readers to take a more careful approach to data compilation. Chapter 13 builds on this discussion by considering relational databases in comparative research.

A common criticism of comparative studies is that they ignore intraspecific variation (e.g., Struhsaker 2000; Strier 2003; see also chapter 11). Another potential problem is that some data are inaccurate or outdated, and as such, they could obscure the comparative pattern under investigation. For example, studies of primate ecology would benefit from using data updated from the classic work of Clutton-Brock and Harvey (1977) because problems have been identified with some of these data and better estimates are now available on more species of primates (e.g., Smith and Jungers 1997; Nunn and Van Schaik 2002). Similarly, one can level criticism at how traits are coded in cross-cultural studies (e.g., Dickemann 1982). Last, it is important to realize that different researchers studying different species or human cultural groups often use different methods, and that the desire to build a broad and taxonomically complete database often involves sacrificing detailed measures of, for example, social network connectivity, for rougher measures, such as mean group size.

Another issue concerns taxonomic sampling (Freckleton 2009). What happens if taxonomic sampling is incomplete—as it almost always is—and the pattern happens to be found in the lineages that are included in the sample, but not in other lineages? Taxonomic sampling biases arise in biological data because some species are easier to study than others, and extinct lineages often will be excluded from the comparative sample (unless of course they are the focus of the study). Similar issues arise in cross-cultural studies of humans. For example, European introductions of infectious diseases often decimated groups of people—such as American Indians—before their languages and cultures could be described adequately. This can affect the results of comparative studies by producing lower levels of language diversity in an area simply because the linguistic data were not collected in time (e.g., Nettle 1999b). In biology, a recent simulation study found that when sampling of species is random, spurious results are unlikely, but taxon sampling can become a problem if species are chosen for inclusion based on characteristics related to the comparative pattern (Ackerly 2000).

In many cases, it is reasonable to expect that data errors will obscure real effects rather than create spurious ones (Köbben 1967; Clutton-Brock and Harvey 1984). In statistical terms, type II errors might be more likely than type I errors. Related to this, we often want to ensure a large sample size: if the dataset is large enough and the pattern is real, increasing the sample size could help to offset issues of poor data. While this might generally be true, it will not be true in all cases and so needs to be addressed for each comparative study on a case-by-case basis. For example, Harmon and Losos (2005) showed that intraspecific variation has no effect on the outcome of nonphylogenetic tests but can negatively impact the results of some phylogeny-based tests when sample sizes per species are low. The reason for this is that measurement error can increase the apparent magnitude of change in closely related species, leading to elevated rejection of null hypotheses of no effect (see also Purvis and Webster 1999; Ives et al. 2007). Harmon and Losos (2005) provided diagnostic tests to investigate this potential problem, and these tests have been applied in at least one study of primate behavior (Kutsukake and Nunn 2006).

Despite the importance of the underlying data for testing adaptive hypotheses, few studies have investigated the effects of using different data compilations on the results of comparative analyses (Purvis and Webster 1999). One study on primate range size investigated how different values influenced the results (Nunn and Barton 2000). In this study, we obtained comparative data based on all studies we could find with information on range size and group composition. We first averaged data across studies, producing average values for the comparative analysis without regard to the populations from which the data actually came. We then repeated the analyses using only data in which a range size and group composition were "matched" from the same study site. In this case, there were very few differences in results among the data sets.

A recent study on sleep patterns across mammals found that lab conditions impacted the sleep durations that were measured by different studies (Capellini et al. 2008). The authors therefore restricted the data to studies in which key aspects of laboratory procedures were held constant. This produced differences from another study that did not implement similar procedures; for example, rapid eye movement (REM) sleep durations exhibited a non-significant negative tendency to associate with relative brain size in a simple regression ($t = -1.61$; $d.f. = 41$; $R^2 = 0.06$; $p = .114$), while a previous phylogenetic study without the same data control measures (Lesku et al. 2006) found the opposite effect when looking at the percentage of REM sleep in a path analysis (although it is difficult to say whether this is due to different measurements or different statistical methods).

A final example involves a study of testes mass in birds. Calhim and Birkhead (2006) found that previous studies of sperm competition using testes mass had often settled for larger numbers of species at the cost of including some erroneous values, and that a smaller sample of higher quality data would have altered conclusions in four of ten studies that they examined.

The Nonindependence Problem

When I play a game of poker with a new deck of playing cards, I take special care to shuffle the deck so that the cards are random, specifically, so that runs of a single suit or number do not occur. In other words, I prefer that the cards occur in a random order, with the suit that is next drawn being statistically independent of the suit in the previous draw. In a similar way, if I were to study clutch size in birds, I might prefer to have a single estimate from each breeding pair rather than multiple estimates from some pairs and not others; to do otherwise would potentially bias the results towards a few prolific breeders. Or, if I was conducting experiments with a fertilizer on different plots of agricultural land, I might expect that spatially adjacent plots would produce similar outcomes because they experience greater similarities in sunlight, slope and soil quality (Purvis and Webster 1999). Such potential effects would need to be controlled in statistical tests of the resulting data.

These examples all involve issues of statistical nonindependence. Statistical tests make many assumptions, such as the assumption that the data are normally distributed, but few of these assumptions are as important as the statistical independence of the data points. Whereas many assumptions of statistical tests can be met through transformation of the data or through use of nonparametric statistics, this is not usually true in the case of nonindependence, where more sophisticated methods are needed to deal with this assumption violation. And the effects of nonindependence can be substantial and unpredictable. In some cases, for example, nonindependence can produce spurious results by suggesting that a pattern exists when in fact two variables are unrelated (i.e., a type I error). In other cases, use of nonindependent data can cause the researcher to miss true patterns (i.e., a type II error).

In biological systems, the critical issue is that more closely related species tend to share traits through common ancestry, yet we can be more confident that a trait is an adaptation if it has evolved repeatedly, rather than once, in association with some other trait or environmental attribute (Harvey and Pagel 1991; Pagel 1994b). Thus, it may be incorrect to consider a trait shared by multiple extant species as independent if it is shared among species through common descent rather than independent origin (Harvey and Pagel 1991). While selection might be the cause of sister species sharing traits through common descent, we cannot be confident of this because, even in the absence of selection, sister species also are expected to share trait values. A stronger test of comparative patterns would be the finding that origins of one trait are consistently associated with origins of another trait, rather than the finding that a single evolutionary origin of the two traits was shared through common descent among a group of organisms.

Nonindependence of species data can be tested explicitly using methods described in the previous chapter for quantifying phylogenetic signal (Freckleton et al. 2002; Blomberg et al. 2003). What causes nonindependence among species? Statistically speaking, nonindependence will occur when traits evolve under a Brownian motion model on any non–star-like phylogeny (Blomberg and Garland 2002; Blomberg et al. 2003). At a more biological level, phylogenetic niche conservatism may play a role in nonindependence, with species being more likely to invade a niche that is similar to the niche of their ancestors (Harvey and Pagel 1991; Harvey and Purvis 1991). Imagine, for example, that a new feeding niche arises, such as the evolution of a new fruit that ripens in tree crowns rather than falling to the ground when ripe. We might expect that an arboreal species that is already adapted to eat fruits in the forest canopy is more likely to take advantage of this new feeding niche, compared to a species that specializes in terrestrial locomotion and is less frugivorous. Similarly, if speciation often occurs through barriers to gene flow (allopatric speciation), then we might expect that new species will remain adapted to their current niches (or at least to similar niches) as the populations become genetically distinct. Other potential drivers of phylogenetic nonindependence include genetic and developmental constraints, correlated change in other traits, time lags in selection, niche conservatism, and phenotype-dependent selective pressures (Harvey and Pagel 1991; Harvey and Purvis 1991; Blomberg and Garland 2002; Desdevises et al. 2003). As noted in chapter 5, these different evolutionary mechanisms can be difficult to disentangle.

The problems of nonindependence are not limited to analyses of different species. Comparison of different biological populations raises similar problems, because populations of biological organisms typically exhibit nested patterns of relatedness and adaptation that also lead to nonindependence among populations (Ashton 2004; e.g., in baboons: Newman et al. 2004; Wildman et al. 2004; Kamilar 2006; in birds: Edwards and Kot 1995). The same issue exists in comparative studies of human genetic traits in different populations (e.g., Dediu and Ladd 2007).

In human cross-cultural studies, researchers have long appreciated that

Figure 6.6. Cross-cultural data and statistical nonindependence. The map shows geographical distribution of sampling for marriage transfer patterns in European societies. Open triangles indicate populations that practice bride-price; filled triangles indicate populations that practice dowry. Notice the geographic structuring of the character states, which is likely to reflect shared ancestry. (Redrawn from Fortunato et al. 2006.)

data points representing different human populations are not statistically independent of one another (figure 6.6), causing statistical tests of "raw" cross-cultural data to be incorrect (Naroll 1961; Dow 1984). The assumption in this case is that cultural behaviors represent a body of socially learned traditions that are inherited vertically. One study found that χ^2 tests of cross-cultural data have a vastly inflated type I error rate (Dow 1993), and recent studies have confirmed that trait values from cross-cultural datasets lack statistical independence (Eff 2004; Dow and Eff 2008). This nonindependence probably arises from shared history—that is, because two societies shared a common ancestor and inherited the traits from that ancestor—through many of the same mechanisms just discussed for biological traits. Indeed, the significance of this problem for comparative anthropology was first pointed out by Sir Francis Galton in his 1889 evaluation of E. B. Tylor's comparative work and has since become known as "Galton's problem" (Naroll 1961; see also chapter 10)—although it would seem to be more appropriate to call it Tylor's problem!

Importantly, nonindependence in human cultural and linguistic data sets also can arise through contact with other cultural groups. This contact can include actual borrowing of linguistic traits or cultural behaviors, conquest of one group by another, or the migration of people from one area into another area. Contact-related transmission of cultural traits can make phylogenetic approaches invalid, thus requiring new methods to control for both historical relatedness and patterns of contact among cultural groups (Dow 1984; Nunn et al. 2006). Linguists have even developed methods for identifying "loanwords," which are words that are borrowed between two distinct language groups (Campbell 2004; McMahon and McMahon 2005; Nakhleh et al. 2005). It is possible that data on loanwords between pairs of societies could be used to estimate the probability that other cultural traits are also borrowed, and hence to control for nonindependence that arises through borrowing (Borgerhoff Mulder et al. 2006; Nunn et al. 2006).

Phylogenies and statistical nonindependence. The issue of nonindependence was the driving force behind the development of many innovative tools in comparative biology. As noted above, phylogeny helps to deal with nonindependence in several ways. First, phylogeny provides a way to identify independent data points—in this case, independent evolutionary origins of biological (or cultural) traits. Second, the usual aim of a comparative study is to document correlated trait evolution (Harvey and Pagel 1991; Garland et al. 2005). Hypothesizing that a dependent variable Y, such as neocortex size, is adaptively linked to an independent variable X, such as social group size, implies that the two variables have evolved in a correlated way, with changes in group size covarying with changes in neocortex size. Many phylogenetic methods allow us to test this hypothesis of *correlated evolution* directly. Finally, incorporation of phylogeny may reduce the effects of unmeasured confounding variables, which are problematic when shared through common descent (see Price 1997).

Felsenstein (1985b) provided a hypothetical example to illustrate how ignoring phylogeny can lead to spurious results in comparative studies (figure 6.7). As described in more detail in the figure legend, the essence of the problem is that when data from two clades are combined, it can appear that a significant association exists, yet the "significant" effect can exist entirely as a difference between the clades rather than patterns of correlated evolution within clades. Without evidence of correlated evolution within clades, we are skating on thin ice when we make statements about the traits being correlated.

Although his example may seem to be a somewhat extreme caricature of "the problem" (as Felsenstein called it), figure 6.7 captures the essence of the potential negative impacts of nonindependence on nonphylogenetic comparative tests. In fact, effects such as these are common. *Clade shifts*, for ex-

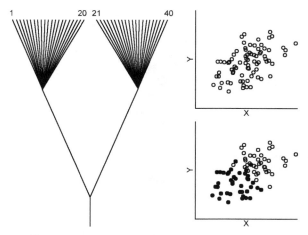

Figure 6.7. The phylogenetic nonindependence "problem." Imagine that we studied a group of organisms for traits *X* and *Y*, and this produced a bivariate relationship shown in the upper right plot. One would be tempted to conclude that a significant relationship exists between the traits. Further suppose, however, that the data come from two clades, as illustrated on the phylogeny, and that within each of these clades no significant relationship exists between *X* and *Y*; instead, there has been a change in *X* and *Y* in the split between the clades (shown as filled and open circles for each of the two clades on the lower right plot). Thus, a cross-species test that ignores phylogeny might conclude that a correlation exists, when in fact the traits are uncorrelated over most of their evolutionary history. (Redrawn and slightly modified from Felsenstein 1985b.)

ample, occur when some variable, shared through common descent, produces a shift in the relationship between the variables with no change in their slopes (Nunn and Barton 2001; Martin et al. 2005). Thus, the slope is the same in the two groups, but the intercept differs, as illustrated, for example, in studies of primate brain size (figure 6.8), ranging patterns (Nunn and Barton 2000), and mammalian development and life history (Martin et al. 2005). In such a case, Felsenstein's (1985b) logic applies, and without phylogenetic control, the regression statistics can easily produce erroneous estimates of the slope (i.e., they would be biased upward in figure 6.8 when placing a single line through all the data points). It is important to appreciate that clade shifts can even cause real patterns to be obscured (figure 6.9; see also Purvis and Rambaut 1995). Thus, ignoring this issue has the potential to mislead research in significant ways, including by missing interesting biological patterns. As we will see in chapter 9, clade shifts and other clade-specific effects are also often useful for testing adaptive hypotheses (Garland et al. 1993) and thus are not simply an annoyance to tame with statistical methods.

Computer simulation studies have decisively shown the magnitude of statistical errors that result when phylogenetic information is ignored (Anthro-

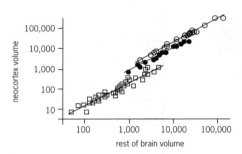

Figure 6.8. Clade shift in brain size. Clade shifts in neocortex size relative to the rest of the brain in insectivores (squares) and primates (prosimian primates, filled circles; anthropoid primates, open circles). For a given body mass, prosimians tend to have larger neocortex size than insectivores, while anthropoids have larger brains than prosimians. Because of these clade shifts, a slope estimated across all species would be greater than slopes within the different clades. (Redrawn from Barton and Harvey 2000.)

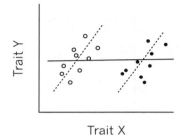

Figure 6.9. Clade shifts can obscure evolutionary patterns. This plot shows hypothetical data. Traits X and Y exhibit correlated evolution within each of two clades, shown as dashed lines and open and filled circles for the species in different clades. When looking across all species combined, however, the association is not significant. Use of a phylogeny would tend to reveal the true association in this case, while species data would not. (Redrawn from Huey 1987 and Gittleman and Kot 1990.)

Tree 6.1; Martins and Garland 1991; Purvis et al. 1994; Nunn 1995; Diaz-Uriarte and Garland 1996, 1998; Harvey and Rambaut 1998; Martins et al. 2002; Rohlf 2006). The effects are staggering. One simulation test, for example, showed that type I error rates can be as high as 44 percent when phylogeny is ignored, compared to an expected error rate of 5 percent (Harvey and Rambaut 1998). A more recent study of paired *t* tests revealed type I errors of over 70 percent when phylogeny is ignored (Lindenfors et al. 2010). In other words, a naïve comparative analysis, such as one that used data straight from figure 6.1, can be nine to fourteen times more likely to indicate a significant pattern than expected by chance, when in fact no relationship exists. The elevated type I error rates in analyses of species data are caused by a wider distribution of statistics than expected if the data were statistically independent (figure 6.10; see also AnthroTree 6.1).

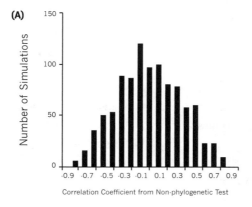

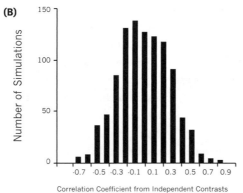

Figure 6.10. Statistical distributions and phylogenetic nonindependence. Data were simulated along a phylogeny with a correlation coefficient of zero, and the resulting data were subject to different analysis procedures (Martins and Garland 1991; see also AnthroTree 6.1). (a) When using standard statistical tests without controlling for phylogeny, the distribution of statistics was wider than expected. (b) When using independent contrasts based on the tree used to simulate the data, however, the statistical distribution narrowed to what would be expected. It is the wider distribution of statistics in (a) rather than bias that causes the elevated Type I error rates in cross-species tests that ignore phylogeny. (Redrawn from Martins and Garland 1991.)

As if that was not bad enough, the statistical power to detect associations can be reduced when phylogeny is ignored (Martins and Garland 1991; Purvis et al. 1994; Harvey and Rambaut 1998). This latter conclusion runs counter to the commonly expressed opinion that nonindependence is simply a "degrees of freedom" problem in which phylogenetic comparative methods are thought to reduce the number of data points for analysis and therefore result in lower statistical power. The next chapter shows, for example, that analyses based on the most commonly used comparative method—independent contrasts—has the same degrees of freedom as one would have in nonphylogenetic analysis (provided that the phylogeny is fully resolved, see Purvis and Garland 1993). In any case, as noted by Pagel (1993), the degrees of freedom analogy can be misleading, as the critical issue is how the variance is partitioned among species rather than the degrees of freedom available for the analysis.

Computer simulation studies also clearly illustrate an important assumption of nonphylogenetic studies that was discussed in the previous chapter. When using species values with no control for phylogeny, the investigator is assuming that the species are linked by a star phylogeny, with all branches emanating as a burst from a single common ancestor (see figure 5.4; Felsenstein 1985b). Many populations and species exhibit more phenotypic similarities to their close relatives than to more distant ones—mouse lemurs tend to be more similar to other species of mouse lemurs (and other lemurs) than they are to monkeys, and gibbons tend to be more similar to one another than they are to other apes. Hence, the assumption of a star phylogeny seems to be violated in many cases, and we can test this possibility using methods described in the previous chapter (Freckleton et al. 2002; Blomberg et al. 2003). Importantly, even imperfect phylogenetic information can provide better statistical performance than ignoring phylogeny (Purvis et al. 1994), as the star phylogeny assumed in a nonphylogenetic test is often a much worse representation of the true phylogenetic structure of the data.

Summary and Synthesis

The comparative method has played a major role in uncovering adaptive trait evolution in biological systems at the species level (Ridley 1983; Harvey and Pagel 1991; Martins 2000; Garland et al. 2005) and for cultural traits and morphological variation in human populations (Mace and Pagel 1994; Mace and Holden 2005). In fact, with the large datasets offered by primatological studies, some of the first statistical applications of the biological comparative method were applied to primate behavior, ecology, and morphology (Milton and May 1976; Clutton-Brock and Harvey 1977; Clutton-Brock et al. 1977; Mitani and Rodman 1979). Likewise, human cultural traits have long been examined comparatively (e.g., Naroll 1965; Cowlishaw and Mace 1996; Ember and Ember 1998; Borgerhoff Mulder 2001; Mace et al. 2003). Phylogenetic methods have been used less frequently to study aspects of language change, but these studies are gaining favor (McMahon and McMahon 2005; Lieberman et al. 2007; Pagel et al. 2007; Atkinson et al. 2008).

From this strong foundation, where does the future of comparative methods lie in evolutionary anthropology? A major area of research will involve incorporating intraspecific variation into comparative studies, especially in studies of nonhuman primate behavior (e.g., Struhsaker 2000; Strier 2003; see chapter 11) and to use this variation to interpret cross-species trends (Martins and Hansen 1997; Martins and Lamont 1998; Ives et al. 2007; Felsenstein 2008). Similarly, the informatics revolution is providing us with larger datasets, including data that can record details on intraspecific variation over time and space (i.e., among different groups and populations). Methods for organizing and accessing such large data sets have yet to be applied to the fullest in studies of primates and humans (see chapter 13).

Last, I wish to end on a cautionary note concerning the "adaptationist program" (Gould and Lewontin 1979). As described by Gould and Lewontin (1979) over twenty-five years ago, we should beware of falling into the pan-adaptationist trap. Some traits may reflect byproducts of selection on other traits, they may be evolutionary relics with no current adaptive function, or they could be the result of processes other than natural selection, such as founder effects. Thus, rejection of one adaptive hypothesis should not lead automatically to a new and different adaptive scenario for a trait, and multiple adaptive solutions may be discovered by evolution for the same environmental or social challenge. Similarly, a trait may have been acted on by many selective pressures, thus requiring multiple, non–mutually exclusive hypotheses for a full explanation of the cross-species variation. Other researchers also have addressed the limitations of the comparative method (Coddington 1988; Reeve and Sherman 1993; Leroi et al. 1994). As noted in the previous chapter, for example, it can be difficult to infer evolutionary process from comparative patterns (Leroi et al. 1994; Hansen and Martins 1996; Blomberg and Garland 2002; Revell et al. 2008). More generally, the comparative method should be applied to adaptive questions with a cautious enthusiasm—an enthusiasm that considers alternative hypotheses, takes care when constructing comparative datasets and treats phylogenetic programs as more than mere black boxes to generate p values.

Pointers to related topics in future chapters:

- Clade shifts and independent contrasts: chapters 7 and 9
- Adaptation, convergence, and single evolutionary origins: chapter 12
- The association between neocortex and group size: chapter 12
- Compiling comparative data: chapter 13

7 Comparative Methods to Detect Correlated Evolutionary Change

The previous chapter demonstrated that both cultural and biological data suffer from serious problems of nonindependence, and this can have major consequences for comparative studies (Felsenstein 1985b; Martins and Garland 1991). If the data from different species and populations are not statistically independent, how can we study the correlated evolution of traits?

In this chapter, I review methods for controlling for the nonindependence of species data points in comparative tests of adaptive hypotheses. Although a wide variety of methodological approaches have been taken to study correlated evolution, I focus in particular on two methods for investigating correlated evolution of continuous characters—independent contrasts (Felsenstein 1985b; Garland et al. 1992) and GLS (Grafen 1989; Martins and Hansen 1997; Pagel 1997; 1999a; Garland and Ives 2000; Lavin et al. 2008)—and one method for investigating correlated evolution of discrete characters, known as Pagel's discrete test (Pagel 1994a; Pagel and Meade 2006a). In a final section, I discuss approaches for analyses of datasets that include both discrete and continuous traits. Throughout, I provide examples of how these methods have been applied in evolutionary anthropology. Additional methods not covered in this book are summarized briefly in AnthroTree 7.1.

Phylogenetically Independent Contrasts

Most traits used in comparative tests of adaptation are quantitative: they vary continuously, including traits such as body mass, latitude, and home range area. The method of phylogenetically independent contrasts is designed to study correlated evolution of continuous characters and is one of the most widely used methods in comparative biology. Indeed, if we take the number of citations of the paper that proposed the method (Felsenstein 1985b), we find that this paper is cited more than three thousand times on the Web of Science database as of the writing of this book—an impressive accomplishment in its own right—and the number of citations appears headed on a positive tra-

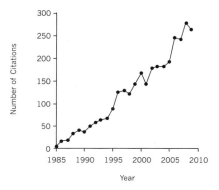

Figure 7.1. Citations per year of Felsenstein's "Phylogenies and the Comparative Method" (1985b). Data come from the bibliographic database Web of Science (http://www.isiwebofknowledge.com/) for the years 1985–2009.

jectory into the future (figure 7.1). Clearly this is a method that has had a huge impact on evolutionary studies and is worth getting to know. I therefore spend much of the first part of this chapter covering independent contrasts and how this method can be applied to evolutionary questions. It is worth noting at the outset, however, that independent contrasts is mathematically equivalent to GLS approaches, and that GLS offers a more flexible approach to studying correlated evolution. Thus, while the method of independent contrasts is well worth the effort to learn, GLS is too, and it may even overtake independent contrasts as the favored method in the future.

Independent contrasts represent differences in trait values between closely related pairs of lineages. These lineages can include those ending in species on the tips of the tree, or among ancestors deeper in the tree. As such, *contrasts represent amounts of evolutionary divergence since two lineages last shared a common ancestor.* By examining evolutionary change rather than the endpoints on the tree (i.e., species), this method overcomes the problem of statistical nonindependence of comparative data, and it more directly gets to the heart of comparative hypotheses that focus on evolutionary change. Figure 7.2 provides an example of independent contrasts involving the scaling of metabolic rate with body mass (Nunn and Barton 2001), which is known as Kleiber's law (Kleiber 1961; Schmidt-Nielsen 1984). This plot can be interpreted as showing that increases in body mass—that is, positive changes in X—are correlated with increases in basal metabolic rate—positive changes in Y. In this plot, body mass contrasts have been forced to be positive (i.e., *positivized*; see Garland et al. 1992). As we will see shortly, independent contrasts

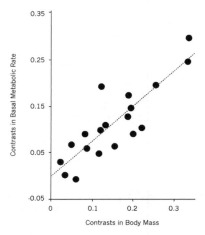

Figure 7.2. Association between basal metabolic rate and body mass in primates. Contrasts in basal metabolic rate are positively correlated with contrasts in body mass ($r^2 = 0.92$, $t_s = 14.2$, $p < .0001$). The slope of the line and the deviations from it (residuals) can be interpreted in the conventional way (see chapter 9). For example, Kleiber's law states that basal metabolic rate increases to the 0.75 power of mass (Kleiber 1961). The 95% confidence intervals for the slope (95% CI, 0.64–0.86) encompass this predicted value. (Taken from Nunn and Barton 2001.)

are typically standardized for evolutionary time (i.e., branch lengths), and in this sense can also be considered rates of evolution (Garland 1992).

On first exposure to independent contrasts, some people find the basic concept difficult to comprehend—how is it that differences in trait values can be equated to evolutionary divergence? If the goal is to study correlated evolutionary change, it might seem more intuitive to investigate correlated evolutionary change along each branch, for example by reconstructing the ancestral nodes of the two traits on a phylogeny, calculating change from the deeper to shallower nodes (including to species values at the tips of the tree), and then testing whether these changes are correlated. Upon further reflection, however, a major problem becomes evident with this alternative branch-by-branch approach: for a fully resolved phylogeny with n measured species, there are $2n - 2$ branches, therefore giving more degrees of freedom in a statistical test than there are actual measured data points (see also Martins and Garland 1991 and Pagel 1993). By comparison, a fully resolved tree (i.e., one with no polytomies) produces $n - 1$ independent contrasts. The method of independent contrasts uses ancestral states as an intermediate step in the calculations, but it examines differences among nodes (or tips) rather than along individual branches. Hence, the method does not inflate the degrees of free-

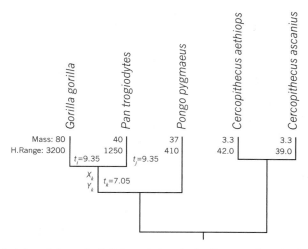

Figure 7.3. Calculating phylogenetically independent contrasts. Shown are five primate species, their phylogenetic relationships, and data on body mass and home range area along the tips of the phylogeny. X_k and Y_k represent ancestral values for body mass and home range area, respectively, and t gives branch lengths.

dom above the number of measured data points—a reasonable requirement for any statistical test.

Calculating unstandardized contrasts. The example used here is a simple one involving calculation of contrasts in body mass and home range area in five primate species (three apes and two Old World monkeys, data are available in AnthroTree 7.2). We might expect that larger bodied primates require a larger home range, and this pattern has indeed been demonstrated in previous work (Milton and May 1976), including studies that used independent contrasts (Nunn and Barton 2000; Nunn and Van Schaik 2002). Clearly the sample size used here is too small to re-test the association between these variables in a conclusive way, but it serves as a tractable, real-world example of how contrasts are calculated. Data were taken from Rowe (1996) and shown with the phylogeny in figure 7.3. Garland et al. (2005) also provide a worked example.

The algorithms for obtaining independent contrasts start at the tips of the tree. Let us begin with the contrast between the gorilla and chimpanzee (i.e., *Gorilla gorilla* and *Pan troglodytes*). The contrast in body mass, which we will identify as the independent variable X, is calculated as $80 - 40 = 40$, and the contrast in home range, Y, is calculated as $3{,}200 - 1{,}250 = 1{,}950$. Two

Table 7.1. Independent contrasts example: body mass.

Contrast label	Contrast description	Unstandardized contrast	Sum of branch lengths	Standardized contrast
A	*Pan—Gorilla*	40	18.7	9.25
B	Ancestor(P + G) – *Pongo*	23	28.23	4.33
C	*C. aethiops – C. ascanius*	0	9.6	0
D	Ancestor(P + G + P) – *Cercopithecus*	47.1	49.93	6.67

Note: Results may differ slightly from values shown here due to rounding during calculations.

Table 7.2. Independent contrasts example: home range area.

Contrast label	Contrast description	Unstandardized contrast	Sum of branch lengths	Standardized contrast
A	*Pan—Gorilla*	1950	18.7	450.9
B	Ancestor(P + G) – *Pongo*	1815	28.23	341.6
C	*C. aethiops – C. ascanius*	3	9.6	0.97
D	Ancestor(P + G + P) – *Cercopithecus*	1430.3	49.93	202.4

Note: Results may differ slightly from values shown here due to rounding during calculations.

important points should be mentioned at this stage. First, the contrasts are *unstandardized* because they have not been corrected for branch lengths (more on that soon). Second, the direction of subtraction is arbitrary and requires only that the direction remain the same for all contrasts calculated at a particular node. Thus, it would be equally valid to subtract the value for the gorilla from the chimpanzee in calculating the contrast for body mass, giving 40 − 80 = −40. By doing so, we must calculate the contrast for home range size in the same direction: 1,250 − 3,200 = −1,950. Many comparative analysis programs choose the direction of subtraction such that one of the independent variables is *positivized*, as was done for body mass in figure 7.2.

Tables 7.1 and 7.2 summarize the contrasts for body mass and home range area, respectively, where letters refer to equivalent contrasts between the same pairs of lineages in the two tables. The tables also record information on the sum of branch lengths, which is necessary for generating *standardized contrasts*. The description and calculations here are for a fully bifurcating tree. Polytomies require additional calculations or assumptions about the nature of the polytomy as hard or soft, with a variety of methods available that include setting some branch lengths to zero, randomly resolving the polytomy in different ways, or taking only a single contrast at polytomous nodes (see Grafen 1989; Pagel 1992; Purvis and Garland 1993; Purvis and Rambaut 1995).

The next step is to estimate the nodal values for the most recent ancestor of the two species that were just contrasted. The algorithm used for this has several special properties. First, it uses only values "above" the node leading towards the tips of the tree (in this case, values from the gorilla and chimpanzee, respectively). This differs from the approach to reconstructing ancestral states of continuous characters by using all surrounding nodes (see chapter 4), and it is essential for maintaining the independence of the contrasts. Second, the average is not a simple average of descendants; instead, the nodal estimate is weighted by the branch lengths leading to each of the descendants. As shown in figure 7.3, if the ancestor is labeled k, the two descendent nodes/species are i and j, and X and t are trait values and branch lengths, respectively, we would calculate the nodal value as:

$$X_K = \frac{(1/t_i)X_i + (1/t_j)X_j}{1/t_i + 1/t_j}$$

Last, a small amount of error is added to the branch below the estimated nodal value (i.e., the branch leading to the ancestor of the node), based on the assumption that branch length reflects the expected variance in trait change. This addition can be viewed as an extra "burst" of evolution (Felsenstein 1985b, 10) to reflect that nodal values are estimates rather than "real data." Thus, the branch is lengthened by this amount:

$$t'_k = t_k + \frac{t_i t_j}{(t_i + t_j)}$$

To make this process more concrete, consider the body mass contrast for A. One would reconstruct the estimated value of the most recent ancestor of the gorilla and chimpanzee by using only data on these two species (not other species or reconstructed nodes). Using the formula from above, we obtain estimates for mass and home range (X and Y) at this node:

$$X_K = \frac{(1/9.35)80 + (1/9.35)40}{1/9.35 + 1/9.35} = 60$$

$$Y_K = \frac{(1/9.35)3200 + (1/9.35)1250}{1/9.35 + 1/9.35} = 2225$$

In this case, the nodal values are simple averages of descendants because the branch lengths are equal along each of the descendent lineages. We then need to lengthen the branch leading to this new node (i.e., to its ancestor) to reflect the statistical uncertainty in its estimation. Thus:

$$t'_k = 7.05 + \frac{9.35 \times 9.35}{(9.35 + 9.35)} = 11.73$$

With these values, we can prune the branches i and j that lead to gorillas and chimpanzees, respectively, resulting in the new tree in figure 7.4.

The algorithm works in a recursive manner through the tree by calculating contrasts at each node, covering each evolutionary path from tips to root such that each branch is used only one time. Thus, another contrast can be calculated between the ancestor of *Pan* and *Gorilla* and the species value for *Pongo* (see tables 7.1 and 7.2). The ancestor would then be calculated for this node, the branch to this node from the root lengthened, and the branches descending from this node to the tips deleted. Notice in this case that the estimated nodal value is not a simple average, due to the fact that the branch lengths differ in this contrast. Thus, the value for X_k and Y_k tends to be more similar to the value for the reconstructed ancestor of *Pan* + *Gorilla* because this ancestor is evolutionarily closer to the node being estimated.

This process continues until no more contrasts can be calculated, with the last contrast thus calculated at the root node. We have produced a set of contrasts that need to be standardized and then analyzed, but before this is done, three points require attention. First, recall that $n - 1$ contrasts are possible when we have n species and a fully bifurcating phylogeny—a fact that can be confirmed by noting that our five species of primates produced four contrasts in tables 7.1 and 7.2. As will be discussed shortly, that one "lost" degree of freedom (i.e., $n - 1$ contrasts compared to n species) is given back in the statistical analysis. Second, some contrasts involve comparison of extant species (*Pan-Gorilla*), others involve comparison of two reconstructed ancestors (contrast D: ancestor of apes vs. ancestor of *Cercopithecus* monkeys), and yet others involve comparison of one extant species and one reconstructed ancestor (e.g., contrast B: *Pongo* versus ancestor of *Pan* and *Gorilla*). It is often valuable to distinguish among these three different types of contrasts, and

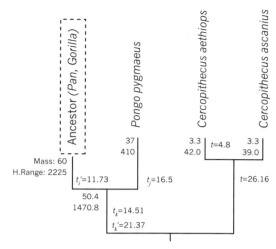

Figure 7.4. Ancestral states and independent contrasts. This figure shows the ancestral states in *Gorilla* and *Pan* along the branch that led to their ancestor in figure 7.3. Estimates are based only on descendants from the node, and that the branch length to the new node is lengthened from t_i to t_i'. This estimate is needed to calculate contrasts deeper in the tree, for example, between the ancestor of *Pan + Gorilla* and *Pongo*. Because these estimates are not based on information above and below the node (e.g., in squared change parsimony described in chapter 4), some users avoid the term "reconstructions" for these values.

more importantly, to use a computer program that allows easy identification of the species involved in a particular contrast (see AnthroTree 7.2). Last, some contrasts can be equal to zero when there are no differences in trait values, as shown in the body mass contrast between the two species of *Cercopithecus* monkeys (contrast *C* in table 7.1). This is not a problem, and in fact it is quite common; it simply reflects that closely related lineages will often have the same (or very similar) trait values, as expected when species share traits through common descent (Blomberg et al. 2003).

Standardized contrasts. The association between the *standardized contrasts* is shown in figure 7.5, which reveals that contrasts in body mass are associated with contrasts in home range size for this very small selection of primate species ($t_2 = 5.58$, $p = .01$). Typically, the standardized contrasts are used to test a hypothesis, and the aim of this section is to explain why standardization is necessary and how it is achieved.

An important issue in any statistical test is that the data should have the same underlying variance, or *homoscedasticity*; violation of this assumption is referred to as *heteroscedasticity* (Sokal and Rohlf 1995). Heteroscedasticity can be an issue with independent contrasts, because a contrast between

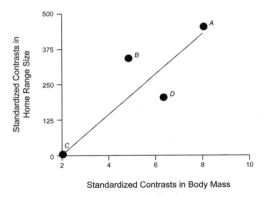

Figure 7.5. Simple example of independent contrasts. Contrasts in body mass covary with contrasts in home range size. These contrasts are standardized for evolutionary time.

two lineages will tend to be larger in magnitude when the lineages shared an ancestor in the more distant past. Thus, we might expect that the contrast between the two ape species is larger than the contrast between the two *Cercopithecus* species, as the last common ancestor of *Pan* and *Gorilla* existed 9.35 million years ago, while the ancestor of the two species of *Cercopithecus* is about half of that age, occurring 4.8 million years ago. This is indeed the case (see tables 7.1 and 7.2). This difference might also partly reflect that larger mean values of body mass and home range—which characterize the apes more than the *Cercopithecus* monkeys—are more variable. Thus, log transformation of the raw values can be useful when calculating contrasts, just as it is useful for analysis of species values.

The problem of heteroscedasticity is where standardization comes into play. Under a Brownian motion model of evolution, the variance accumulates proportional to time. Using this assumption, we can scale the contrasts so that they have the same expected variance. Standardization is relatively straightforward: *based on the Brownian motion model, we simply divide by the square root of the summed branch lengths used for each contrast.* Because the branches represent variance under Brownian motion, the square root of the total time represented by the contrast is the standard deviation. The standardized contrasts are shown in tables 7.1 and 7.2.

In the process of standardizing contrasts, it is important to test the assumptions of the method, especially the assumption that Brownian motion adequately describes the evolutionary process for a particular phylogeny and comparative dataset (Harvey and Pagel 1991; Garland et al. 1992; Purvis and

Table 7.3. Some assumption checks for independent contrasts.

Assumption	Description of Test
Brownian motion	Absolute value of standardized contrasts should be independent of square root of the sum of branches for each contrast (regression not through origin). In addition, the absolute value of standardized contrasts should be independent of the estimated value at the node for that contrast (regression not through origin).
Age	Absolute value of standardized contrasts should be independent of the age of the node for each contrast (regression not through origin). (See also Freckleton and Harvey 2006.)
Residuals	Check for heterogeneity in the variance of the residuals; specifically, calculate residuals from regression of contrasts in Y on contrasts in X (through the origin), and plot these residuals against contrasts in X (all contrasts are standardized).
Outliers	Removal of outliers can often improve the fit of the assumptions. However, outliers may also be biologically interesting (e.g., in the context of clade shifts) and so should be examined carefully in light of possible biological explanations. Use statistical procedures to objectively identify outliers.
Measurement error	When number of samples per species is low or intraspecific variation is known to be high, it is useful to estimate the variation that can be accounted for among species and to carefully examine whether terminal contrasts with short branches are outliers (Harmon and Losos 2005). (See also Ives et al. 2007 and Felsenstein 2008.)

Source: These and other checks are summarized from descriptions for the programs "Comparative Analysis by Independent Contrasts" (CAIC; Purvis and Rambaut 1995) and the "Phylogenetic Diversity Analysis Programs" (PDAP; Garland et al. 1993; Garland et al. 1999; Garland and Ives 2000) and other references (Harvey and Pagel 1991; Garland et al. 1992; Nunn and Barton 2001; Freckleton 2009).

Rambaut 1995; Garland et al. 2005). The importance of assumption checks has been demonstrated in computer simulations (especially by Diaz-Uriarte and Garland 1996, 1998). Simulation studies also have shown that incorrect phylogenetic information can negatively impact the statistical performance of independent contrasts (Purvis et al. 1994; Symonds 2002). In most cases, however, simulation studies have revealed that completely ignoring phylogenetic information has substantially greater negative effects on error rates than assumption violations when using independent contrasts. Thus, a failure to meet the assumptions is not a default excuse for conducting nonphylogenetic tests.

The key assumptions of independent contrasts are given in table 7.3. One of the most important and best understood of these assumptions is that the absolute value of the standardized contrast should be independent of the square root of the sum of the branches involved in the contrast (Garland et al. 1992; Diaz-Uriarte and Garland 1996). As shown in figure 7.6, this assumption appears to be adequate for this very small dataset (i.e., the statistical test

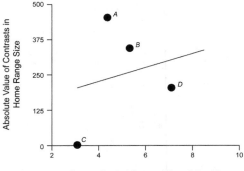

Figure 7.6. Testing the "standardization" assumption of independent contrasts. If the contrasts are properly standardized under the Brownian motion model of evolution, then the magnitude of the contrasts will be independent of the square root of the summed branch lengths used to standardize them (Garland et al. 1992). Because the sign of the contrast is arbitrary, the absolute value is taken. Here, I tested the assumptions for the contrasts in home range size (hectares, not log-transformed). In this very simple case, the assumption is not violated ($t_2 = 0.40$, $p = .73$), suggesting that standardization is adequate. The same was true of body mass ($t_2 = 0.82$, $p = .50$).

of the assumption is not significant), although as revealed in AnthroTree 7.2, other assumptions are violated. Another assumption not mentioned in table 7.3 is that the tree topology and branch lengths are correct. This assumption cannot be tested, but this important issue can be addressed by using a Bayesian sample of trees (e.g., Huelsenbeck et al. 2000; Pagel and Lutzoni 2002; Huelsenbeck and Rannala 2003) or, if such a sample is unavailable, running the analysis on multiple inferences of phylogeny (e.g., Pagel 1994a; Mitani et al. 1996b).

Perhaps the most common question from first-time users of independent contrasts is, "I have some assumption violations, what should I do?" First of all, as it says in *The Hitchhiker's Guide to the Galaxy*, don't panic! Assumption violations are common, and it is usually possible to transform the branch lengths on the phylogeny to meet the assumptions. In addition to transforming branch lengths, transformations of the data, such as taking the logarithms of the values, can often help to meet the assumptions (Garland et al. 1992). And as just mentioned, even when assumption violations occur or cannot be fully rectified, the method of independent contrasts generally exhibits superior performance relative to nonphylogenetic tests of species data (Diaz-Uriarte and Garland 1996). In my own experience analyzing nonhuman primate data, I have found that log transformation of the raw data typically—but not always—helps to meet the assumptions. Branch lengths often need to be transformed and again, log transformation has worked well in my analyses of primate data. It is important to emphasize, however, that the assumptions need to be checked for each dataset on a variable-by-variable basis.

When branch lengths are unknown, users often assume equal branches. However, some other branch length transformations are available and often better meet the assumptions (AnthroTree 7.3). One, called the Grafen transformation (not to be confused with ρ in chapter 5), sets the ages of the nodes equal to one less than the number of species arising from that node; from these "dated" nodes, it is possible to infer the branch lengths (Grafen 1989). One can also assign ages based on the log number of descendants, which is known as the Nee transformation (see also Pagel and Harvey 1992; Purvis 1995). This transformation has more biological validity, as it is the expectation under some models of clade diversification (i.e., the birth-death model; see chapter 8).

In summary, one should try a transformation of data or branch lengths, determine if that helps to meet the assumptions, and continue with the analysis once the assumptions are adequately met. If the assumptions cannot be met, then one can investigate whether outlier contrasts are causing problems with assumptions violations, with subsequent removal of the outliers to better meet the assumptions.

Testing for correlated evolution. After completing these steps, we are finally at a point to test whether contrasts in the two (or more) traits show evidence of correlated evolution. One can simply import the contrasts into a statistical package and use standard analytical methods, including bivariate correlations, multiple regression, and even principal components analysis (PCA; Ackerly and Donoghue 1998; Revell 2009). One critical issue arises, however: *regressions, correlations, and even PCA should be forced through the origin* (Harvey and Pagel 1991; Garland et al. 1992; Ackerly and Donoghue 1998). The reason for this follows from a point that was discussed earlier, namely that the direction of subtraction is arbitrary when calculating contrasts. Because of this, the expected value of a contrast is zero (e.g., in the example above for body mass contrast A, the value was either 40 or −40, giving an expectation of 0). Because the regression line must pass through the mean of X and Y, the line must go through the expected values of zero for each trait—i.e., the origin, at 0,0. It is possible to estimate the phylogenetically valid intercept in the raw state space (Garland et al. 1993, 1999; Garland and Ives 2000). This is often important in studies of allometry, which is covered in chapter 9.

Regression through the origin causes much confusion and distress, and sometimes even resistance from authors and reviewers. In fact, this requirement is valuable to the method in a way that is not widely appreciated. As noted above, a phylogeny with n species has $n - 1$ contrasts; thus, it would appear that an independent contrasts analysis loses one degree of freedom, which is unfortunate, especially for small datasets where we need as many degrees of freedom as possible. However, by forcing the line through the origin, that degree of freedom is, in effect, "refunded" to the user because the intercept is not estimated (assuming a bifurcating tree; Purvis and Garland 1993). More specifically, by not estimating an intercept in a regression model, the degrees of freedom are calculated as the number of contrasts minus 1, rather than number of contrasts minus 2 (Garland et al. 1992). Thus, when using a bifurcating tree, the degrees of freedom for analyses of contrasts and species values are the same.

Regression through the origin is also confusing to some users in the context of clade shifts. As noted in chapter 6, clade shifts represent changes in the intercept for a given relationship between two traits using species values (see figures 6.8 and 6.9; Nunn and Barton 2001; Martin et al. 2005). This information on intercept is effectively discarded when using independent contrasts forced through the origin. As will be discussed in chapter 9, however, a clade shift that occurs between two clades will represent only a single contrast, which may (or may not) show up as an outlier on a plot of the contrasts (Garland et al. 1993; Purvis and Webster 1999). Thus, analyses based on independent contrasts are usually less affected by clade shifts than are analyses of species data, where many more data points typically bias slope estimates (Nunn and Barton 2000, 2001; but see Martin et al. 2005).

Calculating contrasts might seem like a lot of work. Fortunately, a number of easy-to-use computer packages have been developed to calculate contrasts (AnthroTree 7.2 and 7.4). Most of the calculation steps are hidden from the user, except for assumptions checks, which the user typically needs to evaluate for him- or herself.

Issues involved with using independent contrasts. In addition to the assumption checks, some other issues arise when using independent contrasts. One of these involves intraspecific variation (Purvis and Webster 1999; Harmon and Losos 2005; Ives et al. 2007; Felsenstein 2008). Intraspecific variation can result in apparent differences among closely related species—especially on the tips of the tree—when in fact the differences are well within the range of variation expected given our ability to measure the trait in question (see Ricklefs and Starck 1996; Purvis and Webster 1999). This is one case where it is useful to use contrasts calculated at internal nodes (e.g., Nunn and Barton 2000), with the expectation that internal contrasts will represent more biologically meaningful differences. Chapter 11 discusses additional approaches for addressing issues related to measurement error.

Another question that commonly arises is whether to run analyses using both phylogenetic and nonphylogenetic approaches. As noted in the previous chapter, the results of phylogenetic and nonphylogenetic approaches are often congruent, but not always (see figure 6.5; Ricklefs and Starck 1996; Price 1997; Carvalho et al. 2006). In the past, I advocated reporting both sets of statistical results (Nunn and Barton 2001), and I followed this approach in my own research (e.g., Nunn 1999b; Nunn et al. 2003). However, I no longer follow the approach of reporting results based on analyses of both independent contrasts and species data (see also Freckleton 2009). Instead, users should use diagnostics from chapter 5 to assess whether the traits show phylogenetic signal, and if so, undertake a phylogenetic comparative analysis that meets the assumptions based on the level of signal. We will see in the next section that it is relatively easy to integrate measures of phylogenetic signal and evolutionary patterns with comparative analyses based on GLS (which can include a star phylogeny representing a nonphylogenetic test; Pagel 1999a; Spoor et al. 2007; Lavin et al. 2008; Gartner et al. 2010). In fact, many of the evolutionary diagnostics from the previous chapter can be estimated and used to transform the branches for a contrasts analysis.

It is common to hear statements like, "Independent contrasts and related approaches rely too heavily on instances of character change, whereas stasis might also be consistent with an adaptive hypothesis, for example in the context of stabilizing selection." Although methods have been developed to study stabilizing selection (see the next section), these methods have not yet played a prominent role in studying adaptation—or at least not as prominent as methods based on independent evolutionary change, such as independent contrasts. This probably reflects that, when testing adaptive hypotheses, the most convincing evidence comes from repeated origins of a particular trait with another trait (Pagel 1994b; see chapter 6). In other words, correlated evolutionary change is generally more convincing than correlated evolutionary stasis. However, this is an area of active research, and in the future new methods may gain favor for studying correlated evolution in the context of stabilizing selection (Hansen et al. 2008).

Finally, it is worth mentioning that the method of independent contrasts can be adapted to study a wide array of other questions. In chapter 5, for example, I discussed how evolutionary differences can be used as a measure of phylogenetic signal (Blomberg et al. 2003) and to study rates of evolution (Garland 1992). Another example concerns evolutionary lag, that is, whether evolutionary change in one trait lags behind change in another trait (Deaner and Nunn 1999). This approach makes use of unstandardized contrasts by asking whether the amount of change in one trait relative to another trait is a function of the time since pairs of lineages diverged. Lindenfors et al. (2004) used this method to test whether the number of males lags behind the number of females in primate social groups, as predicted if female behavior drives variation in the number of males (Altmann 1990; Nunn 1999b). They found support for evolutionary lag. Thus, an understanding of independent contrasts provides a general framework for addressing a broad array of evolutionary questions.

Examples. The method of independent contrasts is probably the most widely used method in comparative biology, and so it is not surprising that evolutionary anthropologists have also applied the method to a diverse array of questions (see Nunn and Barton 2001). Across primates, for example, a number of studies have investigated whether sexual dimorphism reflects selection on male body size arising from competition over mates (Clutton-Brock et al. 1977), and more recent studies have analyzed this important question using independent contrasts (Kappeler 1997; Plavcan and Van Schaik 1997b; Lindenfors and Tullberg 1998; Smith and Cheverud 2002). Take, for example, a study by Mitani et al. (1996a). These authors reasoned that the best measure of intrasexual competition would come from using the operational sex ratio, that is, the ratio of reproductively active males to reproductively active females. To calculate this ratio, they used data on the sex ratio of adults in groups and days per year that females in groups are sexually active. After controlling for body mass, they found that evolutionary increases in sexual dimorphism tended to occur with evolutionary increases in the operational sex ratio (figure 7.7; note that intermediate steps, such as controlling for body mass using residuals, should also be conducted phylogenetically; see Revell 2009). Thus, species with relatively greater sexual dimorphism tended to have relatively more males per available female.

Chapter 1 discussed some of the work in comparative psychology that has made use of comparison, including independent contrasts analyses by

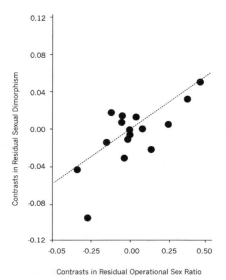

Figure 7.7. Competition for mates and sexual dimorphism. Contrasts in sexual dimorphism and operational sex ratio controlling for body mass by taking residuals. This association is highly significant ($r^2 =$ 0.49, $t_{15} = 3.77$, $p < .01$). Redrawn from Mitani et al. (1996a).

Reader and Laland (2002; see figure 1.2). As the name implies, research in this field often involves comparisons of different species. These comparisons are usually made on a scale that is too limited for application of contrasts approaches, often because only a handful of species are compared. Contrasts approaches have been used, however, in studies of brain size in relation to play (Lewis 2000; Iwaniuk et al. 2001), general cognitive performance (Deaner et al. 2007), and energetic costs of brain size (Martin 1998; Isler and Van Schaik 2006).

Other authors have applied the method of independent contrasts to study human behavioral and life history features. In one study, Holden and Mace (1997) used independent contrasts and other methods described below (Pagel 1994a) to investigate three hypotheses for the evolution of adult lactose digestion capacity in human populations. First, lactose tolerance could be related to consumption of milk, and thus, the presence of dairying. Second, lactose digestion could be related to latitudinal deficiency of vitamin D and the facilitating role of milk in calcium uptake. Third, the ability of adults to digest lactose could be related to the need for fluids in highly arid environments. Holden and Mace's (1997) data came from 7,905 individuals representing 62 cultures, and they used three phylogenies to assess the sensitivity of their results to genetic and linguistic relationships among the human cultural groups. Their analyses supported the link between dairying and lactose tolerance, but found no support for the other two hypotheses; specifically, consumption of raw milk was unrelated to solar radiation or aridity.

Generalized Least Squares Approaches to Correlated Evolution

Another approach for studying correlated evolution of continuous characters makes use of the GLS methods discussed in previous chapters (Grafen 1989; Martins and Hansen 1997; Pagel 1997, 1999a; Garland and Ives 2000; Rohlf 2001). With a GLS framework, alternative models can be compared easily using likelihood ratio tests (LRTs) and other model selection approaches, including comparison of models with and without correlated evolution. It is important at the outset to note that the method of independent contrasts is mathematically identical to GLS, and that GLS approaches provide a more flexible approach to studying correlated evolution (Garland and Ives 2000; Rohlf 2001, 2006; Lavin et al. 2008). It seems likely that GLS-based approaches will become increasingly attractive to comparative biologists in the future. Even so, it is still important to learn independent contrasts, given that so much previous work has been based on the method, and because it will likely remain in the phylogenetic toolkit for many years to come.

In the context of correlated evolution, GLS investigates whether one or more independent traits predict values of another (dependent) trait. The model used for two traits can be represented by the equation $y_i = \alpha + \beta\, x_i + \epsilon$, where x and y are trait values, α is the intercept, β is the regression slope, and ϵ is the error term. This general statistical modeling framework is probably familiar to most readers with a basic understanding of statistical methods, such as regression, where we are interested in estimating the slope and intercept. How does one include phylogeny in this approach? As in the previous uses of GLS discussed in chapters 4 and 5, phylogenetic relatedness is incorporated into the error term using a variance-covariance matrix (see figure 5.5), which is derived from the phylogenetic topology and branch lengths. It is the "generalized" aspect of the method that enables nonstandard assumptions, such as assumptions about the correlations among the data points (i.e., nonindependence).

GLS is less widely used than independent contrasts, yet as a platform it offers some important advantages. As noted above, for example, GLS approaches provide more flexibility than standard independent contrasts approaches. This flexibility is perhaps most obvious in terms of using maximum likelihood methods to probe the underlying evolutionary pattern of trait change. For example, the parameter λ (and some other parameters discussed in chapter 5) can be estimated and taken into account by scaling the variance-covariance matrix (Freckleton et al. 2002; Revell 2010). Thus, while independent contrasts effectively assumes that $\lambda = 1$ and users must transform the data and phylogeny to try to meet this assumption, the variance-covariance matrix used in GLS provides a way to more flexibly and systematically incorporate scaling parameters into the analysis of correlated evolutionary change (including $\lambda = 0$, which is equivalent to a star phylogeny; see figure 5.4). Often it is important to use the maximum likelihood estimate of λ when studying correlated evolution, for example when λ differs significantly from both 0 and 1 (e.g., Capellini et al. 2010; see chapter 5). In addition, it is possible to implement a model of evolution based on stabilizing selection (the OU model; Hansen 1997; Martins and Hansen 1997; Hansen et al. 2008; Lavin et al. 2008; Gartner et al. 2010) or to adjust parameters to make the statistical model appropriate to the ecological niche-filling model (e.g., g and δ; Pagel 1999a; Blomberg et al. 2003). The flexibility of GLS also carries over easily to address other questions, such as phylogenetic meta-analysis (Adams 2008; Lajeunesse 2009) or integrating directional models of trait evolution into studies of adaptation (Pagel 1997, 1999a; see also chapter 4).

As already noted, results from independent contrasts represent a special case of GLS (Garland and Ives 2000; Rohlf 2001, 2006). When using GLS with the assumptions of Brownian motion evolution, GLS will return results that are identical (or nearly so) to independent contrasts. However, GLS offers additional advantages over standard implementations of independent contrasts. When analyzing independent contrasts, for example, the intercept is forced to equal zero, and explicit estimation of an intercept in the raw data requires additional steps in the analysis (i.e., positioning the regression line through value at the root node; see Garland et al. 1993, 1999; Garland and Ives 2000). Intercepts, standard errors, and many other statistical parameters can be extracted more directly from a GLS analysis (Pagel 1997). Analyses and assumption checks might be more intuitive when using independent contrasts, but the flexibility of the GLS method is likely to make this the method of choice in future comparative research.

While calculation of independent contrasts can be done by hand (although this is not recommended!), investigation of correlated evolution by GLS requires a computer. Luckily, several recent software developments have made this method easier to implement. An example dataset and instructions are provided in AnthroTree 7.5.

Examples. Spoor et al. (2007) addressed the question, "What factors have influenced variation in the size of the semicircular canals in mammals?" This is an important yet neglected question, because the semicircular canals play a major role in balance, vision, and coordination of body movements. In addition, the semicircular canals can be measured from fossils, thus providing a means to investigate locomotor patterns in extinct species—provided of course that we understand variation in living species to make inferences for fossils (Silcox et al. 2009). Spoor et al. (2007) compiled a dataset of 210 mammalian species, which included 91 living primates and subfossil lemurs. They tested the hypothesis that mammals exhibiting higher levels of locomotor agility will have larger semicircular canals. Using a GLS model that included body mass as a covariate and incorporated the degree of phylogenetic signal, the authors found support for this hypothesis (figure 7.8).

Pagel (1997, 1999a) provided an example of the GLS approach involving two well-studied comparative patterns in mammals: the association between brain size and body size and the association between metabolic rate and body mass. In an analysis of how brain size scales with body mass among twenty-three species of mammals, estimates of δ and λ were not significantly different from one (Pagel 1999a). Thus, rates of evolution have been relatively constant, they are consistent with Brownian motion, and the traits show phylogenetic signal. Remarkably, however, the GLS regression slope is much lower than previous estimates—that is, 0.59 with 95% confidence intervals of 0.52 to 0.67 (Pagel 1999a). When applied to fifty-nine species of primates, a similar result emerges (b = 0.48; 95% confidence interval, 0.39–0.57). By comparison, the association between metabolic rate and body mass in fifteen species of mammals returns a value close to the expected "three-quarters rule": Kleiber's law (0.72; 95% confidence interval, 0.60–0.84) (Kleiber 1961; see chapter 9 for more recent comparative studies that strongly challenge the three-quarters scaling of metabolic rate proposed under Kleiber's law).

Hansen et al. (2008) developed a comparative method consistent with adaptation toward a fitness peak in the framework of an OU process. As an example, they applied the method to study the association between dimorphism and body mass in primates (i.e., Rensch's rule; Abouheif and Fairbairn 1997). Using data on the ratio of male-to-female body mass and female body mass for 105 primate species, they tested whether a significant association is

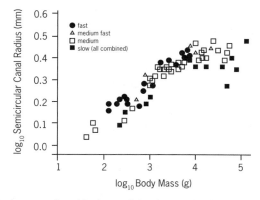

Figure 7.8. Agility, body mass and semicircular canals in primates. Plot shows species values and reveals an association between body mass and the size of the semicircular canals ($n = 91$). Analyses were conducted using phylogenetic generalized least squares, and further identified locomotor mode as a predictor of semicircular canal size. (Redrawn from Spoor et al. 2007.)

found among these traits. They found support for the predicted association, but it explained only 13 percent of the interspecific variation.

At a worldwide scale, women exhibit a tendency to produce slightly more male than female offspring, yet different human populations exhibit intriguing variation in this trait. Mace et al. (2003) used GLS to examine variation in biased sex ratios in modern human populations. For example, African populations tend to exhibit a weaker male bias in the sex ratio, while European populations show a stronger male bias. Few researchers have investigated this variability in a systematic way. Controlling for ancestry using genetic information, Mace et al. found that sex ratio at birth covaries strongly with sex ratio of children that survive to age 14 among Old World populations. This suggests that sex ratios at birth are not correlated with sex differences in childhood mortality, as might be expected if the slight male bias is due to higher male mortality. They found additional effects on the sex ratio involving fertility rates and the risk of death during childbirth.

To wrap up, it is fair to say that on the whole, GLS approaches represent a valuable improvement over analyses based on phylogenetically independent contrasts. While independent contrasts is a tried and true method and is unlikely to mislead, it can actually take more work to run a contrasts analysis than GLS, especially given the assumption checks and associated transformations of data and branch lengths needed to meet the assumptions of independent contrasts (although GLS approaches are also associated with assumption checks; see Freckleton et al. 2009; Revell 2010). Moreover, GLS provides a

significantly more flexible framework for investigating correlated evolutionary change under different evolutionary models, and for taking into account
scaling parameters of interest to those models, such as using λ as a measure of
phylogenetic signal. Recent advances in phylogenetic GLS offer great promise
for the future, specifically with regard to assessing phylogenetic signal, correlated evolution, and multiple predictors in a single analytical framework
(Spoor et al. 2007; Lavin et al. 2008; Gartner et al. 2010; Revell 2010).

Correlated Evolution of Discrete Traits: Pagel's Discrete Method

Pagel (1994a) developed a maximum likelihood approach to statistically
compare two models of evolution: one in which two traits are evolving independently of one other, and another in which the traits exhibit correlated
evolution, such that states of one trait affect the probability that the other
trait changes (and vice versa). The traits, X_1 and X_2, are binary, meaning that
they take only one of two values (e.g., absence or presence, usually coded as
0 and 1, respectively). An advantage of this method over other approaches for
the analysis of discretely coded data (e.g., Ridley 1983; Maddison 1990) is that
it makes use of branch lengths, thus allowing for the possibility that changes
are more likely to occur on longer branches. In addition, Pagel's (1994a)
method does not rely on explicitly reconstructing ancestral states for internal
nodes. Other methods typically treat these nodes as if they were actual data
and, by performing statistical tests using the reconstructions, the results are
conditioned on our ability to accurately piece together the evolutionary history of the traits (e.g., Maddison 1990). AnthroTree 7.6 provides an example
dataset for running an analysis using Pagel's discrete method.

Pagel's (1994a) approach characterizes evolutionary change along branches
based on a continuous-time Markov model. In effect, it makes use of all possible character states at internal nodes, and explicit statistical comparison of
reconstructions is not required. The method compares two models of trait
change: one in which the traits are evolving independently of one another (the
independent model), and another in which the traits exhibit correlated evolution (the *dependent model*). The likelihoods of these two models are then
compared, with the expectation that the *dependent model* has a significantly
higher likelihood if the traits are evolving in a correlated fashion.

The *independent model* is based on the maximum likelihood approach discussed in chapter 3 for reconstructing ancestral states in discrete characters,
which characterizes transitions from states 0 to 1 and 1 to 0 on branches of

Table 7.4. Transition matrix for dependent evolution model.

	00	01	10	11
00	$1-(q_{12}+q_{13})$	q_{12}	q_{13}	0
01	q_{21}	$1-(q_{21}+q_{24})$	0	q_{24}
10	q_{31}	0	$1-(q_{31}+q_{34})$	q_{34}
11	0	q_{42}	q_{43}	$1-(q_{42}+q_{43})$

Note: Shows character states for traits X_1 and X_2 separated by commas (i.e., X_1, X_2), with transitions from the initial state shown down the left side, and ending states across the columns. For example, q_{13} represents the transition from 00 to 10 (see also fig. 7.9). These rates are instantaneous rates over infinitesimally small amounts of time. It is assumed that two states do not change simultaneously, although dual changes are possible along a branch. Note that the model requires estimation of eight parameters.

the tree. Thus, it uses a transition matrix as described previously (see table 3.3), where evolutionary changes are assumed to occur independently on different branches, rates of change are constant, and the rate applies to an infinitesimally small period of time (instantaneous rates). Transition rates in this matrix are solved as the values that maximize the likelihood of the data. With these rates, it becomes possible to compute the likelihood of the character states of trait X_1 in the *independent model* based on every possible set of ancestral states and summed across different reconstructions (and thus not conditioned on any particular reconstruction). This process is repeated for trait X_2. The likelihood of the *independent model* is simply the product of the likelihoods for the two traits.

Estimation of the likelihood for the *dependent model* involves more parameters, with the transition rates of one trait estimated for particular states in the other trait. This requires a 4×4 transition matrix (table 7.4) to determine if transitions in one trait are a function of the state in the other trait (which, as we will see, is identical to the independent model when some transition parameters are forced to be equal). In the case of X_1 changing from 0 to 1, for example, we need to obtain a transition rate for cases in which X_2 is in state 0, and another transition rate when X_2 is in state 1. These transitions correspond to transition parameters q_{13} and q_{24}, respectively, in table 7.4. Given that these are instantaneous rates of change, the model assumes that only one trait changes at a time. Thus, zeros are placed in cells that involve "dual transitions" (e.g., state 0,0 $\rightarrow$ 1,1, or 0,1 $\rightarrow$ 1,0), indicating that such transitions are not allowed (see table 7.4). Note that this assumption does *not* restrict change to occur in only one trait per branch; it only means that traits are assumed to not change at the same instant.

This sounds very abstract, but think for a minute about what the transi-

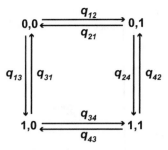

Figure 7.9. Trait transitions for correlated evolution. The dependent model investigates how the state of one trait influences the probability that the other trait changes. The subscripts identify the beginning and ending states of a particular transition, where the values 1, 2, 3, and 4 correspond to the states (0,0), (0,1), (1,0), and (1,1), respectively. Thus, q_{24} refers to beginning state 2, which is (0,1), and ending state 4, which is (1,1). In other words, the first trait changes from 0 to 1 when the second trait is in state 1. The figure also reveals the importance of "rate pairs" in the dependent model, which are parallel arrows pointing in the same direction (e.g., q_{13} and q_{24}). If estimates for rate pairs differ, this is consistent with correlated evolutionary change.

tion parameters of the dependent model represent in the context of correlated trait change. Imagine that the traits are correlated such that when the second trait is in state 1, the first trait also experiences selection for state 1. In such a case, q_{24} will be greater than q_{13}, and the same will be true of other pairs of transition rates. This is illustrated in figure 7.9. The relevant pairs for comparison are opposite to one another on the figure, with arrows pointed in the same direction (i.e., q_{13} and q_{24}, q_{12} and q_{34}, etc.). When one or more of these "rate pairs" differ substantially, the dependent model will produce a higher likelihood than an independent model in which the rates are the same. Thus, a higher likelihood for the dependent model is consistent with correlated evolutionary change.

Under the assumptions that the rows in table 7.4 sum to 0 and that two traits cannot change simultaneously, only two transition parameters are needed for each row, giving a total of eight transition rates for the dependent model (see also figure 7.9). In contrast, only four transition parameters are needed for the independent model—that is, two parameters for each of the two traits. The independent model can be viewed as nested within the dependent model, because within the matrix describing transitions in the dependent model it is possible to design an independent model of evolution by setting certain transition parameters to be equal (see Pagel and Meade 2005, 2006a). Referring to figure 7.9, for example, if X_2 is independent of the state of X_1, then the transition parameters q_{12} and q_{34} should not be significantly different, that is, change in X_2 is independent of the state of X_1. Thus, the 4×4 dependent model is equivalent to the independent model when $q_{12} = q_{34}$, $q_{21} = q_{43}$, $q_{13} = q_{24}$, and $q_{31} = q_{42}$.

Once the likelihoods of independent and dependent models are obtained, it is relatively easy to conduct an LRT to determine if the dependent model gives a statistically higher likelihood than the independent model (see Anthro-Tree 3.5 and 4.7; Pagel 1994a). In the case of correlated evolution, one would compare an eight-parameter *dependent model* to its reduced four-parameter *independent model*, thus giving four degrees of freedom for the statistical test.

One of the appealing aspects of this framework is that it can flexibly test a wide range of evolutionary scenarios. Thus, in addition to testing for correlated evolution of two traits, it is possible to investigate evolutionary scenarios in which change in one trait is contingent upon particular values for other traits, for example, to test whether one trait changes before the other trait (i.e., the temporal order of trait changes), and to build restricted models in which some types of transitions are not allowed (which are then compared to an unrestricted model; see Pagel 1994a). Note that by providing a means to investigate which of two traits evolves first on a tree, we can address issues of causality raised in chapter 6. "Flow diagrams" can be used in this regard to describe the history of evolutionary change for a pair of traits, as demonstrated by Pagel and Meade (2005, 2006a) and illustrated in an example in the next section. A final advantage of the method is that one can fix ancestral states when the fossil record provides evidence for a particular ancestral state.

Examples. Holden and Mace (1997) applied Pagel's (1994a) method to investigate a fundamental question in recent human evolution—the evolution of adult lactose tolerance. As with their analysis using independent contrasts described above, the analysis using Pagel's discrete method also found that adult lactose tolerance has arisen more often in human societies that keep cattle for milk production. Holden and Mace (1997) further demonstrated that evolutionary origins of milking occurred *before* the evolution of lactose digestion, supporting hypotheses that invoke culturally driven genetic change (i.e., gene-culture coevolution), rather than an alternative hypothesis that a population with the gene for lactose digestion began herding cattle and drinking milk. These results have been supported by subsequent genetic studies (Tishkoff et al. 2006), and the evolution of adult lactose tolerance is generally considered a classic case study in gene-culture coevolution (Durham 1991).

Another cross-cultural study used a phylogeny of Bantu languages (Holden 2002) to investigate the hypothesis that the spread of cattle results in the loss of matriliny, where matriliny refers to group membership identity, altruism, and inheritance of wealth primarily along female lines of descent (Holden and Mace 2003; Holden 2006). Patriliny occurs when group membership flows to a greater extent through the paternal line, while a mixture of maternal and paternal descent patterns is also possible (and common among extant societies). This pattern might be expected to switch if cattle confer greater fitness benefits on males than females; such an effect is plausible, given the great value of cattle to many African societies and its link with polygyny. Holden and Mace (2003) found support for this hypothesis, with evolutionary origins of cattle keeping associated with transitions to patrilineal descent (or a mixture of matrilineal and patrilineal descent). The analysis further revealed that having matrilineal descent and cattle keeping was an unstable state, resulting in a high rate of movement towards patrilineal descent or loss of cattle.

Pagel's (1994a) method also has been applied to investigate primate behavior. For example, I used the method (Nunn 2000) to investigate predictions from economic models of collective action and free riding (see also Heinsohn and Packer 1995; Van Schaik 1996; Nunn and Lewis 2001). In the case of males defending access to females, I specifically predicted that individuals in more seasonally breeding macaque species would not exhibit loud calls; this prediction was based on (1) the assumption that loud calls function in extra-group defense and (2) empirical findings that seasonality increases scramble competition among males (Paul 1997; Nunn 1999b), such that females within the group are not easily monopolizable (and all males would therefore benefit from extra-group defense). Using discrete categorizations of loud calling and seasonality in macaque monkeys, I found support for the prediction (Nunn 2000). Pagel's (1994a) method was particularly helpful in this analysis, as parsimony reconstructions of loud calls and seasonality could not be reconstructed unambiguously on the phylogeny of macaques—a requirement for some other methods of analyzing discrete traits (Ridley 1983; Maddison 1990).

A Bayesian approach. Pagel's (1994a) discrete method can be implemented in a Bayesian framework to control for uncertainty in the evolutionary model and/or phylogenetic history (Pagel et al. 2004; Pagel and Meade 2005, 2006a). Across humans, for example, Pagel and Meade (2005) investigated whether transfer of resources at marriage covaries with mating system (see also Hartung 1982; Gaulin and Boster 1990; Fortunato and Mace 2009). Using a dataset of Indo-European language groups and associated cultural trait data, the authors found support for this prediction. Pagel and Meade's (2005) analyses further suggested that the derived state of polygyny with bride-price occurs through a first transition in mating system (monogamy to polygyny), and then a transition in wealth transfer (dowry to polygyny).

Pagel and Meade (2006a) provided another example of the Bayesian implementation to examine whether exaggerated sexual swellings covary with multi-male mating systems in Old World primates (Clutton-Brock and Harvey 1976). Importantly, a previous test using two different phylogenetic methods failed to find statistical support for the linkage between exaggerated swellings and multi-male mating systems (Nunn 1999a). In contrast, Pagel and Meade (2006a) found overwhelming support for a linkage between sexual swellings and multi-male social systems (figure 7.10). In their Bayesian MCMC analysis across models and phylogenies, an independent model of no association was "visited" on less than 0.009 percent (!) of the samples, while dependent models were sampled approximately thirty times more often than expected by chance. Pagel and Meade (2006a) further investigated the order of evolutionary events and concluded that mating system evolved first, which then created selective pressure for sexual swellings to evolve (figure 7.11).

In summary, Pagel's (1994a) discrete method offers a flexible, rigorous framework to test adaptive hypotheses, including tests of correlated evolution and the order of evolutionary events. The method also provides a way to incorporate fossil data into the analysis—a feature that could be useful for studies in evolutionary anthropology that examine paleontological data.

Combining Continuously and Discretely Varying Traits

The methods discussed so far consider cases where both traits are either discrete or continuous. What should one do when the analysis involves a mixture of discrete and continuous variables? This is a common situation in real-world comparative datasets where, for example, we might be interested in whether classifications of mating promiscuity covary with immune system parameters (Nunn et al. 2000; Nunn 2002a), whether dietary and activity period categories covary with brain structures (Barton 1998), or whether extinction risk categories covary with measures of sexual selection (Morrow and Pitcher 2003; Morrow and Fricke 2004).

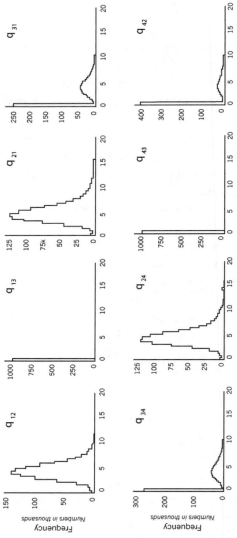

Figure 7.10. Sexual swellings and multi-male social systems. Plots show values of rate parameters from Bayesian analyses that controlled for uncertainty in both the estimation of the evolutionary model and the phylogenetic history of anthropoid primates (Pagel and Meade 2006a). The figure is arranged to ease comparison of "rate pairs" by examining a plot and its corresponding rate pair below the plot. When these rate pairs differ, this provides evidence for correlated evolution, as illustrated with these data (e.g., for pairs q_{13} and q_{24}).

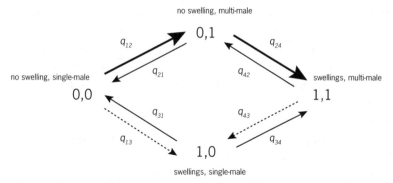

Figure 7.11. **Flow chart summarizing the evolution of exaggerated sexual swellings.** Chart reveals the likely paths involving correlated evolution of exaggerated sexual swellings and multi-male social systems in anthropoid primates. For each pair of numbers, the first represents exaggerated sexual swellings (0 = absent, 1 = present), and the second represents the social system (0 = single-male, 1 = multi-male). The thickness of the lines represent support for particular transition probabilities, ranging from lowest support (e.g., 97.7% posterior probability of being zero for q_{13}) to strongest support (q_{12}). The analysis reveals that multi-male systems are likely to evolve before the evolution of exaggerated sexual swellings. (Redrawn from Pagel and Meade 2006a.)

One approach is to use the Brunch algorithm in a comparative analysis program known as CAIC, which stands for Comparative Analysis by Independent Contrasts (Purvis and Rambaut 1995, see AnthroTree 7.7). Briefly, this method identifies the discrete binary trait as the independent variable and codes the states as 0 or 1 (nonbinary data can be used, but it is less straightforward). The program then calculates contrasts with the direction of subtraction set such that the discrete variable is positive, retaining the same direction of subtraction for contrasts in the continuous variable. One then tests whether contrasts in the continuous dependent variable are consistently positive or negative over evolutionary "increases" in the discrete independent variable (see also Purvis and Webster 1999; Nunn and Barton 2001).

The Brunch algorithm is probably a conservative approach to analyzing a mixture of continuous and discrete data; thus, significant results can be trusted but lack of significance could reflect low statistical power. Another approach is to treat the discrete independent variable as if it is continuous and thus similar to using a "dummy" variable in regression (see AnthroTree 7.7 and documentation for the PDAP Package of Mesquite; Lavin et al. 2008; Gartner et al. 2010). This seems to imply a non-Brownian motion model of evolution, however, and while probably statistically valid, this is an area in which future simulation research would be valuable.

Last, one can investigate the data using a generalized least squares approach. If the discrete variable is one of the predictors, it is valid to include it in the GLS model as a dummy variable (e.g., Lavin et al. 2008). If the discrete variable is the response, one can incorporate an appropriate link function into the GLS model (Martins and Hansen 1997). More recently, Ives and Garland (2010) developed an approach for phylogenetic logistic regression. This method can also be used to quantify phylogenetic signal in discrete traits and to examine the association between two discrete traits (see also Paradis and Claude 2002). In general, it is becoming much easier to perform analyses involving mixtures of continuous and discrete variables. Be sure to visit Anthro-Tree 7.0 for updates as new methods and simulation studies become available.

Summary and Synthesis

Most work in comparative biology has focused on developing methods to test adaptive hypotheses. As discussed in the previous chapter, a primary way to investigate adaptative hypotheses is to examine associations between traits or between traits and environmental features (Harvey and Pagel 1991).

It is important to emphasize that this chapter covered only three methods relevant to studying correlated evolution; additional methods are available (see AnthroTree 7.1 and Ridley 1983; Cheverud et al. 1985; Maddison 1990; Harvey and Pagel 1991; Garland et al. 1993; Martins and Hansen 1996; Rohlf 2001; Felsenstein 2004). In one important comparative study of primate ecology, for example, Clutton-Brock and Harvey (1977) examined variation at the genus level (with some genera split if the species therein varied in key ecological traits). This decision was validated quantitatively using nested analysis of variance (ANOVA), which demonstrated that significant variation existed at the genus level for seven of eight variables examined. As discussed in Harvey and Pagel (1991), however, nested ANOVA only shifts the nonindependence problem from one level to another—in this case from species to genera—because the higher levels themselves also exhibit hierarchically structured relationships. This can be seen in figure 7.12, which shows how grouping taxa at the generic level still produces a hierarchical pattern of relatedness. This "higher node," or "node averaging," approach has the additional problem of reducing the degrees of freedom, thus diminishing the statistical power of the comparisons. On all these accounts, this method and others similar to it (e.g., Smith 1994) should be avoided.

The approach of using traits that are sampled widely and assumed to be independent has also been used in cultural anthropology; therefore, these

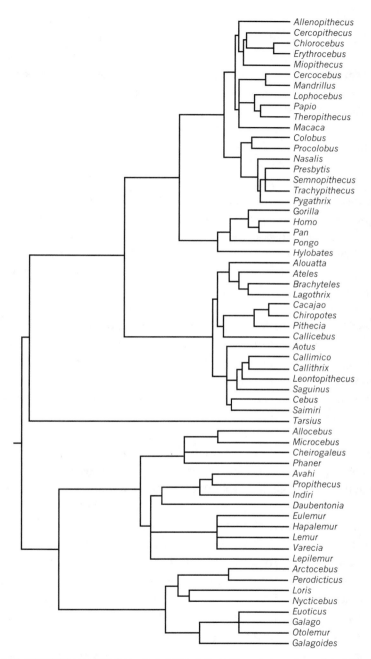

Figure 7.12. Using "higher nodes" does not eliminate phylogenetic nonindependence. Calculating mean values for various higher taxonomic groups, such as the genera shown here for primates, reduces sample sizes, and fails to reduce nonindependence arising at deeper nodes because the genera continue to exhibit variation in relatedness. In other words, a hierarchical structure remains across genera.

approaches are similar to the higher nodes approach in biology. One example involves Murdock and White's (1969) standard cross-cultural sample (SCCS), derived from the impressive collection of cultural traits known as the *Ethnographic Atlas* (Murdock 1967). As with biological data, such datasets fail to provide a principled way of dealing with statistical nonindependence (Mace and Pagel 1994). While these compilations of cultural data are absolutely essential for cross-cultural research, the statistical errors associated with applying them without statistical corrections are likely to be substantial. Indeed, Dow (1993) demonstrated higher type I error rates when using the SCCS (see also chapter 10; Eff 2004; Dow and Eff 2008).

Other approaches to deal with nonindependence make use of autocorrelation methods, and these methods have been applied to diverse questions in biological and cross-cultural research (White et al. 1981; Cheverud et al. 1985; Edwards and Kot 1995). For cultural traits, Dow and colleagues (Dow 1984; Dow et al. 1984) developed autocorrelation approaches to control for the nonindependence of population data. One method, for example, controls for the nonindependence of data in cross-cultural studies by using two matrices: one for linguistic or genetic similarity and another for spatial proximity (Dow 1984). Autocorrelation methods can also be used to assess phylogenetic signal (Gittleman and Kot 1990; Nunn 2002a) and more generally have been shown to have appropriate statistical performance (Martins 1996b). As a generalization, however, they offer fewer advantages than the newer approaches described above.

In conclusion, a wide array of tools is available for investigating whether two or more traits exhibit an evolutionary association. Indeed, there is more than one way to "skin the cat" in phylogenetic comparisons, just as is true for statistical options for most other scientific questions. In the future, we can expect greater integration of methods under one statistical roof, probably within a GLS framework that can accommodate flexible evolutionary modeling and a mixture of discrete and continuous characters (e.g., Lavin et al. 2008; Gartner et al. 2010). In addition, it should become easier to implement Bayesian methods to control for phylogenetic uncertainty and to routinely incorporate intraspecific variation into comparative analyses (Ives et al. 2007; Felsenstein 2008, see also Chapter 11). As in the past (e.g., Clutton-Brock and Harvey 1977; Pagel 1994a), comparative studies of primates are likely to play a major role in the further development of these methods (e.g., Pagel and Meade 2006a).

Pointers to related topics in future chapters:

- Clade shifts and incorporating their effects into GLS: chapter 9
- Estimating the intercept when using independent contrasts: chapter 9
- Integrating ancestral state reconstruction and functional relationships: chapter 9
- Incorporating intraspecific variation: chapter 11
- The "higher nodes" approach of averaging variation at higher levels: chapter 11
- Standardizing contrasts: chapter 12

8 Using Trees to Study Biological and Cultural Diversification

In the early 1970s, a group of paleontologists assembled for a series of meetings at Woods Hole Biological Station. The scientists included Stephen Jay Gould, David Raup, and Thomas Schopf, and their goal was to assess the role of chance in evolutionary history (Raup et al. 1973; Gould et al. 1977). Their conclusions were revolutionary for the times: they proposed that some patterns observed in the fossil record might have resulted from random events rather than adaptive evolution. For example, rather than postulating "key innovations" leading to the rise of a clade, a large clade might just have been "lucky," with a few speciation events early on producing a strong foundation for subsequent growth of the clade. From this, the "Woods Hole Group" proposed that observed evolutionary patterns should be viewed within a framework that examines null patterns of diversity. They were among the first to use computer simulations to investigate how speciation and extinction influence macroevolutionary patterns and to compare real-world data to simulated null models of diversification.

The findings of the Woods Hole Group had reverberations beyond paleontology. In fact, much of what will be covered in this chapter can be traced, either directly or indirectly, to the philosophical perspective of the Woods Hole Group and to earlier scientists that provided the theoretical shoulders for the Woods Hole Group to stand upon (Yule 1925; Kendall 1948; Moran 1958). Remarkably, many of the questions asked by this group can now be addressed in the absence of fossil data; these methods will be the focus of this chapter. When fossil data are available, it obviously makes sense to incorporate those data into the analysis. Indeed, an important area for future work is to unify data on extant diversity and paleontological-archaeological data in ways that can offer the strongest insights to biological and cultural diversification (e.g., Renfrew 1992; Diamond and Bellwood 2003; Nee 2004; Paradis 2004; Etienne and Apol 2008; Rabosky and Alfaro 2010).

What kinds of questions can we ask concerning primate or human diversification? We might want to know, for example, if one primate lineage has undergone a more rapid rate of diversification than other lineages, or if some linguistic groups expanded at the expense of others. We could also investigate whether the branching of lineages reveals evidence for adaptive radiation, which would be indicated by a more rapid rate of lineage splitting early in the history of a clade, or to investigate (and even estimate) the rate at which lineages have gone extinct over some time period. Adaptive radiations could also be important for understanding the rise of cultural innovations. For example, the domestication of livestock, the development of agriculture, and innovations related to long-distance oceanic travel are thought to have spawned human radiations, and these questions can be investigated using phylogeny-based methods (e.g., Gray and Jordan 2000; Gray and Atkinson 2003; Gray et al. 2009). Last, have certain characteristics of organisms produced higher rates of diversification in lineages with those traits? As one example, sexual selection is thought to increase the rate of speciation (Lande 1981; Turelli et al. 2001), predicting a higher accumulation of species in lineages characterized by greater sexual selection (e.g., Barraclough et al. 1995).

In previous chapters, we were largely concerned with how traits map onto a phylogeny or covary with other traits over phylogenetic history. In all of the examples just given, however, it is the phylogeny itself that is analyzed, and thus the tree topology and branch lengths provide the data for hypothesis testing. In what follows, I review some of the ways that a phylogeny can be used to make inferences about the evolution of biological species, populations, and languages. Some of these methods have been applied to study primate evolution (e.g., Purvis et al. 1995; Gittleman and Purvis 1998; Nunn et al. 2004; Matthews et al. 2010) and linguistic evolution (e.g., Atkinson et al. 2008; Gray et al. 2009).

Background

An important first step in all that follows concerns the definition of diversification. Diversification refers to the rate at which new lineages arise on a phylogeny. In some cases, the tree includes fossil taxa, but more often it is based only on the relationships among living (extant) species. Many people equate diversification with speciation (or cladogenesis). *However, speciation is only part of the picture, with diversification actually a reflection of two processes: speciation and extinction.* A higher speciation rate will increase the number of lineages on a phylogeny, but this increase will be eroded by extinctions that cull these lineages. As we will see, extinction produces a signature on a

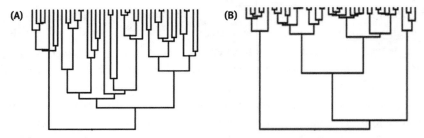

Figure 8.1. The signature of extinction on a dated phylogeny. The tree in (a) was generated with a lower extinction rate than the tree in (b), holding the speciation rate constant in both trees. Specifically, $d = 0.02$ in (a) and $d = 0.16$ in (b), with $b = 0.2$ for both simulations. In both cases, the simulation was allowed to run until forty lineages had accumulated. Trees were simulated in Mesquite, version 2.5.

dated phylogeny that can be detected even in the absence of fossil data. For example, figure 8.1a shows a clade that was simulated using a lower extinction rate than the clade in figure 8.1b (all other parameters were identical).

The study of diversification has a long history, but only recently have statistical tools become available to investigate patterns of diversification and its components. In this chapter, I will be concerned with four basic questions:

1. **Given an observed pattern of diversification for a clade, what were the speciation and extinction rates?** The speciation rate should be higher than the extinction rate among living clades—otherwise, all of the lineages in the clade are expected to go extinct—but by how much does speciation exceed extinction?

2. **Are some lineages more diverse than others?** This requires an appropriate null model for the accumulation of lineages over time, and a way to compare observed and expected patterns in a statistically rigorous way (Nee et al. 1994a).

3. **Did the evolution of some trait cause shifts in diversification rates?** This question relates to key innovations. We also might want to know if a correlation exists between a continuously varying trait, such as a body mass, and the number of species in a clade (e.g., Gittleman and Purvis 1998; Conroy 2003; Freckleton et al. 2008).

4. **Does diversification rate covary with the rate of molecular or morphological evolution?** In particular, do lineages characterized by higher rates of evolution produce more descendent lineages?

To begin, an important foundation for many of these questions is the Yule model (Yule 1925). This model is a useful simplification in which extinction does not occur, and thus it is also known as a pure birth process, where birth refers to the origin of new lineages. Nee (2001) provides an overview of this model. A common variant on this model is the birth-death model (Harvey et al. 1994; Nee et al. 1994b; Nee 2001), which requires separate estimates for the speciation rate (i.e., lineage *births*) and the extinction rate (i.e., lineage *deaths*). In one special case of the birth-death model, the number of lineages is held constant (Hey 1992), resulting in what might be called density dependent cladogenesis or a Moran process (Nee et al. 1994b; Nee 2004). In what follows, I will consider a model in which no constraints are placed on the total number of lineages in the clade.

Evolutionary biologists investigated the birth-death model in the early 1990s when the advent of molecular methods began to produce an avalanche of phylogenies (Nee et al. 1992; Sanderson and Bharathan 1993; Kubo and Iwasa 1995). A constant-rate version of the model starts with a single lineage at time zero. New lineages arise at a speciation rate b and lineages disappear with an extinction rate d, corresponding to birth and death rates, respectively. These rates are assumed to be constant over time and across branches of the phylogeny. It is worth noting that previous authors have often used the Greek symbols λ as the speciation rate and μ as the extinction rate, reflecting that these are biologically relevant parameters that need to be estimated. Here, I follow Nee (2006) and use b and d throughout, because it is easier to re-member that they refer to birth and death rates, respectively, and to avoid confusion with the now widely used λ as a measure of phylogenetic signal (al-though note that Blomberg et al. 2003 also use a parameter called d in their OU model-fitting approach, which should not be conflated with d used in this chapter).

The primary goal is to use a dated phylogeny of living species to recon-struct aspects of the evolutionary process. Thus, referring back to an earlier figure, we have the phylogeny in figure 3.1b, but wish to estimate the rates of speciation and extinction in the actual evolutionary history of this hypotheti-cal clade, shown in figure 3.1a (see also Nee et al. 1994b).

Most people are surprised to learn that extinction leaves a fingerprint on a dated phylogeny of extant species—this fingerprint is intriguing and often challenging to comprehend, at least when first exposed to it. A starting point for describing the signature of extinction is to plot the number of lineages in a phylogeny versus time; this has become known as a *lineages through time plot* (AnthroTree 8.1; Nee et al. 1992; Harvey et al. 1994; Mooers and Heard 1997). Figure 8.2 provides lineages through time plots for the trees shown in

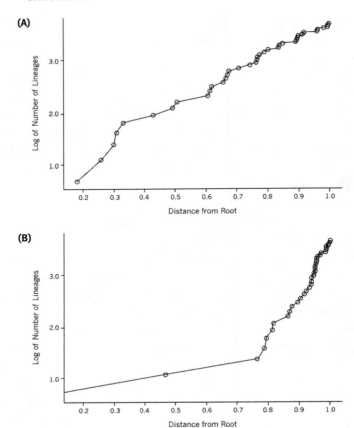

Figure 8.2. Lineage through time plots. Plots correspond to phylogenies in figure 8.1, with (a) showing an apparently more steady increase in lineages through time, as compared to the plot in (b). This corresponds to the higher extinction rate in (b) and illustrates the "pull of the present."

figure 8.1. By comparing the two examples in figure 8.2, it is easy to see that the signature of extinction involves an *apparent increase in the rate of lineage accumulation closer to the tips of the tree*, which has been called "the pull of the present." The word *apparent* is important; it refers to the fact that speciation rates *appear* to increase, but in fact the speciation rate remained constant through time in the simulation that produced the lineages represented in figure 8.2 (see AnthroTree 8.2 for programs that can simulate the birth-death model). This apparent increase is the signal that we can use to estimate extinction rates.

What causes the pull of the present? As shown previously with figure 3.1, it is helpful in this case to examine the "actual" phylogeny of a group of liv-

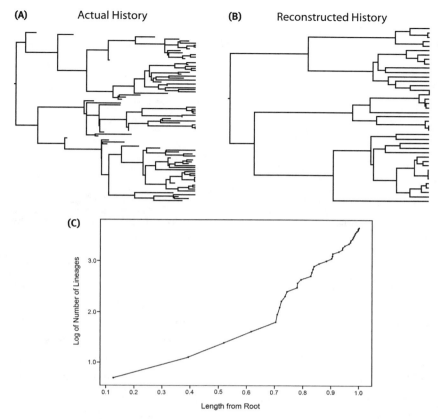

(A) Actual History

(B) Reconstructed History

(C)

Figure 8.3. The reconstructed evolutionary process. (a) The true history of living and extinct lineages, based on $b = 0.2$ and $d = 0.1$. (b) The phylogeny as it would appear when reconstructed from living species. (c) The lineage through time plot for the reconstructed phylogeny. The phylogeny and lineage through time plots were generated in Mesquite, version 2.5.

ing and extinct organisms (figure 8.3a), and to then look at what is "reconstructed" when only the living species are included (figure 8.3b). The lineages through time plot for this example reveals the pull of the present (figure 8.3c), and comparison to the actual tree in figure 8.3a reveals why. Many of the lineages that arose early in the history of this clade have gone extinct and thus are missing from the tree reconstructed from extant species. By comparison, the lineages with more recent origins have had less time to go extinct, resulting in the appearance of a faster rate of diversification as one moves toward the tips on the reconstructed phylogeny.

Another feature of lineages through time plots is the "push of the past," which is observed on the actual history of diversification rather than the re-

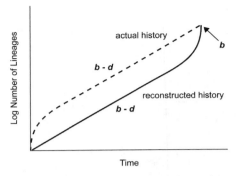

Figure 8.4. Expected number of lineages over time. Lineages are expected to accumulate at the constant rate of $b - d$ over most of the reconstructed and actual histories. On the reconstructed history, the pull of the present can be seen at the right of the plot; the curve is expected to approach the birth rate, b. The push of the past can be seen on the left of the curve for the actual evolutionary history. (Redrawn from Harvey et al. 1994 and Nee et al. 1994a.)

constructed history (figure 8.4). The push of the past occurs because, when looking at the clades that persist to some later sampling point (e.g., the present), the clades that are successful are likely to have had a strong start at the beginning, and this strong start can occur simply by chance (Harvey et al. 1994). Figure 8.4 provides an overview of these expected patterns and is useful for the next section, which reviews methods for estimating speciation and extinction rates from dated phylogenies.

Before proceeding, it is important to mention some assumptions that underlie many—but not all—of the methods for studying diversification. First, we are assuming that we have a perfectly known dated phylogeny. In reality, minor errors in the tree are unlikely to have a major impact on the analysis. To more confidently rule out such effects, however, it is possible to use the Bayesian methods described in previous chapters to control for phylogenetic uncertainty (Huelsenbeck et al. 2000; Pagel and Lutzoni 2002), as was done in a recent analysis of primate diversification (Matthews et al. 2010). Second, we are assuming that all of the living species in a clade are included in the reconstructed phylogeny; in other words, we assume complete sampling of extant lineages. As we will see, violations of this assumption can have a profound impact on the results because missing species can cause the tail of the lineages through time plot to peter out at the end (Nee et al. 1994a). Last, it is worth amplifying a point made earlier, namely that b and d are typically assumed to be constant through time and across lineages. Again, this can have an impact on lineages through time plots and estimates of b and d (Nee et al.

1992; Harvey et al. 1994; Zink and Slowinski 1995; Rabosky 2010), which is important because some studies of primates have suggested that lineages vary in their rates of speciation and extinction (e.g., Purvis et al. 1995; Paradis 1998; Pybus and Harvey 2000).

What is the Speciation and Extinction Rate?

As shown in figure 8.4, the patterns that emerge on lineages through time plots are directly related to the rates at which lineages speciate (b) and go extinct (d). In particular, we can see on the linear parts of the lineages through time plots that the log number of lineages accumulates at a rate that is proportional to the difference in birth and death rates ($b - d$). In addition, the curvature of the line at the end points—where the *push of the past* and the *pull of the present* occur—provides insights to b. More specifically, the slope of this curvature asymptotically approaches the birth rate; in other words, at the very end of the curve, the slope will become very close to the actual birth rate, b. Under a pure birth process, with $d = 0$, the number of lineages is expected to accumulate proportional to time, with slope equal to b, and the distance between the lines on figure 8.4 increases as d approaches b (Harvey et al. 1994). Thus, a higher probability of extinction (relative to speciation) produces a greater divergence between the actual and reconstructed histories.

With this estimate of b and $b - d$, it is possible to estimate d. Because we rarely if ever have true knowledge of the actual evolutionary history of a group of organisms, these estimates are based on the reconstructed rather than the actual phylogeny. The relative values of b and d are often as interesting as the absolute values (Nee et al. 1994a). For example, $b - d$ can be used as a description of the rate at which species accumulate, the diversification rate. Similarly, the ratio $d \div b$ describes the magnitude of the pull of the present on reconstructed phylogenies and captures the degree to which small clades are vulnerable to extinction.

The graphical representation of this process is obviously helpful for understanding how one can estimate b and d from the evolutionary tree reconstructed from living species. The actual estimates are made using maximum likelihood (Nee et al. 1994a, 1994b). While these estimates have become ever easier to compute, one must beware of treating the programs as black boxes, and thus forgetting the key assumptions that underlie this approach. In particular, a failure to completely sample all the extant lineages in the tree will produce a flattening of the reconstructed evolutionary process and thus an

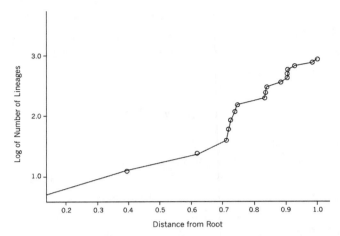

Figure 8.5. Missing lineages. When extant lineages are missing from a phylogeny, this can cause the lineage through time plot to flatten out at the end. Such patterns can provide information on sampling, for example, when estimating the population size of an infectious disease or the number of species in an understudied clade of organisms (Nee et al. 1994a).

apparent slow-down in the rate of diversification as one approaches the present (Nee et al. 1994a). An example of this is given in figure 8.5, where a random selection of 50 percent of the lineages in figure 8.3 was removed. Notice in particular how the lineages through time plot in figure 8.5 becomes a bit "limp" at the end, compared to figure 8.3, which can result in nonsensical—or at least lower than expected—estimates of b. Simulation tools can be used to investigate different evolutionary and taxonomic sampling scenarios, including effects of mass extinctions, imperfect sampling and varying rates of evolution (see AnthroTree 8.2 and Harvey et al. 1994; Rambaut et al. 1996).

Let us apply what we have just learned to the simulated trees in the figures from this chapter, and then turn to analyses of real data in primates. Returning to the example in figure 8.1, panel a was generated with $b = 0.2$ and $d = 0.02$, while panel b was generated with $b = 0.2$ and $d = 0.16$. How well can we estimate these rates from the phylogenies? I used the program Mesquite, version 2.5 to estimate speciation and extinction rates, which produced estimates of $b = 0.204$ and $d \approx 0$ for figure 8.1a, and $b = 0.196$ and $d = 0.159$ for figure 8.1b (see AnthroTree 8.3). Similarly, for figure 8.3b, the actual values of b and d were 0.2 and 0.1, respectively, while estimated values were 0.197 and 0.103. Last, for the random sample of 50 percent of the lineages shown in figure 8.5, the estimates of birth and death rates were biased downward, as expected, with estimates of $b = 0.089$ and d close to 0.

As single cases, these examples are anecdotal, and they actually offer undue optimism about the ability to accurately estimate speciation and extinction rates, especially since these examples were selected among several simulation runs to illustrate the approach. Only a properly designed simulation procedure, with multiple runs of the simulation, can assess how well the methods estimate b and d. Indeed, several such tests have been conducted, with sobering results. Some studies found, for example, that d is more difficult to estimate than b (Kubo and Iwasa 1995; see also Maddison et al. 2007). Another simulation study found that estimates of d tended to be biased downward, and b and d based on extinct lineages (i.e., the actual history) performed substantially better than those based only on the reconstructed phylogeny (Paradis 2004). Perhaps even more worrisome, a recent study found that estimating b and d is particularly problematic when rates of diversification vary throughout the tree (Rabosky 2010). Because these rates are likely to vary across clades and through time in real-world data, the author concluded that his "study does not paint a promising picture of fossil-free extinction estimates" (Rabosky 2010, 1821).

It is worthwhile to briefly review some previous applications of these methods to primates. Purvis et al. (1995) applied the methods to estimate birth and death rates for strepsirrhines, New World monkeys, Old World monkeys, and apes (table 8.1). Maximum likelihood estimates of b for these clades ranged from 0.093 to 0.342. Interestingly, in two of the clades the authors could not detect extinction (i.e., d was estimated to be 0), and both of these clades involved monkeys. It seems very unlikely that extinction rates are actually zero in these clades, and in fact the fossil record reveals evidence for extinct anthropoid lineages (e.g., Fleagle 1988; Jablonski 1993; Simons 1995). Similarly low rates of extinction were obtained in a more recent study that used a Bayesian sample of phylogenies (Matthews et al. 2010). These results likely reflect the bias in estimating d identified by Paradis's (2004) simulation study.

Last, it is worth considering whether lineages through time plots can be used to investigate mass extinctions. It might seem that the number of lineages through time gives some insights to mass extinctions; surely an extinction event would cause a decline in the number of lineages, and if large enough, this should be detectable on the phylogeny reconstructed from extant species. Unfortunately, mass extinctions are not typically easy to discern on a lineages through time plot (Harvey et al. 1994). While mass extinctions leave some clues on the reconstructed phylogeny (figure 8.6), the methods

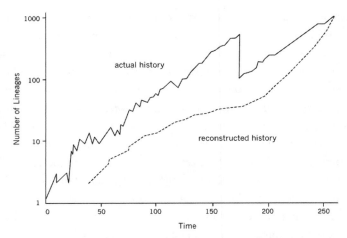

Figure 8.6. Detecting mass extinctions. Use of a dated phylogeny can also be used to detect mass extinctions, but the power to do so is relatively more limited than one might expect. In this simulation, Harvey and colleagues (1994) simulated evolution with a birth rate (*b*) of 0.1 and death rate (*d*) of 0.075. Upon reaching five hundred lineages, a one-time catastrophe was simulated, resulting in the loss of four hundred of the lineages, and the simulation continued until a thousand lineages were generated. The top line shows the actual number of lineages through time, while the bottom line reveals the number of lineages that would be reconstructed from the lineages that persisted to the end of the simulation. (Redrawn from Harvey et al. 1994.)

discussed here probably detect these clues less powerfully than examining a well-sampled fossil record.

Are Some Lineages More Diverse than Others?

It is one thing to estimate the rate at which lineages arise and go extinct and remarkable that such information can be extracted from a dated phylogeny of extant species. But what do these numbers really mean? For example, is a speciation rate of 0.342 exceptionally high? Often, it is more useful to examine patterns of diversification comparatively. Thus, instead of simply quantifying rates, we might wish to assess heterogeneity in diversification rates across different clades, and if variation exists, to identify which lineages show significantly higher or lower diversification rates.

Questions concerning differences in diversification rates are not new (Sanderson and Donoghue 1996). Here I briefly review three broad sets of methods involving analyses of tree balance (or shape), sister clade and related comparisons, and estimating shifts in diversification rates from dated phylogenies. Measures of tree balance give insights to whether patterns of di-

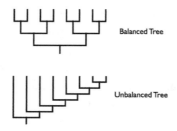

Figure 8.7. Tree balance. The tree on the top is perfectly balanced, with each lineage producing two descendent lineages, while the tree on bottom is unbalanced, with lineages producing variable numbers of descendent lineages.

versification vary in the tree, while the other approaches provide insights to where the shifts occur.

Measures of tree balance. A broad set of methods have investigated *tree balance* (or *symmetry*) as measures of differential diversification rates (Anthro-Tree 8.4; Shao and Sokal 1990; Kirkpatrick and Slatkin 1993; Mooers and Heard 1997; see also the special issue of *Systematic Biology* on tree shape, edited by Mooers and Heard 2002). Tree balance refers to the degree to which nodes on the tree result in subgroups of equal size. We can imagine two extremes of tree balance—a perfectly balanced (or symmetrical) tree, shown at the top panel of figure 8.7, and a perfectly unbalanced (or asymmetrical) tree, shown in the bottom panel. Variation in diversification rates across a tree is expected to produce a more unbalanced tree. The goal is to identify whether the observed pattern is statistically different from a null model in which all lineages have the same probability of speciation and extinction (i.e., random branching, which is not illustrated in figure 8.7).

A number of tree balance statistics have been proposed (Mooers and Heard 1997). Kirkpatrick and Slatkin (1993), for example, presented six different measures of tree balance. Importantly, they also provided expected values under random models of diversification and confidence intervals for up to fifty species; this information can be used to determine whether observed measures of tree balance differ significantly from null expectations for each of the six measures (see also Shao and Sokal 1990). In another paper, Chan and Moore (2002) generated measures of tree balance that are generalizations of a single-node approach (Slowinski and Guyer 1989) to the whole tree. They showed how statistical significance can be obtained using a simulation approach and provided a computer program to generate null distributions (Chan and Moore 2005).

Heard (1992) calculated measures of tree balance for real and simulated phylogenies. He demonstrated that real trees are more unbalanced than expected by chance and that smaller trees are more unbalanced than larger trees. In other empirical research, Chan and Moore (2002) applied their method to study diversity across all primates and in major groups of primates (see also Moore et al. 2004). They found that for two of their statistics, the apes and Old World monkeys exhibited significant degrees of asymmetry, indicating that shifts in diversification rates occurred in these two clades. Variation at the whole-primate clade level, however, was not convincingly asymmetric (for additional analyses of primate tree shape, see Isaac et al. 2005 and Heard and Cox 2007).

Kirkpatrick and Slatkin (1993) used computer simulations to investigate the statistical power of several measures of tree balance, and Agapow and Purvis (2002) extended their study with additional simulations and measures. These simulations revealed that many of the statistics have low power. Similarly, Losos and Adler (1995) used simulations to investigate the degree to which the assumption of "instantaneous" speciation affects measures of tree balance. The methods generally assume that speciation is an instantaneous process, rather than requiring time for reproductive isolation to take place, yet a lengthier period of speciation could impact patterns of tree balance, specifically resulting in more symmetric trees (Losos and Adler 1995).

Sister clade comparison and related approaches. The method of sister clade comparison is a simple way to investigate whether two clades differ in the number of lineages they contain (Slowinski and Guyer 1989; Jensen 1990). The method is based on testing whether two clades of the same age have significantly different numbers of extant lineages; as such, it focuses on relative differences among clades of the same age, compared to variation in the observed versus expected number of lineages under a given birth-death model (Slowinski and Guyer 1989). The methods focus on a single node and compare the number of taxa on either side of that node.

Slowinski and Guyer (1989) provide formulae for calculating the significance of an observed difference in taxon numbers based on a null model of lineage accumulation (a randomly branching Markov process). This method has low statistical power, but it has other advantages. For example, it is not necessary to have detailed information on the full topology and branch lengths; one only needs to be sure that the clades compared are sister clades and that all the species are assigned to the correct clade. In addition, if applied across multiple sets of taxa in which comparisons are made for each set, the power of the test can be increased, and stronger inferences about the causes of the differences can be made (see Mitter et al. 1988; Slowinski and Guyer 1993). When conducting multiple comparisons, the sets of clades should be statistically independent of one another, with no branches shared in common among the compared sets of taxa.

Another method examines the distribution of the number of descendent lineages per taxon (e.g., Nee et al. 1994a). Thus, this approach can be applied using taxonomic rather than explicit phylogenetic information (Dial and Marzluff 1989), although it assumes that taxonomy accurately reflects the phylogenetic history of the groups involved. A more phylogenetically informed approach examines a "slice" of phylogeny at an early point in the radiation of a clade to ask, how many subsequent lineages were produced by each of the lineages that were present at the slice point? Plotting this frequency can provide insights to the lineages that experienced higher rates of diversification (Purvis et al. 1995). Notice, for example, the large number of Old World monkeys present in the extant clades at twenty-five million years ago in primates (figure 8.8). This plot strongly suggests a higher speciation rate (or lower extinction rate) in this clade. Similarly, one can plot the frequency of the number of subtaxa per taxon at the phylogenetic slice of choice. It is then possible to assess the probability of seeing particularly speciose clades by comparing the observed distribution to that expected under a geometric distribution (Nee et al. 1992, 1994a; Nee 2004).

Quantifying variation in diversification rates. More powerful statistical methods can be used to investigate variation in diversification rates using phylogenetic information (Sanderson and Donoghue 1994, 1996; Paradis 1998). In a study of primate diversification patterns, for example, Purvis et al. (1995) tested whether some primate clades exhibit higher rates of diversification than others. The data suggested that the Old World monkeys have in fact radiated at a faster clip (table 8.1). The authors tested for differences among clades by forcing clades to take values from other clades and comparing the likelihoods using an LRT. They found that Old World monkeys radiated more rapidly than other clades of primates (table 8.2). In another test, Purvis et al. (1995) statistically compared growth rates ($b - d$) for the four clades, which also revealed significant differences.

Purvis et al. (1995) further investigated variation in diversification rates on primate phylogeny. Specifically, they wished to identify where the change in

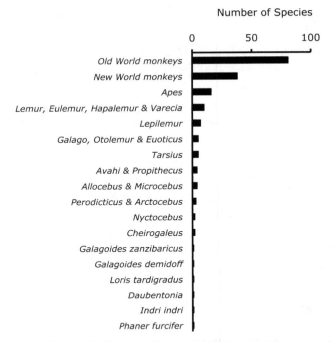

Figure 8.8. Time-window analysis of primate radiation. A "slice" of primate phylogeny was taken at twenty-five million years, and the number of species arising from each lineage at this time slice was obtained, based on the Bininda-Emonds et al. (2007) phylogeny. This analysis reveals that some clades, such as the two clades of monkeys at the top, have given rise to an extraordinary number of lineages, while other clades, including many strepsirrhines, have produced few descendent lineages in the past twenty-five million years. (Updated and redrawn based on the original analysis by Purvis et al. 1995.)

"vital rates" occurred on the tree, and whether the results were common to all Old World monkeys or only to particular clades within this group. They mapped the significant increases in diversification rate onto primate phylogeny using parsimony to identify where on the tree changes in rates occurred. This revealed that all major groups of Old World monkeys contained rapidly evolving lineages, implying that the increase is common to the entire clade. A similar result was found for the galagos within the strepsirrhines. In this case, however, the authors caution against strong conclusions because the age of this clade was questionable in the phylogeny that they used.

In another analysis, Moore et al. (2004) applied a maximum likelihood approach to infer shifts in diversification rate across primates. Their analysis was based on Purvis's (1995) supertree for primates, allowing them to compare their results to Purvis et al. (1995). They detected three statistically significant diversification rate shifts in primates: at the origin of anthropoids; within New World

Table 8.1. Birth (*b*) and death (*d*) rates of primate lineages.

Clade	*b*	*d*
Strepsirrhines	0.093	0.023
New World Monkeys	0.118	0.000
Old World Monkeys	0.342	0.000
Apes	0.134	0.037

Source: Purvis et al. 1995

Table 8.2. Comparing diversification rates in major clades of primates.

Clade	Strepsirrhines	New World Monkeys	Old World Monkeys	Apes
Strepsirrhines	—	2.8	64.5**	2.2
New World Monkeys	3.7*	—	53.7**	0.7
Old World Monkeys	47.7**	32.0**	—	29.3**
Apes	0.6	0.0	10.0**	—

Source: Purvis et al. 1995
Note: Values show likelihood ratio statistics indicating support for each clade's own values of birth rate (*b*) and death rate (*d*) (maximum likelihood estimates), with the support obtained with the maximum likelihood estimates of *b* and *d* from other clades (listed at the top). Significance was based on a χ^2 distribution with d.f. = 2. Strepsirrhines include lemurs, galagos, and lorises.
* *p* < .05
** *p* < .001

monkeys (the clade containing *Callithrix, Cebuella, Leontopithecus,* and *Saguinus*); and within Old World monkeys (the clade containing all *Presbytis* species, except *P. entellus*). They also identified several marginally significant rate shifts, including two in the Old World monkeys and two in the strepsirrhines. These results are similar, but not perfectly congruent, with those reported by Purvis et al. (1995). Importantly, however, the results of Moore et al. (2004) also were not perfectly congruent with the "whole-tree" statistics used to detect variation in diversification rate across primates in a different paper (Chan and Moore 2002). For example, the apes were indicated as having variation in diversification rate with the whole-tree measure, but the maximum likelihood method revealed no evidence for shifts in diversification rate in this clade.

Paradis (1998) developed methods from survival analysis to investigate shifts in diversification rates on a phylogeny. The approach uses likelihood ratio tests (or the Akaike information criterion, AIC) to assess the fit of models with equal or variable diversification rates among specific clades. Paradis (1998) applied the method to Purvis's (1995) inference of primate phylogeny. As demonstrated by Purvis et al. (1995) using different methods, Paradis (1998) found significant heterogeneity in diversification rates among primates.

A last example comes from a recent study of language diversification in- volving the spread of people through Polynesia. Using a linguistic phylogeny for Austronesian languages, Gray et al. (2009) compared branch lengths along hypothesized "pauses" in the colonization of Polynesia to all other branches on the tree, with the prediction that these pauses should be characterized by longer branches. This pattern was generally upheld, but was only marginally significant. The authors also performed a Bayesian analysis of diversification rate changes on the phylogeny and found strong support for expansion pauses and pulses in the settlement of Polynesia (see chapter 10).

What Factors Influenced Diversification Rates?

One of the most famous sayings in evolutionary biology is attributed to J. B. S. Haldane. When asked what could be inferred about the "creator" from the works of nature, Haldane reportedly replied, "He had an inordinate fond- ness for beetles." What is it about beetles that made them so successful or, in an example from mammals, why are bats and rodents so diverse, yet mono- tremes and sloths are so rare? As revealed in the example above, similar kinds of questions can be asked about human population splitting and linguistic di- versification (Gray et al. 2009). For example, why are there so many languages in New Guinea, or so many language "stocks" in the New World? Many such questions in linguistics have yet to be tackled at a large scale with phyloge- netic approaches. In an example from biology, Barraclough et al. (1995) com- pared sister taxa of birds that differed in the number of species showing di- chromatism. They found that more dichromatic lineages have more species, consistent with a link between sexual selection and speciation.

Identifying differences in diversification rates among clades—whether by increased speciation rates or decreased extinction rates—again only pushes the question to another level. Why have diversification rates differed in dif- ferent clades? This question lies at the core of research on *key innovations*, which refers to the origin of a new trait that gives one group of organisms an evolutionary advantage over its sister group without the trait, resulting in higher rates of speciation, lower rates of extinction, or both (see Heard and Hauser 1995). It also could be that species with particular traits can divide up niche space more finely, experience geographical divergence (vicariance) events more commonly, or undergo ecological or genetic isolation more read- ily. In other words, it might have more to do with the environmental circumstances of the species in a clade than an intrinsic feature of their biology that gives more speciose clades an advantage in competition with other clades.

Many of the methods described in the previous section can be applied to study key innovations. For example, one can use sister clade comparison to test whether a particular trait or environmental characteristic consistently affects the size of clades (Slowinski and Guyer 1993). However, several caveats apply. First, as with "convergence approaches" to study correlated evolution, it is important to find repeated evolutionary origins of the hypothesized innovations; otherwise, when investigating whether change in a trait along a single branch influenced rates of subsequent diversification, it is difficult to rule out the possibility that some other trait influenced rates of diversification. Second, the methods apply best to situations where entire clades in the larger tree have (or do not have) the trait that is hypothesized to influence diversification rates. What should be done when the trait is not so strictly shared through common descent (Ree 2005), as in our example of exaggerated sexual swellings in Old World monkeys in chapter 3, where we saw that swellings were probably lost in some clades? One possibility is to compare the proportion of taxa in pairs of clades that generally have or do not have the trait, or to use a "sliding scale" threshold to assign presence or absence of a characteristic to a clade (e.g., Barraclough et al. 1995).

In an elegant approach based on maximum likelihood, Maddison et al. (2007) described a statistical model that allows inference of speciation and extinction rates in the context of a binary (discrete) trait (AnthroTree 8.5). Thus, one can ask, is the rate of speciation or extinction elevated in lineages that have a particular character state, such as diurnal activity period? Or, for cultural traits, do origins of agricultural practices increase speciation rates and reduce extinction rates? The model contains six parameters to be estimated—two speciation rates (one for each of the two binary states), two extinction rates (again, for each of the two binary character states), and a model of character state change as the independent variable (transition rates from 0 to 1 and 1 to 0). The approach thus builds on an original proposition made by Pagel (1997) and others (Paradis 2005; Ree 2005) by including extinction rates in addition to speciation rates. This is important, because a character that influences diversification (speciation and extinction) might also impact the distribution of character states on the resulting tree (Maddison 2006). Maddison et al. (2007) provided a simulation test of the estimation procedure, which showed that the method is better able to detect differences in speciation rates (b) than extinction rates (d). However, more simulations are needed to fully evaluate this and related approaches.

A related set of questions applies to continuously varying traits. For example, some authors have proposed that body mass influences diversification rates, with smaller-bodied species better able to fill niche space (e.g., Hutchinson and MacArthur 1959; Brown 1995; Gittleman and Purvis 1998). One approach is to conduct a contrast-based analysis that investigates how contrasts in a continuous character covary with contrasts in the number of lineages that descend from a node (AnthroTree 8.6, Agapow and Isaac 2002; Isaac et al. 2003). This sister clade comparison test can be conducted for nodes with three or more descending lineages, and ideally such a test requires a fully resolved and complete phylogeny (i.e., with no polytomies or missing taxa). Thus, one asks whether contrasts in some trait, such as body mass, correlate with contrasts in the number of taxa that descend from that node. A more recent approach makes use of the number of nodes along root-to-tip path lengths in relation to traits of interest, under the prediction that traits leading to higher diversification should be associated with more nodes along the path from the root to the tips (Freckleton et al. 2008).

Fewer methods analogous to the approach of Maddison et al. (2007) have been developed for continuous traits. In one study of primates, however, Matthews et al. (2010) binned a continuous distribution of body mass into binary categories of "large" and "small" body mass to apply Maddison and colleagues' discrete approach. They failed to find a significant association between body mass and rates of speciation or extinction. Recently, a new method has been developed to investigate how a continuously varying trait covaries with speciation and extinction rates (FitzJohn 2010). In an application of the method to primates, it also failed to find a consistent association between body mass and speciation or extinction rates.

Sister clade comparisons of continuously varying traits have been applied several times to questions in primate evolution. Unlike the methods just discussed, which can estimate speciation and extinction rates separately, these methods examine overall diversification rate. For example, Gittleman and Purvis (1998) tested whether smaller-bodied primates diversify more rapidly but failed to find significant support for this hypothesis (see also Conroy 2003; Isaac et al. 2005). Another approach, based on maximum likelihood estimation of diversification rates, did find that larger bodied primate species exhibit lower rates of diversification, and the magnitude of the effect varied among primate lineages (Paradis 2005). In a more comprehensive analysis of multiple variables that might influence diversification rates across mammals, Isaac et al. (2005) found that population density and group size correlate with diversification rates in primates. They proposed that a larger population could buffer a species from extinction risk while also potentially favoring the origin of peripheral isolates that favor the creation of new species.

Last, it is worth emphasizing again the challenges of determining causality (chapter 6). If we find that a character is associated with speciation, extinction, or overall diversification, we cannot be sure that this character is the causal factor influencing these processes. It could be, for example, that some other character covaries with the measured one and is the true cause. As noted by Maddison et al. in the context of their method, "the correct conclusion given a significant result . . . is that the character examined *or a codistributed character* appear to be controlling diversification rates" (2007, 708). Thus, we should keep in mind the spurious correlation between ice cream sales and murder rates that was mentioned in chapter 6.

Along similar lines, sometimes we cannot even determine whether the trait influenced diversification, or whether diversification influenced the trait (i.e., a reversal of causality). An example from my research illustrates this difficulty in the context of parasitism. We investigated whether primate lineages with more parasites exhibit a higher rate of diversification (Nunn et al. 2004). We found support for this association, but we were unable to discern with great confidence the underlying cause: does increased parasitism cause higher rates of host diversification through some sort of coevolutionary arms race, or does a higher rate of host diversification create better "habitat" for parasites, for example, by providing more opportunities for generalist parasites to establish in overlapping host populations of closely related species? Or, returning to the issue raised in the previous paragraph, does some other factor drive both parasitism and host diversification rates? The evidence we were able to muster suggested that host diversification influences parasitism by creating more opportunities for host sharing and host shifting, but not all predictions for this causal link were statistically significant.

Diversification and Evolutionary Rates

A last important question arises from our previous discussion concerning evolutionary models, rates of evolution, and directional models of evolution (chapters 4 and 5). Are rates of molecular evolution associated with rates of diversification? Using a molecular phylogeny, one prediction is that the distance from the root to the tips of the tree will correlate positively with the number of nodes along this path; such an association would indicate that divergence events are associated with increased rates of molecular change. Webster et al. (2003) investigated this prediction using forty-three published molecular phylogenies and found support in 35 percent of the trees.

A related question involves whether the evolution of a trait leads to higher rates of genetic or linguistic evolution. Lutzoni and Pagel (1997) considered this possibility in a study of the evolution of mutualism among fungi, where they found elevated rates of evolution in nuclear ribosomal DNA following the evolution of mutualism. Similarly, Wlasiuk and Nachmann (2010) and Garamszegi and Nunn (2011) found links between measures of molecular evolution and parasitism. For languages, Atkinson et al. (2008) found that the splitting of languages is associated with greater amounts of linguistic change.

In studies of evolutionary rates of molecular evolution, it is important to be aware of the *node-density artifact* (Sanderson 1990; Venditti et al. 2006; Venditti and Pagel 2010). The basic idea is that rates of evolution will be better estimated on parts of the tree that are characterized by a denser taxon sampling. Conversely, on longer branches, multiple changes in a character (including reversals) are often hidden to the investigator, who will perceive fewer changes at a particular genetic locus. Collectively, this effect can generate artifactually shorter summed lengths along root-to-tip paths that have fewer nodes. This artifact biases estimates of branch lengths especially in parsimony analyses, and also in maximum likelihood and Bayesian approaches for inferring phylogeny (Venditti et al. 2006; Hugall and Lee 2007).

The node density artifact could also impact the results of comparative tests, for example, in independent contrasts analyses that make use of branch lengths (chapter 7) or tests of directional models using GLS (chapter 4; see figure 4.7). In addition, phylogenetic studies that compare phenotypic and molecular change (e.g., Omland 1997a) also are susceptible to the node-density artifact (Bromham et al. 2002). Luckily, methods are available to detect the node-density artifact (Venditti et al. 2006), and further research will undoubtedly help to develop additional evaluation diagnostics and solutions for correcting the effect.

Summary and Synthesis

As noted by Agapow and Purvis, "Phylogenies are valuable troves of information concerning the trajectory of evolution, and their shapes carry the fingerprints of the processes that have formed them" (2002, 866). In this chapter, I reviewed some of the questions that can be asked, and specifically how the fingerprints of evolutionary history can be deciphered to answer these questions. Many of these questions can be addressed using null models in which speciation and extinction are treated as stochastic factors. Such null models, however, do not mean that actual speciation and extinction events are random, but rather, as noted by Slowinski and Guyer, "that a group's size, no matter how great, is not *prima facie* evidence that the group arose from nonrandom speciation and/or extinction" (1993, 1020).

More work is needed to assess the strengths and weaknesses of methods to investigate speciation and extinction rates (Kubo and Iwasa 1995; Paradis 2005; Maddison et al. 2007; Rabosky 2010; Matthews et al. 2010; Rabosky and Alfaro 2010). One conclusion from simulation studies thus far is that the power to infer differences in diversification rates is low. Thus, it is important to run analyses of tree balance and estimate speciation and extinction rates from phylogenies that are as large (and complete) as possible. In addition, when doable, research should incorporate information on extinct lineages. A more worrying issue concerns the difficulty and possible bias in estimating the extinction rate. This means that the methods may do a better job of testing hypotheses involving the factors that influence speciation (e.g., that small body mass leads to greater division of niche space) or overall diversification, compared to hypotheses involving extinction (e.g., that larger-bodied primates experience higher extinction because their populations are less able to rebound from ecological disturbances).

Studies of primate diversification have been relatively common, whereas these models have yet to be applied widely in studies of cultural traits, including language (Gray et al. 2009). At a smaller scale, extinction rates are likely to vary among human populations. One study found, for example, that cultural group extinction in war-faring populations of New Guinea ranged from 1.6 percent to 31.3 percent per generation (Soltis et al. 1995). It would be interesting to link these micro-level patterns of extinction to broader variation in extinction rates in reconstructed linguistic trees, although data are presently too limited to do so.

Pointers to related topics in future chapters

- The birth-death model applied to languages: chapter 10
- Extinction risk: chapter 11

9 Size, Allometry, and Phylogeny

A critical issue in many studies of morphology, behavior, and physiology is how traits covary—or *scale*—with some measure of size, particularly body mass (Schmidt-Nielsen 1984; LaBarbera 1989; Martin 1990). Understanding scaling is necessary to test hypotheses that make specific predictions for how a trait changes with changes in size. With detailed estimates of how a trait scales with body mass, it is also possible to make specific predictions for how other, possibly related traits also scale with mass. For example, the scaling of metabolic rate and body mass (Kleiber 1961) predicts that other traits involving energy acquisition, such as home range size or surface area of the gut, will exhibit the same scaling relationship (Gittleman and Harvey 1982; Martin et al. 1985; Nunn and Barton 2000).

Understanding scaling is also essential when statistically controlling for the effects of size as a confounding variable in comparative research. Statistical control is often needed, for example, in studies of life history evolution (Harvey and Clutton-Brock 1985; Ross and Jones 1999) or brain size (Martin et al. 1985; Harvey and Krebs 1990; Gittleman 1994; Martin 1996; Deaner et al. 2000). Body mass typically covaries strongly with life history traits, brain measurements and many other characteristics, yet we are often interested in testing whether other traits like diet influence life history or brain size. Thus, it is necessary to control for these size-related effects to rule out the possibility that variation in the trait of interest is driven simply by variation in size.

In morphological studies, scaling is also essential for investigating *shape*, where shape refers to the information in a morphological trait after scale, location and rotational effects are removed. Shape has been investigated, for example, in studies of cranial and dental morphology in relation to diet in mammals (Dumont 1997) and in studies of forelimb morphology of tamarins and marmosets (Falsetti et al. 1993). Understanding scaling is also important in biomechanics where, for instance, dimensionless ratios provide a framework for studying the effects of various forces, such as gravity or viscosity, in animals of different size or to create models that capture real aspects of the question at hand, but at a scale that enables easier experimental investigation. In these and other situations, the association between shape and size is often the interesting pattern to examine, rather than to simply "control" it away (see Falsetti et al. 1993; Jungers et al. 1995). Alberch et al. (1979) provide a framework for investigating morphological change in "size-shape" space, thus explicitly embracing how shape and size covary.

These examples reveal some of the reasons to investigate whether (and how) a trait covaries with size. In my own work on behavioral and life history traits, I have been more concerned with statistically controlling for size (Nunn 2002b; Lindenfors et al. 2007) or making predictions for how some trait, such as home range size, covaries with other traits, such as metabolic needs of a social group (Nunn and Barton 2000). However, research in morphology or biomechanics is often focused on different issues involving variation in shape and discerning exactly how shape covaries with size. Using the example of dimorphism in the human corpus callosum, Smith (2005) provides an excellent overview of issues involved with "relative size versus controlling for size." In all comparative studies, it is important to incorporate phylogeny when controlling for confounding effects of body mass (Garland et al. 1992, 1993; Revell 2009) and when examining the evolution of size and shape (e.g., Cannon and Manos 2001).

Several books and articles provide introductions to the fundamentals of scaling (Peters 1983; Calder 1984; Schmidt-Nielson 1984). Therefore, in what follows, I provide only a brief overview of these topics and other general statistical issues, keeping the focus on how phylogenetic methods can be used to study a wide array of topics involving size, including allometry, clade shifts, controlling for size, and even reconstructing size (and variables related to size) in the fossil record. As we will see, phylogenetic comparative methods are appropriate for investigating scaling questions, and in most cases are expected to outperform analyses that ignore phylogeny.

Because much of this chapter involves morphological variation, let us first examine a case that reveals the importance of incorporating phylogeny when investigating the association between body mass and a morphological trait—the intermembral index (IMI), which was discussed earlier in this book (see chapter 4).

Example: IMI, body mass, and phylogeny. Primate morphology is commonly studied using comparative methods, including studies of sexual dimorphism in body mass and canine size (Clutton-Brock et al. 1977; Plavcan et al. 1995;

Figure 9.1. Suspensory locomotor adaptation in a white-handed gibbon (*Hylobates lar*). This female gibbon from Khao Yai National Park, Thailand, is hanging from a branch by one hand. (Photo courtesy of Ulrich Reichard, Southern Illinois University, 2008.)

Mitani et al. 1996a; Plavcan 2001; Smith and Cheverud 2002), aspects of dental morphology and development (Smith 1983; Kay 1984; Holly Smith et al. 1994; Ravosa and Hylander 1994; Plavcan and Daegling 2006), and locomotor morphology and physiology (Napier and Walker 1967; Martin 1990; Demes and Jungers 1993). Morphological research on primates has a long and rich history (e.g., Schultz 1930; Fleagle 1988; Martin 1990, and chapters in Tuttle 1972; Jenkins 1974), yet many morphological patterns studied prior to the availability of phylogenetic comparative methods have yet to be examined in an explicitly phylogenetic context.

Consider, for example, the IMI, which plays a major role in discussions about locomotor adaptations and limb morphology in primates (Napier and Walker 1967; Napier 1970; Martin 1990). The IMI quantifies relative limb length; it is calculated simply as the ratio of forelimb to hindlimb length, multiplied by 100. A higher IMI corresponds to longer relative forelimb length, as found in species with suspensory locomotion like gibbons (figure 9.1). Indeed, the IMI corresponds closely to different locomotor categories in primates; it approximates 70 in species that exhibit vertical clinging and leaping, 70–100 in quadrupedal primates, and 100–150 in primates that exhibit suspensory locomotion (Martin 1990).

The IMI is easily calculated, it is intuitively simple, and it provides a way to investigate locomotor categories in a quantitative, comparative framework. Given the tight theoretical and empirical linkage between IMI and locomotor adaptations, this index has played a prominent role in reconstructing the behavior of extinct primates (Napier and Walker 1967; Martin 1990). Jungers (1978) calculated the IMI for several different species of *Megaladapis*, a subfossil giant prosimian from Madagascar. He documented values between 110 and 129, which is well outside the range of variation found in extant Malagasy prosimians. Jungers therefore concluded that *Megaladapis* made use of vertical supports but did not possess hindlimb morphology consistent with leaping behavior (see Cartmill 1974).

While the IMI is a useful descriptor of different locomotor categories, is this index associated with other variables related to locomotion, such as body mass? Jungers (1978, 1984) and Martin (1990) noted that the IMI correlates with body mass across species, with larger primates exhibiting higher values of the IMI. This scaling relationship played a role in Jungers's (1978) biomechanical analysis of locomotion in *Megaladapis*. Using data on IMI and body mass from Rowe (1996), I confirmed that a significant relationship exists between body mass and IMI using the raw species data (figure 9.2a; b = 0.08; $R^2 = 0.36$; $F_{1,128} = 73.1$; $p < .0001$).

As discussed in previous chapters, however, such "nonphylogenetic" tests violate a critical assumption of most statistical analyses: closely related species tend to share similar trait values, rendering "species-level" data to be nonindependent and producing invalid statistical tests of whether two traits covary (Harvey and Pagel 1991; Martins and Garland 1991; Nunn and Barton 2001). Indeed, if we examine the species that make up the data points in figure 9.2b, we find that species within major primate clades are clumped together in morphological space. Apes are at the far right side of the figure, for example, and gibbons cluster together at the left edge of the "cloud" of ape data. In addition, the relationship between these traits varies among primate taxa, with a relatively flat line in prosimians and even a negative slope in apes. Clearly, the links between IMI and body mass are not as straightforward as one might expect from a simple nonphylogenetic test, and the relationship may even vary among primate clades.

To explore this association further, I undertook an analysis of the IMI relative to body mass using phylogenetically independent contrasts (chapter 7). In a contrasts analysis, the association between body mass and the IMI is substantially weaker than suggested by the nonphylogenetic test; evolutionary

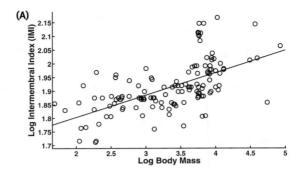

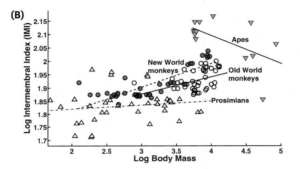

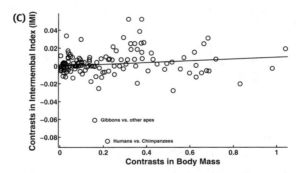

Figure 9.2. The allometric scaling of the intermembral index. Plots show how the IMI varies with body mass for (a) all primates treated as independent data points, (b) for primates in four major clades (apes [inverted triangle], New World monkeys [filled circle], Old World monkeys [open circle], and prosimians [triangle]), and (c) when using phylogenetically independent contrasts. Two outliers occur among contrasts involving apes, as indicated on the plot. Phylogenetic tests used log-transformed data and branch lengths based on S. Nee's transformation (described in Purvis 1995) for body mass and Grafen's transformation (1989) for IMI, with the topology following Bininda-Emonds et al. (2007). Independent contrasts were calculated using the PDAP Package in Mesquite, version 1.08.

rates of change in body mass are not as consistently associated with evolutionary rates of change in IMI (figure 9.2c). The relationship remains statistically significant ($t_{128} = 2.20$; $p = .03$), but body mass explains very little of the variation in IMI ($R^2 = 4\%$). Thus, one would conclude that the evolutionary association between mass and IMI is, at best, a weak one, suggesting that the association is not very useful for making predictions for IMI based only on body mass. (For a similar analysis that compares the IMI in the largest and smallest species among closely related taxa, see Jungers 1985. Such an analysis fails to make use of all the variation, but before the development of independent contrasts, it was one of the few approaches available for dealing with phylogenetic nonindependence.)

The IMI therefore provides several lessons as we embark on the material covered in this chapter. First, it reveals the importance of controlling for phylogeny in cross-species comparative studies of morphology, even for traits that appear to have a tight biomechanical linkage with other morphological measures. Second, it illustrates a common step in comparative studies in which investigators assess whether the trait of interest covaries with body mass. In this case, however, the association with body mass weakened considerably when phylogeny was taken into account. Last, the IMI provides an example of how variation in extant species can be used to make inferences about behavior in extinct species, subject to the strength of the evolutionary association that exists among the characters of interest. The specific functional linkage between IMI and locomotor behavior was not examined here, but could be tested using phylogenetic comparative methods in a future study.

Allometry

Interspecific *allometry* investigates how quantitative change in Y covaries with change in size across species, where size is often measured as body mass (M). This can be represented by an equation of the form

$$Y = \alpha\, M^\beta$$

where α is the *allometric coefficient* and β is the *allometric exponent*. This equation reveals that when $\beta \neq 1$, the ratio of Y to M changes with changes in M (Huxley 1932; Harvey and Pagel 1991; Jungers et al. 1995). For example, if $\beta > 1$ and Y is spleen mass, the ratio of spleen mass relative to body mass will increase with increasing body mass. For primate spleens, I previously found that $\beta = 1.07$, with confidence intervals encompassing a slope of 1 (Nunn 2002b, see also Stahl 1965). This suggests that the ratio of spleen mass to body mass is relatively constant across primates that vary in body mass.

Allometric relationships are typically examined in log transformed space, which gives the following equation:

$$\log Y = \log \alpha + \beta \log M$$

Thus, β becomes the slope of the relationship between the trait of interest and body mass and $\log(\alpha)$ is the intercept, which provides a more convenient way to investigate the relationship statistically (e.g., using linear line-fitting methods). A number of references provide a more comprehensive overview on allometry and issues to consider when conducting allometric studies (Gould 1966; Smith 1980, 1993; Sweet 1980; Harvey and Pagel 1991; Niklas 1994).

These equations emphasize that scaling relationships represent an example of correlated trait evolution discussed in chapters 6 and 7. In the case of allometric studies, however, we are expressly interested in the actual form of the relationship—that is, the slope estimate in logarithmically transformed data space—rather than simply whether the association is positive, negative or nonsignificant. In terms of the slope, it is often interesting to investigate whether a trait is *isometric* with respect to body mass, meaning that the slope equals 1. Or more specifically, isometry applies when the trait exhibits geometric similarity at different sizes, which means that the slope can differ from one when, for example, comparing a volume to a linear measure (see Jungers 1985; LaBarbera 1989; Spence 2009); thus, the units on the axes are critical for interpreting scaling relationships.

Under isometry, the trait varies directly in proportion to changes in size, as with the example of spleen mass given above. Alternatively, traits can exhibit *positive allometry* ($\beta > 1$) or *negative allometry* ($\beta < 1$), with proportionally more or less change, respectively, relative to increases in body mass (assuming that the traits are not already expressed on a per-gram basis and are in the same units). As noted by Harvey and Pagel, these terms "probably do more to obfuscate than to clarify" (1991, 172) as negative allometry does not require that the slope (β) is less than 0; rather, it indicates that its ratio to mass declines with increasing mass, which only requires that $\beta < 1$.

As with any question involving correlated evolution, determination of the precise form of a scaling relationship requires that we consider phylogeny (Harvey and Pagel 1991; Garland and Ives 2000; Nunn and Barton 2001; O'Connor et al. 2007). Incorporating phylogeny is especially important because body mass typically exhibits phylogenetic signal (Blomberg et al. 2003; Purvis et al. 2005a; Capellini et al. 2010), even in intraspecific datasets (Ashton 2004). Unfortunately, one can find suggestions in the literature that independent contrasts are not appropriate for investigating allometric relationships (Smith 1994; Martin 1996; Martin et al. 2005). Some original critics have reversed course in later papers due to methodological and other developments (e.g., Smith and Cheverud 2002), yet it is worthwhile to clarify some misconceptions that still arise when using independent contrasts to study allometry (see also Nunn and Barton 2001). For example, how can we estimate the intercept? As was discussed in chapter 7, the intercept is forced to equal zero in an analysis of independent contrasts. In studies of allometry, however, researchers often wish to estimate the intercept in order to assess whether a species falls above or below the regression line. Similarly, it might seem that by using contrasts we are moving into a different "trait space" that is not comparable to the raw species data. Is it possible to estimate a biologically meaningful slope when using independent contrasts, which are essentially amounts of evolutionary divergence rather than raw data?

Let us consider first the issue of the intercept. For slopes estimated with independent contrasts, it is straightforward to place the resulting phylogenetically correct regression line through the raw data. Specifically, the line obtained using independent contrasts would go through the *phylogenetic mean* of traits X and Y for the species data; this phylogenetic mean is estimated as the ancestral states for X and Y at the root node in a contrasts analysis (see chapter 4, Garland et al. 1993, 1999; Garland and Ives 2000). With the phylogenetically valid slope thus placed on the raw data, the intercept is the predicted value of Y when X equals 0 (see AnthroTree 9.1). Other approaches, such as GLS, provide a way to estimate allometric slopes in a phylogenetically correct way, and output from such models typically includes an estimate of the intercept without any further computational steps (see chapter 7, AnthroTree 7.5, and Pagel 1997).

In terms of estimating a biologically meaningful slope with independent contrasts, again, this is not a problem (Garland and Ives 2000; O'Connor et al. 2007). Slopes estimated from analyses of independent contrasts are perfectly comparable to those obtained using raw species data (see Harvey and Pagel 1991; Pagel 1993, 1999a; Martin 1996; Nunn and Barton 2001; Rohlf 2006). Thus, if we estimate a slope with independent contrasts and also using the raw

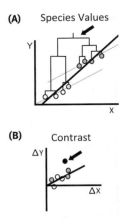

Figure 9.3. Independent contrasts and clade shifts. (a) Different ecological conditions are indicated by gray-filled and open data points that are shared through common descent in two clades. In this hypothetical case, the different conditions represent a clade shift where the allometric exponent (slope) remains the same within clades, but the allometric coefficient (intercept) differs between them. In species analysis, phylogenetic clumps of data may therefore bias the estimated slope across all species. Here, the overall slope will be biased upward (black line), but in other cases, it can be biased downward (i.e., when the grade is shifted "down" with mean increases along the x axis; see figure 6.9). (b) Contrasts represent evolutionary change, and so clade shifts between the different ecological categories should be represented by a larger than expected contrast in the branch linking the two subclades (filled point indicated by arrow). The slope is less likely to be biased upward.

species data, the expected value of these two slopes is identical. It is important to remember, however, that type I error rates generally will be elevated and power generally reduced in nonphylogenetic tests, because the distribution of statistics are wider than expected (see figure 6.10). Similarly, in real datasets, clade shifts can create artifacts that can severely obscure true relationships when phylogeny is ignored—a point we turn to next.

What about estimating the allometric exponent (slope) when phylogenetic effects, such as clade shifts, are present? Recall from chapter 6 that a clade shift occurs when a set of species shares a confounding character state through common ancestry that produces a shift in the relationship between the main variables with no change in the slope (see figure 6.8 and Martin 1990). The slopes are the same in the two groups, but the coefficient differs. In analyses of species data, a clade shift has the potential to severely bias the slope estimate because multiple data points are treated incorrectly as independent. Imagine, for example, that we estimate a regression line across the data in figure 9.3a without incorporating phylogeny; the line would be biased upward, relative to the true association indicated with the light gray lines within each of the two clades (which is the true relationship between X and

Y). With the method of independent contrasts, however, only a single contrast is calculated across the clade shift (figure 9.3b), which might be expected on average to produce less bias than the multiple data points among clades in a species analysis.

Hence, an independent contrasts analysis is likely to be more appropriate than a species-level analysis when clade shifts exist, especially when clade-specific effects can be identified and contrasts across the grades are "flagged" or even eliminated from the analysis (Nunn and Barton 2000, 2001). However, we often can do better than this by incorporating the clade shift directly into the analysis, that is, by including a variable in the statistical model that reflects the different phylogenetic groups (see Lavin et al. 2008; Fritz et al. 2009; Gartner et al. 2010). In addition, we will see later in this chapter that phylogenetic methods can be used to investigate the clade shift itself, including testing specific hypotheses that predict a shift in the intercept between two (or more) groups of organisms.

Example: Kleiber's law. One of the best-known examples of allometry involves what has become known as *Kleiber's law*. Max Kleiber evaluated how basal metabolic rate scales with body mass across mammals, with the resulting plot known the "mouse-to-elephant" curve for the wide range of organism sizes that it encompassed. The relationship between body mass and basal metabolic rate exhibits a scaling exponent of 0.75 (Kleiber 1961; Schmidt-Nielsen 1975; McNab 1986), and primates generally fall along this line (figure 9.4; see also figure 7; Martin 1990; and Genoud 2002). Kleiber's law thus refers to this "three-quarters" scaling rule.

As noted earlier, a scaling relationship between metabolic rate and body mass predicts that other features of an organism's biology related to energy acquisition should exhibit the same relationship. In primates, for example, Martin (1985) investigated how the surface area of the small intestine scales with body mass, which he predicted would exhibit a slope of 0.75 based on Kleiber's law. He found support for this prediction in a nonphylogenetic test, which was generally upheld in a later analysis by Lavin et al. (2008) across mammals that controlled for phylogeny (a slope of 0.703; $R^2 = 0.94$, in a regression model that fit an OU evolutionary model). Thus, the existence of one relationship makes predictions for similar relationships in other traits, and these predictions can be tested quantitatively.

Underpinning predictions such as these is the assumption that metabolic rate scales with body mass to the 0.75 consistently across taxa, or, to put it

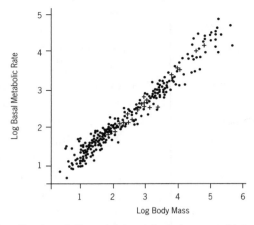

Figure 9.4. The scaling of basal metabolic rate in primates (and other mammals). Basal metabolic rate co-varies strongly with body mass and exhibits a slope close to 0.75. Primates are indicated with a "+," while other mammals are indicated by dots. Data kindly provided by M. Genoud (2002).

differently, that Kleiber's law really is a "law." This is active an area of research where allometric scaling is flourishing, and it is worth briefly taking stock of what researchers are discovering when using these methods. In particular, recent work has shown that it is difficult to distinguish between an allometric exponent of 0.75 and 0.67 for the scaling of basal metabolic rate and body mass, with the latter expected based on surface-to-volume ratios and loss of heat (i.e., with surface having two dimensions and volume having three dimensions, which predicts scaling of two-thirds). One phylogenetic study found, for example, that different phylogenies produce different results, with slopes falling between 0.65 and 0.77 (Symonds and Elgar 2002). More recent studies—many of them phylogenetically based—also have called into question crucial aspects of Kleiber's law, including the consistency of the exponent across clades (e.g., White et al. 2007, 2009a; Sieg et al. 2009; Capellini et al. 2010; Kolokotrones et al. 2010). Clearly this is an area of research where the latest phylogenetic methods will be needed to provide statistical power to detect fine-scale differences among clades. In fact, with the recent findings from applying phylogenetic and more appropriate statistical methods, it seems fair to suggest that Kleiber's "law" has been overturned (and needs a new name).

Line-fitting methods and evolutionary models. An important (and often overlooked) issue in allometric studies involves the statistical method used to estimate the slope (Harvey and Mace 1982; Harvey and Pagel 1991; Warton et al.

2006). Least-squares (LS) regression is the most widely used and readily available line-fitting technique. However, LS regression assumes that the independent X variable is measured without "error." Error in this case refers to deviations between the observed and true values for a trait; it can arise through measurement error, sampling error, temporal fluctuations, or other sources of variation (see Ives et al. 2007). When this assumption is violated—as it probably is for most estimates of body mass (Pagel and Harvey 1988; Smith and Jungers 1997)—LS estimates will consistently underestimate the true slope, and this bias will increase with decreasing correlation between the variables (Harvey and Pagel 1991). For this and other reasons, some authors have discussed the importance of using traits other than body mass to control for the effects of size, for example, by using "the rest of the brain" in studies of the neocortex and other brain parts (Barton 1998; Deaner et al. 2000; Lindenfors et al. 2007).

Methods for incorporating intraspecific variation into phylogenetic analyses are considered in chapter 11. Here, I focus on specific line-fitting methods that can account for error in the independent variable, including methods, such as reduced major axis (RMA) and major axis regression (Harvey and Mace 1982; LaBarbera 1989; Harvey and Pagel 1991; Niklas 1994; Warton et al. 2006). In general, these other methods will tend to produce a steeper slope estimate than LS regression, although differences will converge on the same slopes when the correlations among the traits of interest are very high. Moreover, before giving up on LS regression entirely, it is important to appreciate that LS residuals are obtained as vertical deviations of data points from the expected values. Hence, these residuals are independent of the variable on the x axis, which is essential for some methods of controlling for mass (see Harvey and Pagel 1991).

Harvey and Pagel (1991) provide an introductory overview of line-fitting methods. They also review how least squares, RMA and major axis line-fitting relate to a more general framework called the structural relations model (SRM), which provides a way to estimate slopes that incorporate variation in both the independent and dependent variables more directly (see also McArdle 1988; Pagel and Harvey 1988; Berrigan et al. 1993; Ives et al. 2007). The critical variable in the SRM is the ratio of the error variance in Y to the error variance in X. In major axis regression, the error variances are assumed to be equal, and hence the ratio equals one. RMA assumes that the ratio can be estimated as the ratio of the variances in the actual data, and LS regression assumes no error variance in X, giving an undefined (infinite) value for the ratio of error variances (McArdle 1988). Thus, rather than making assumptions about the nature of error, the SRM provides a way to incorporate actual estimates of variance into the analysis, giving a more general framework for estimating lines of best fit. Indeed, if the error variances of the independent and dependent variables are known, then calculations based on the SRM should provide unbiased slope estimates (Rayner 1985; Harvey and Pagel 1991). In species analyses, the error ratio can be estimated from intraspecific variation in the two traits. Previous authors have suggested that these estimates of the error ratio from the raw data can then be used when analyzing contrasts data (Pagel and Harvey 1989; Harvey and Pagel 1991; Berrigan et al. 1993; but see Nunn and Barton 2000 for a discussion of challenges in applying this approach and Ives et al. 2007 for more recent perspective).

In research that I conducted with Robert Barton (2000), we compared different line-fitting methods in both phylogenetic and nonphylogenetic analyses. We focused specifically on the scaling of home range size relative to the metabolic needs of the group, which was based on body mass, group composition, and the assumption that metabolic rate exhibits an allometric exponent of 0.75 (figure 9.5). More specifically, we tested the hypothesis that home range size scales with group metabolic needs with a slope of 1, where group metabolic needs was measured as the summed metabolic needs of individuals in a typical group for a primate species (see also Martin 1981; Gittleman and Harvey 1982). In nonphylogenetic analyses of raw species data, we found that home range scales with body mass as predicted, that is, with confidence intervals that encompassed a slope of 1 (Nunn and Barton 2000). This analysis also revealed several potential confounding effects that arise when ignoring phylogeny. We noticed, for example, steeper-than-predicted slopes among terrestrial primates and larger than expected home ranges among some small-bodied diurnal arboreal primates. Indeed, our phylogenetic analyses revealed that the slope was significantly greater than one when using RMA to fit the line. Thus, we would have drawn the wrong conclusion had we failed to incorporate both phylogeny and a more appropriate line-fitting method.

Another overlooked issue in allometry involves the evolutionary model. As described in chapter 5, it is now possible to fit parameters that represent, for example, an ecological niche-filling model or a model based on stabilizing selection (the OU model). Recently, these scaling parameters have been incorporated into scaling analyses (Spoor et al. 2007; Lavin et al. 2008). By fitting the parameter κ, for example, one can test whether a speciational model fits the data better than a gradual model (or some combination thereof if κ falls

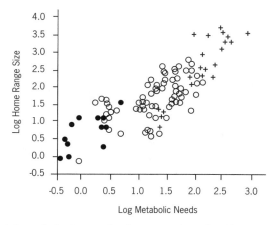

Figure 9.5. Kleiber's law and primate range size. Home range size scales with group metabolic needs with a slope close to or greater than the expected value of 1, depending on whether phylogeny is taken into account and the regression model used (ordinary least squares versus reduced major axis). Diurnal arboreal species are indicated with an open circle, diurnal terrestrial species indicated with a "+," and nocturnal arboreal species indicated with a filled circle. (Data from Nunn and Barton 2000.)

between zero and one; see Pagel 1999a, 2002) and then take this into account in a study of allometry. Similarly, recent work has applied an OU model when investigating correlated evolutionary change (Hansen et al. 2008; Lavin et al. 2008; Gartner et al. 2010).

Last, it is important to correct for logarithmic transformation bias when making predictions from allometric relationships estimated from logarithmically transformed data (Smith 1993). Specifically, after transforming the data to logarithms, the predicted values from the regression model are estimates of the geometric mean rather than the arithmetic mean, which results in an underestimated prediction when de-transformed back to the original unlogged units. This bias can be substantial yet fortunately can be corrected (Hayes and Shonkwiler 2006).

Clade Shifts: Testing Hypotheses and Controlling for Body Mass

Clade shifts thus far have been treated as mere annoyances. In many cases, however, we are interested in using variation around an allometric relationship to test an adaptive hypothesis that controls for size. Consider an example involving jaw morphology in Old World monkeys. This group includes a clade of organisms that tend to eat foliage and seeds (colobines) and another clade that is more frugivorous (cercopithecines). Ravosa (1996) investigated

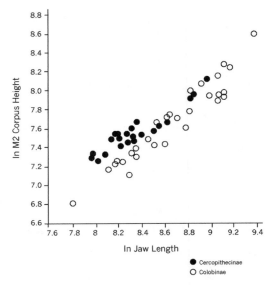

Figure 9.6. The height of the mandibular corpus relative to jaw length. Plot shows data separately for colobines (filled circles) and cercopithecines (open circles). For a given body mass, colobines tend to have taller jaws (under the second molar) than do cercopithecines. Both slopes and intercepts differ. (Redrawn from Ravosa 1996.)

whether the folivorous colobines have more robust dental characteristics, such as the vertical dimension of the lower jaw, compared to cercopithecines. Because the jaw also scales with body mass, Ravosa investigated this question in a way that controls for size (more specifically, for the length of the jaw). In a nonphylogenetic test, he found that colobines have more robust jaws than similarly sized cercopithecines along several key dimensions (figure 9.6). Other examples of using deviations from an allometric line to test hypotheses can found elsewhere in this book, including the example in chapter 1 of testes mass and mating systems (controlling for body mass, see figure 1.1).

Here, we are interested in two related issues: controlling for size and investigating whether organisms with particular features show different intercepts on a plot that controls for size (i.e., consistent deviations from an underlying relationship with size). In many cases, these features involve the phylogenetic position of the group of species; in other words, the factor of interest represents different clades. It is possible to take into account phylogenetic history and models of evolution in such studies (e.g., figure 7.8; Spoor et al. 2007). As a starting point, I first consider a classic solution to these issues developed by Ted Garland and colleagues (1993)—phylogenetic analysis of covariance—

and then I move on to more recent approaches to both investigate clade shifts and control for body mass more generally.

Phylogenetic analysis of covariance. A statistical method commonly used to address clade-related differences involves analysis of covariance (ANCOVA; Garland et al. 1993). ANCOVA provides a way to assess whether the intercept in a regression model differs relative to some other qualitative factor; as such, it controls for variation in the independent continuous variable(s). This is exactly the situation that we face when assessing whether some categorical variable, such as diet, mating system or taxonomic affiliation, impacts variation in another trait, while also controlling for body mass.

Garland et al. (1993) provide a discussion of ANCOVA in a phylogenetic context. These authors used the example of how home range size scales with body mass in forty-nine species of carnivores and ungulates, with the prediction that carnivores, as organisms on a higher trophic level, should have larger home ranges than ungulates, many of which are consumed by carnivores (see also McNab 1963). Thus, in a bivariate plot of home range size regressed on body mass, carnivores should fall in a "cloud" of values that is generally distinct from (and above) the values for ungulates. This is what the authors found in the raw species data (figure 9.7). In this case, the functional trait (i.e., trophic level) is perfectly confounded with phylogenetic affinity (i.e., carnivore versus ungulate), thus giving only one evolutionary transition that represents a difference between carnivores and ungulates.

It is straightforward to apply ANCOVA to the species values to test for a difference in intercepts (although it is important to ensure that the slopes are equal across groups, see Sokal and Rohlf 1995). If we want to conduct this analysis in a phylogenetic context, however, how should we proceed? Several options have been proposed. Garland et al. (1993) developed a Monte Carlo simulation approach, which is flexible, intuitive, and relatively easy to implement (for a recent example of its application to primate biology, see Vinyard and Hanna 2005 and Plavcan and Ruff 2008). The analysis starts by performing an ANCOVA on the species data. To assess statistical significance, one then generates a distribution of the same statistics under the null hypothesis of no difference between the groups; this null distribution incorporates evolutionary history and a specific evolutionary model by simulating traits on the hypothesized phylogeny for the group of organisms with a user-defined model of evolutionary change (see also Garland et al. 2005).

Garland et al. (1993) applied this method to assess whether carnivores have

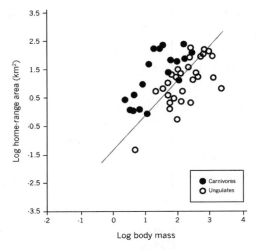

Figure 9.7. Home range area and body mass in carnivores and ungulates. Carnivores (filled circles) tend to have larger home range areas for a given body mass than do ungulates (open circles), but this difference was not statistically significant in a phylogenetic test (Garland et al. 1993). The line is a phylogenetically correct mean estimated using independent contrasts and fit back onto the observed values. (Redrawn from Garland et al. 1993.)

larger home ranges than ungulates after controlling for body mass and phylogeny. They found that the difference between carnivores and ungulates was not statistically significant when using the simulation approach, despite the sharp distinction seen in the bivariate plot of raw data and highly significant results in a nonphylogenetic test. This analysis emphasizes an intuitively obvious point that single evolutionary origins are challenging to study statistically; huge effect sizes are needed to overcome the low statistical power of such an analysis, which effectively involves an "n of 1" (see also chapter 12). Studies of traits that have undergone convergent evolution, such as mating systems in relation to testes mass (Harcourt et al. 1995), provide more degrees of freedom for such tests, because the functional trait is not perfectly confounded with phylogenetic affinity. Better incorporation of evolutionary models also can improve the power to detect effects. Thus, in a later study that used an OU model, Garland et al. (2005) did find evidence for a significant difference in ungulate and carnivore home range size using the simulation approach.

Garland et al. (1993) discussed other ways to increase the power of phylogenetic ANCOVA, like using continuous (rather than categorical) measures of the "shift" variable in question—for example, by including in the model the percentage of meat in the diet, rather than binary categories of "carnivore versus non-carnivore"—and by increasing the phylogenetic sampling breadth of species to encompass more independent evolutionary transitions in the traits of interest. Use of continuous variables is statistically more powerful because it typically increases the amount of evolutionary change that can be reconstructed, compared to using discrete classifications. Indeed, many categorical variables represent an underlying continuous distribution of trait values, such as diet, terrestrial substrate use, and locomotor behavior, and as a general rule, it is better to use the continuous measures rather than discrete categories if possible.

Other approaches to investigate clade shifts. As discussed in chapter 7, a more recent approach is to include a dummy variable for the clades in a phylogenetic GLS model (Lavin et al. 2008; Fritz et al. 2009; Gartner et al. 2010; see also AnthroTree 7.7). With this approach, the researcher is testing whether the clade effect is significant independent of body mass (or other variables). Given the ease of running this analysis, it is likely to become the preferred approach to investigate these types of data in a phylogenetic context.

Garland et al. (1993) considered other approaches to assess statistical significance in comparisons across phylogenetic or functional groups while controlling for size. For example, they applied an independent contrasts approach by comparing the single standardized contrast that links carnivores and ungulates to the other contrasts in the data set (i.e., those that are calculated within carnivores or within ungulates, but not between these two groups). If a difference exists between carnivores and ungulates, this single "connecting" contrast is expected to be larger than the other contrasts (or at least to be in the outer 95 percent of values; see also McPeek 1995; chapter 12; and the Brunch algorithm of the computer application CAIC described in Purvis and Rambaut 1995). It is possible to control for body mass in such a test and to implement it as a nonparametric test (see Garland et al. 1993). Garland et al. also applied phylogenetic autocorrelation methods to produce "phylogeny-free" values that can be examined using standard statistical tests (Cheverud et al. 1985; Gittleman and Kot 1990). Both the independent contrasts and autocorrelation methods yielded nonsignificant effects when comparing carnivore and ungulate home range size (Garland et al. 1993).

Using the raw species data, other methods are available to identify clade shifts. For example, Martin et al. (2005) proposed a nonparametric "rotation method" that makes fewer assumptions about the distribution of error and is less sensitive to outliers (Isler et al. 2002). The method can be used to provide an objective assessment of clade shifts, which can then be taken into account in further phylogenetic or nonphylogenetic tests (e.g., by running phylogeny-based tests within clades). Analyses of species data can be useful for giving clues to the existence of clade shifts, but phylogenetic analyses are essential for testing hypotheses in a statistically rigorous way.

Controlling for body mass more generally. Several methods are available to control for mass when the traits are all continuously varying, with the resulting data then analyzed using phylogenetic methods for continuously varying traits (chapter 7). One possibility is to simply divide the trait of interest by body mass. As revealed above in the discussion of allometry, however, for this approach to effectively control for body mass, the trait must scale isometrically with mass and have an intercept of zero on a log-log plot (see discussion in Smith 1984a, 2005; Jungers et al. 1995; Packard and Boardman 1999). If the trait exhibits positive allometry, then the ratio of the trait to mass will increase with increasing body mass. In this case, a ratio would be ineffective for controlling mass, because larger bodied organisms in the comparative sample would have larger values of the trait relative to body mass, for example in the case of primate sexual dimorphism (Smith and Cheverud 2002). If the ratio turns out to be a significant predictor and body mass also is a potential predictor of the dependent variable, the researcher would be unable to rule out the possibility that body mass actually drives the pattern.

Another common approach is to regress the trait of interest on body mass using least squares regression and then take the residuals (figure 9.8). As noted above, it is important to use LS regression techniques when calculating residuals, as this line-fitting method minimizes the vertical deviations from the regression line (Harvey and Pagel 1991). Thus, LS residuals are, by definition, independent of body mass and can be viewed as "size corrected," whereas the same is not necessarily true of residuals obtained using other line-fitting methods. This general approach can use either an empirically fitted line (estimated using phylogenetic comparative methods) or an a priori line based on other information, and is sometimes referred to as the "criterion of subtraction" (Gould 1975a, 1975b; Smith 1984a). However, other researchers have cautioned against using this approach for datasets with many potentially correlated predictor variables (Freckleton 2002, 2009).

A related approach involves multiple regression (and other multivariate statistical modeling approaches). If we are interested in controlling for the effects of body mass, it is relatively straightforward to include body mass as

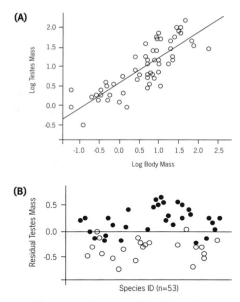

Figure 9.8. Residuals as a way to control for body mass (and other variables). Panel (a) shows how testes mass covaries with body mass across primates. The line is a least squares regression line through the raw data; this line can also be estimated using phylogenetic methods and forced onto the raw data prior to calculating residuals. Panel (b) shows the resulting residuals for each species, with species alphabetically ordered along the x axis (names not shown). Species with multi-male mating systems are shown as filled circles, and single-male mating systems, as open circles (Harcourt et al. 1995; Nunn 1999a; Van Schaik et al. 1999); species with uncertain mating system were eliminated (cf. figure 1.1). Data on testes mass and male mass are from Harcourt et al. (1995).

one of the predictors in a multiple regression model (Freckleton 2009). With body mass as a predictor, any significant effects of the other independent variable would be independent of body mass. Users taking this latter approach may still find it useful to present results graphically using residuals, as this provides a way to represent bivariate relationships. In addition, it is important to be aware of the dangers of collinearity, in which correlated predictor variables can cause unstable regression models (Petraitis et al. 1996). Given that mass covaries with so many traits of interest, collinearity may be especially problematic when body mass is one of the predictors. The first step in dealing with collinearity is to determine if it actually exists; this can be done using variance inflation factors, which are commonly given in statistical packages. The second step is to eliminate the collinearity, either by restricting the analysis to independent variables that are more statistically independent, or by applying methods such as ridge regression (Hocking 1976; but see O'Brien 2007 for a different perspective on this issue).

Reconstructing Body Mass (and Other Traits) in Extinct Species

If we understand the functional links among traits in living species, we can also make inferences about those traits in fossils. For example, if body mass covaries strongly with femur length or tooth size in extant primates, we can use these associations to make estimates of body mass in extinct lineages for which we have data on femur length or tooth size. Similarly, if we document clear functional links between the curvature of the digits and locomotion in living species, we can use this association to make statistically rigorous inferences about the locomotor behavior of an extinct species based on its digit morphology (e.g., Stern and Susman 1983; Jungers et al. 1997).

In this section, the focus is on integrating ancestral state reconstruction (chapters 3 and 4) with studies of correlated evolution and the scaling of body mass (chapter 7). Hence, this section extends and integrates methods from previous chapters, and is placed in this chapter because such methods are often used to estimate body mass (and traits related to body mass) in extinct species. I begin by considering the traditional approach of understanding relationships among various traits and body mass in extant species to make predictions for body mass in extinct species. I then review new approaches that introduce greater phylogenetic rigor, including by considering the phylogenetic placement of the species whose traits are being reconstructed.

Given the importance of body mass to many aspects of mammalian life history, ecology and behavior, body mass is one of the most important traits to estimate in fossil taxa. Thus, many studies have attempted to infer body mass of extinct species. Typically, this is accomplished by documenting relationships in living species between traits that fossilize and body mass (which does not fossilize), and then using these relationships and fossil data to make predictions for body mass in the fossil species. In one study, for example, Gingerich (1977) investigated the relationship between molar size and body mass among living apes. He found that these two traits are highly correlated ($r = 0.94$), and from a larger sample of thirty-eight primate species, he used the scaling of these two traits to estimate the body mass of two fossil hominoids: *Aegyptopithecus zeuxis* (estimated to be 5.6 kg; 95% confidence interval, 4.5–6.5 kg) and *Proconsul africanus* (estimated to be 27.4 kg; 95% confidence interval, 16–34 kg). In another example, McHenry (1975) estimated the mass of hominins based on the scaling of the cross-sectional area of the last lumbar vertebra. From the equation in living humans, he estimated that *Australopithecus robustus* (specimen SK 3981) weighed 36.1 kg (with a range of 25.4 to 46.8 kg).

It is worth noting that many of these earlier studies were not conducted phylogenetically because the comparative methods or the phylogenies were not yet available. Thus, it is important to reinvestigate these and other comparative patterns in a phylogenetically rigorous way (e.g., Spocter and Manger 2007). As illustrated with these two examples, it is also essential to place confidence intervals on the resulting estimates, usually expressed as a plausible range of variation derived from the statistical fit of the model (Smith 1996). Bayesian approaches could be used to place credible intervals on the estimates. As discussed in previous chapters, nonphylogenetic analyses of continuously varying characters are not in general biased, but instead produce wider distributions of test statistics than expected given the number of data points (see figure 6.10). Phylogenetic analyses should create better allometric relationships by sifting out relationships that were type I errors and establishing new relationships that were missed by nonphylogenetic tests due to their reduced statistical power.

The approach just described essentially relies on the statistical association between traits, with phylogeny playing a "supporting" role by ensuring that the association is statistically valid, that is, that the data points are statistically independent of one another. Is it possible to take phylogeny into account more explicitly when inferring body mass (or other traits) for a fossil taxon? New phylogenetic approaches are integrating prediction of unknown character states with studies of correlated trait evolution (e.g., Garland and Ives 2000), which has implications for inferring body mass (and other traits) in extinct species.

In one example of this integrative approach, Organ et al. (2007) investigated genome size in dinosaurs. One goal of their study was to identify the point at which genomes were reduced in size relative to the evolution of birds, which are thought to have smaller genomes to compensate for the energetic expenditure of flight by reducing metabolic costs. Organ et al. (2007) used data on the sizes of bone cells (i.e., osteocytes) as a predictor of genome size, and they investigated this relationship using a Bayesian analysis of correlated evolution that also controlled for phylogeny. First, they confirmed that a significant association exists between osteocyte size and genome size in extant vertebrates. Next, they generated posterior probability distributions of genome size in thirty-one dinosaurs, controlling for uncertainty in the underlying evolutionary correlation between these two traits. They found that for all but one of the theropods within the lineage that gave rise to birds, genome sizes fell within the range of variation found in living birds (figure 9.9).

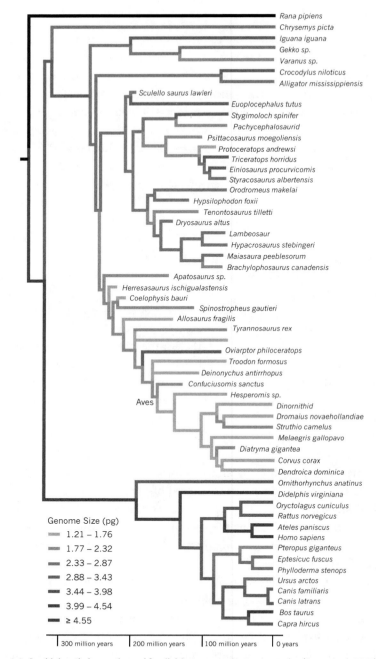

Figure 9.9. Combining phylogenetics and fossil data to reconstruct genome size (Organ et al. 2007). Phylogeny shows genome size reconstructed across a wide range of vertebrates, including dinosaurs. Reconstructions were based on a Bayesian analysis of the association between genome size and bone cells (i.e., osteocytes) in living organisms. The analysis incorporated phylogenetic relatedness using phylogenetic generalized least squares. (Redrawn from Organ et al. 2007.)

Surprisingly, this suggests that the evolutionary reduction in genome size occurred *before* the evolution of flight.

This example shows how one can use evolutionary correlations to make phylogenetically informed predictions about traits that do not fossilize (such as body mass) based on both phylogenetic placement of the lineage and statistical associations with other features that do fossilize (such as dimensions of teeth, long bones or vertebrae). Much potential exists for applying this basic approach to study human evolution (see chapter 12) to reconstruct body mass and other traits throughout primate evolution, and even to investigate the evolution of human cultural behavior on a linguistic tree.

We might call the estimates so far discussed "primary" estimates; we are estimating body mass from molar size, or genome size from osteocyte size. In these cases, the researcher is inferring the value of one trait from another trait that is measurable in the fossil record, as based on comparative studies of the pair of traits in extant primates (and phylogenetic position in the case of osteocyte size). In many cases, however, these primary estimates are then used to make "secondary" inferences about other traits or behaviors, such as group size or life history traits (see examples in Smith 1996).

These secondary estimates can be problematic if the calculations are not done appropriately. Most notably, researchers are "skipping a step" in the sequence of prediction, and this should be reflected by giving wider confidence intervals. For example, we might have an estimate of how some fossilizable trait, such as tooth size, covaries with body mass, and how body mass covaries with another trait, such as lifespan. Thus, in the prediction of lifespan, we really only have data on tooth size, and it is therefore the relationship between tooth size and lifespan that should be examined in living primates to make the inferences about lifespan in the fossil specimen. When the study is done in a way that takes these chains of prediction into account, the confidence intervals often widen considerably, and can even become so broad that prediction is irrelevant (e.g., see Nunn and Van Schaik 2002). Smith (1996) provides a comprehensive overview of this and other potential problems when making inferences in the fossil record based on comparative data.

Summary and Synthesis

Many comparative studies have investigated how traits covary with body mass, and some scaling relationships, such as Kleiber's law, make predictions for how other traits should covary with mass. Other research has used scaling relationships to control for body mass in comparative studies, including comparative studies of nonmorphological characters, such as life history traits, or to compare the shape of morphological traits across species. Finally, it is less widely appreciated that comparison plays an essential role in efforts to infer body mass (and other traits) in extinct lineages. Indeed, many approaches to reconstructing traits in the fossil record require a rigorous comparative perspective based on functional relationships between a trait that fossilizes and some other trait. It is now possible to integrate correlated evolution and phylogeny to make better predictions for unmeasured or extinct species (Garland and Ives 2000; Organ et al. 2007).

In terms of methodological development, greater consideration of appropriate line-fitting methods is needed, especially in the context of using phylogenetic GLS methods and incorporating variation in the data (e.g., Sieg et al. 2009). Indeed, all major line-fitting methods now in use probably make unrealistic assumptions about the distribution of error, and given that most correlations in morphological studies are well below $r = 1$, this is likely to impact estimates of allometric exponents. Similarly, methods are now available for incorporating evolutionary models into studies of scaling, as well as measures of intra-specific variation (and other sources of statistical "error"; Ives et al. 2007). In terms of application, many "classic" comparative studies have yet to be conducted in a rigorous phylogenetic context. As the example of the intermembral index revealed (see figure 9.2), such research is likely to change conclusions from some previous studies, and in the process to generate new insights to morphological evolution in primates.

Pointers to related topics in future chapters:

- Evolutionary singularities and human evolution: chapter 12
- Confidence intervals on fossil reconstructions: chapter 12
- Integrating ancestral reconstruction and correlated evolution: chapter 12

10 Human Cultural Traits and Linguistic Evolution

Why are some human societies characterized by polygyny and others monogamy? What are the environmental and social factors that lead to hierarchical societies, compared to more egalitarian societies? Are residence patterns structured more commonly along male or female lines, and what factors account for whether men or women leave home after marriage? How do languages evolve, and is it possible to build predictive models for how different parts of speech or particular words change over time? More generally, what factors influence the evolution of human cultural diversity, including language?

A wide variety of questions can be investigated through comparisons of different cultures, societies, or political entities, which will hereafter be called "cross-cultural" studies. These questions include fundamental aspects of human behavior at an individual or higher level, characteristics of societal structure and patterns of wealth transfer, the drivers of linguistic variation, and basic questions concerning human health and welfare (e.g., Birdsell 1953; Ember and Ember 1971; Hartung 1982; Burton and White 1984; Bentley et al. 1993; Borgerhoff Mulder et al. 2001; Guernier et al. 2004; Bloom and Sherman 2005; Tehrani and Collard 2009). In addition, cross-cultural comparative research provides a means to understand the distribution of traits geographically, and for linguistic data, to test specific hypotheses about a history that is otherwise not easily studied, such as migration patterns or rates of linguistic evolution (e.g., Nichols 1997; Gray and Jordan 2000; Pagel et al. 2007).

As shown by these examples, comparative anthropologists are interested in documenting differences in the configurations of human cultural traits and in understanding how and why particular sets of traits arise. Comparison has long played a role in this endeavor (Naroll and Cohen 1970; Mace and Pagel 1994; Ember and Ember 1998). The first formalized approach to cross-cultural comparison was developed by Edward Tylor (1889). Tylor was explicitly interested in developing systematic approaches to investigate cross-cultural variation, including correlations—or what Tylor called "adhesions"—among different traits. Lacking possibilities for experiments, he realized that comparison of different societies provides a way to uncover the factors that account for human cultural diversity. Francis Galton criticized Tylor for failing to deal with statistical nonindependence in the cross-cultural data (i.e., Galton's problem; see chapter 6; Naroll 1961). However, Tylor's work spawned a number of followers who developed an empirically and theoretically rich approach that flourishes to this day (Naroll and Cohen 1970; Ember and Ember 1971; Mace and Pagel 1994; Borgerhoff Mulder et al. 2001; Holden and Mace 2003). As we will see, recent studies and new statistical methods offer more principled and user-friendly approaches for dealing with Galton's problem, but a number of fundamental issues remain to be addressed.

In this chapter, I provide a brief overview of some key questions and approaches used in cross-cultural research. To begin, I would like to share a few "snapshot" examples of the comparative approach in modern cross-cultural research. My goal is to show that despite a long history of applying the comparative method, this approach is far from a relic of the past in studies of human behavioral variation. Indeed, the comparative approach in cross-cultural research is flourishing, and it is likely to become even more important as larger-scale datasets and more rigorous statistical methods become available. In the examples that follow to get us started, it is important to note that most of the studies failed to take into account phylogenetic history or other factors that might influence the distribution of data. This partly reflects a failure by most cross-cultural anthropologists to appreciate the importance of controlling for nonindependence, yet it also reflects the real challenges of implementing phylogenetic or other controls when analyzing cross-cultural data (e.g., Billing and Sherman 1998). More generally, this gap presents an opportunity for a new generation of anthropologists, linguists and archaeologists to investigate these questions with statistically valid comparative approaches.

Examples of cross-cultural research. As a starting example, Hartung (1982) tested an adaptive hypothesis for the transfer of wealth through paternal versus maternal lines of descent. He reasoned that parents should pass wealth to sons if it helps them acquire additional mates, and such a pattern should be found in societies that practice polygyny. He tested this hypothesis using data from two large datasets that were compiled independently from the hypothesis he was testing (Murdock 1967; Murdock and White 1969). Hartung found that male biased inheritance is associated with greater polygyny (table 10.1). He also found that polygyny covaries with transfer of resources from the

Table 10.1. Male bias in inheritance and polygyny.

Polygyny	Low or no male bias	High male bias or males only
None	32	44
Limited (<20%)	27	106
General (>20%)	6	196

Source: Hartung 1982
Note: Numbers reflect number of societies with the states from the Ethnographic Atlas (Murdock 1967) (N = 411).

groom's family to that of the bride (i.e., bride-price), providing further evidence for a crucial link between wealth and polygyny. A variety of later studies built on the groundwork laid by this comparative study and other studies on descent mechanisms and residence patterns (Aberle 1961; Ember and Ember 1971), including a recent phylogeny-based comparative study of cattle-keeping in relation to patrilineal descent (Holden and Mace 2003).

The comparative method also has been used to address fundamental questions about "disease risk" using data from the ethnographic record (Low 1987, 1990) and modern societies (Gangestad and Buss 1993; Guernier et al. 2004). In one comprehensive study spanning 150 different countries, for example, Guégan et al. (2001) investigated whether the diversity of infectious diseases corresponded to human life history and socioeconomic variables, including population density and growth rates, mortality and fertility rates, and per capita gross national product. Parasite data reflected the presence or absence of up to 16 different infectious diseases, including malaria, schistosomiasis, and dengue fever. The analyses revealed a positive association between female fertility and the total diversity of infectious diseases (figure 10.1). In poorer countries, greater mortality rates (as might be caused by higher parasite pressure) could lead to earlier female reproduction or higher total numbers of children—although the authors were unable to identify the causal mechanisms responsible for this association (see also Guégan and Teriokhin 2000). In another study, Guernier et al. (2004) found that the diversity of parasites showed a latitudinal gradient with more diverse pathogen communities found closer to the equator (see also Low 1990).

Another area of recent comparative interest involves diet. Human populations exhibit incredible dietary diversity, and since we all take pleasure in eating some foods but not others, it is worthwhile to understand the factors that drive this variation. Billing and Sherman (1998) addressed a basic and fascinating question in the context of dietary diversity: why do some human populations include more spices in food preparation? The authors used the

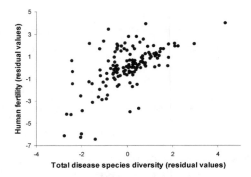

Figure 10.1. Association between female fertility and the diversity of infectious diseases in humans. Data are based on sixteen parasites and pathogens sampled at the country level. Plots show residual values from a multivariate analysis, controlling for key environmental, demographic and socioeconomic factors. (Redrawn from Guégan et al. 2001.)

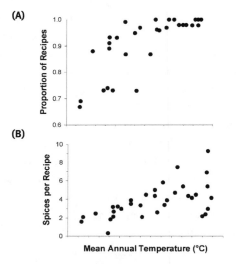

Figure 10.2. Relationship between mean annual temperature and the spiciness of recipes among thirty-six cuisines. Significant positive associations were found between mean annual temperature and (a) proportion of meat-based recipes that require at least one spice and (b) average number of spices used per recipe. Both relationships are statistically significant. (Redrawn from Billing and Sherman 1998.)

resources at Cornell University's School of Hotel Administration to obtain traditional recipes for meat-based dishes from more than thirty countries. From this, they quantified the use of forty-three different spices and food flavorings. Billing and Sherman demonstrated that spices are more common in cuisines from countries characterized by a hotter climate (figure 10.2). Based on additional evidence showing that spices inhibit the growth of a wide diversity of bacteria, the authors concluded that the geographical distribution

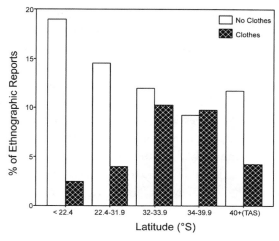

Figure 10.3. Clothing use in relation to latitude among Australian Aborigines. Reports of unclothed aborigines decline with increasing latitude in mainland Australia, with an apparent reversal of the trend in the last category, which involves Tasmania (TAS). (Reprinted with permission from: I. Gilligan, "Clothing and climate in aboriginal Australia," *Current Anthropology* 49 [2008]: 487–95.)

of spicy foods reflects a need to prevent food spoilage caused by bacteria in hotter climates. They ruled out alternative explanations involving nutritional benefits of spices, use of spices to conceal the taste or smell of spoiled food, greater availability of spices at lower latitudes, and use of spicy foods to promote perspiration for evaporative cooling. If their conclusion is correct, enjoying a good curry could owe as much to bacterial pathogens as it does to human ingenuity and cultural transmission.

Another fascinating question concerns the origins of clothing. What factors influence the use of clothing, and when did clothing come into fashion? One obvious explanation for use of clothing is to provide protection from the elements, particularly cold weather. To test this hypothesis, Gilligan (2008) compiled reports of clothing use in Australian aborigines from the ethnographic record and then compared these data to climatic data. The analysis revealed a general association between clothing use and temperature (figure 10.3), with increased use of clothing as one moves away from the equator (from left to right on the figure). Interestingly, however, the areas closest to the equator and farthest from it (i.e., Tasmania) showed some exceptions to the general trend across populations. In the far south, these exceptions may reflect cultural extinction (Henrich 2004) and biological adaptations to colder thermal conditions (Gilligan and Bulbeck 2007), and in the far north, exceptions may have arisen through influence from other cultural groups in the area.

In a different application of the comparative method to the question of clothing, Kittler et al. (2003, 2004) used phylogenetic methods to date the divergence of lice into a lineage that infests head hair of humans and another lineage that infests the rest of the body (and can also live in clothing). They estimated that this evolutionary split occurred around 107,000 years ago, thus providing an estimated date for the origin of clothing in human evolution. Furthermore, the genetic diversity among African lice was greater than that of non-African lice, suggesting an African origin for clothing (Kittler et al. 2003).

Archaeologists commonly compare artifacts across space and over time, and many of the interpretations of variation and temporal change can be placed in a phylogenetic framework (Neff 1992; O'Brien et al. 1994; O'Brien and Lyman 2002; Foley and Lahr 2003). For example, "evolutionary archaeologists" have examined artifact variation using cladistics, including studies of Polynesian ceramics (Cochrane 2004), Middle Eastern and European agricultural crops (Coward et al. 2008), and American Indian projectile points (figure 10.4; O'Brien et al. 2001; O'Brien and Lyman 2002; Buchanan and Collard 2007). Evolutionary archeology takes as a starting point a central tenet of evolution, namely, that cultural variants are differentially transmitted through time. For example, evolutionary archaeologists make use of "clade-diversity diagrams" and other measures to investigate the richness of artifact types over time (Lyman and O'Brien 2000; O'Brien and Lyman 2000, 2002). Archaeologists increasingly use more sophisticated model-based methods to draw inferences about cultural transmission and population histories (e.g., Harmon et al. 2006; Coward et al. 2008; Lycett 2008, 2009), and this is sure to be an area of future methodological application and development.

Can we use variation in the rate at which words undergo change to understand the mechanisms that drive linguistic evolution? In a recent study, Pagel et al. (2007) used phylogenetic methods to estimate rates of word evolution in eighty-seven Indo-European languages (see also Pagel and Meade 2006b). The authors found that words for different meanings exhibit remarkable variability in their rates of evolution. For example, the word "two" had the same cognate (i.e., words with a common origin) in all eighty-seven cultures, whereas the word "dirty" exhibited more than forty different cognates. Using data on word use in four different (and divergent) languages, they made the remarkable discovery that the frequency with which words are used predicts the rate of linguistic evolution: words that are used most frequently evolve at significantly slower rates (see also Lieberman et al. 2007).

In another phylogenetic study of language evolution, Atkinson et al. (2008)

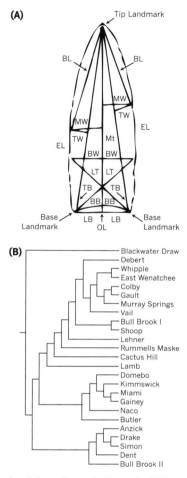

Figure 10.4. Cladogram based on Paleo-Indian projectile points. (a) Using fifteen measurements from projectile points from across the continental United States, Buchanan and Collard (2007) found four equally parsimonious cladograms, including the one shown in (b). The "taxa" at the tips are sites where the projectile points were found. For details on the measurements indicated in (a), see Buchanan and Collard (2007). Characters: EL, edge boundary length; TB, tip landmark to base landmark; TW, width of tip from base length to maximum inflection position; BL, blade length; MW, maximum width; BB, base boundary length; LB, linear measure of base; Mt, midline length; OL, overall length; BW, basal width across first third of point; LT, length from base to one-third along opposite edge. (Redrawn from Buchanan and Collard 2007.)

discovered that rates of word evolution are influenced by the number of diversification events (i.e., nodes) on a linguistic tree, suggesting that newly formed sister languages undergo more rapid linguistic evolution. This is consistent with either smaller population sizes (founder effects), or linguistic change generated by social functions of language, such as strengthening group identity early in the splitting of a linguistic group. Similar speciational effects are found in biological evolution (Venditti and Pagel 2010).

These examples reveal the wide range of questions that can be addressed involving cultural evolution and give a sense of the importance of comparison in anthropological, sociological and linguistic research. In this chapter, I consider the following questions that are relevant to cross-cultural comparisons of ethnographic, archeological and linguistic data:

1. **To what extent are cultural traits shared through common descent versus borrowed from other societies?** This question forms the basis of fundamental research on *phylogenesis* versus *ethnogenesis* in understanding cultural diversity. Moreover, the assumption of common descent underlies phylogeny-based methods, which are increasingly being applied to study cultural variations.

2. **What are the consequences of cultural trait borrowing for comparative methods?** Do cases of borrowing require new comparative approaches to study correlated evolution, and which methods are most appropriate?

3. **What are the prospects for using comparative cultural data, particularly languages, to study human migration patterns?** While linguistic trees are commonly used as a basis for investigating cultural trait evolution using phylogeny based comparative methods (Mace and Pagel 1994; Mace and Holden 2005), these trees also can reveal interesting patterns about language evolution and human history (Gray et al. 2007).

In what follows, I present an overview of some key studies that have addressed these questions and the phylogenetic approaches that are available for cross-cultural comparisons. Interested readers can pursue these and related questions through further reading (Dixon 1997; Nettle 1999b; O'Brien and Lyman 2000, 2003; Shennan 2002, 2009; Campbell 2004; Mace et al. 2005; Lipo et al. 2006b).

To What Extent Are Cultural Traits Shared through Common Descent?

I follow Richerson and Boyd by defining culture as "information capable of affecting individuals' behavior that they acquire from other members of their species through teaching, imitation, and other forms of social transmission" (2005, 5). Cultural traits have many similarities to genetic traits: they exhibit variation, and variants are inherited across generations and can compete with one another (Mesoudi et al. 2004; Mace and Holden 2005). We may lack a "gene" for particular cultural traits, and identifying the cultural "species" to study is challenging. However, we face similar challenges in studies of many (if not most) biological characteristics; for example, we rarely understand the genetic underpinnings of behavioral or morphological traits of interest, many nongenetic traits may show phylogenetic signal (e.g., parasite richness; see chapter 5), and deciding how to operationally define a biological species is far from simple (see chapter 11). It is becoming more widely accepted to examine cultural traits on a phylogeny as evolved traits (Mace et al. 2005).

In contrast to the phenotypic traits that interest most comparative biologists, however, cultural traits can spread "horizontally" among unrelated individuals rather than through common descent along strict genetic, or "vertical," lines of transmission (note, however, that horizontal transmission of genetic elements is probably also widespread in biological systems, including mammals; Tosi et al. 2002; Roca et al. 2005; Burrell et al. 2009). This horizontal transmission of behavioral variants between individuals creates the possibility for horizontal transmission between different recognizable cultural groups, whether these are religious denominations, speakers of the same language, ethnic food traditions, or myriad other definable culture groups (i.e., the "species" in the comparative analysis). It is therefore important to critically evaluate the assumptions of vertical transmission and the appropriateness of different methods when this assumption is violated (Borgerhoff Mulder et al. 2006; Nunn et al. 2006).

Two points are worth emphasizing at this stage. First, group-level horizontal transmission is often termed *ethnogenesis*, while group-level vertical transmission is typically called *phylogenesis*. More specifically, phylogenesis refers to the spread of traits through vertical descent in the splitting of societies, and ethnogenesis refers to the spread of cultural traits through blending and borrowing among contemporaneous societies (Moore 1994; Tehrani and Collard 2009). Second, horizontal transmission requires that we have some reference history for comparison. However, genes, languages, and human cultural traditions may have independent histories, and functionally related subsets of these characters may also have incongruent histories (Boyd et al. 1997; Matthews et al. forthcoming). Hence, we have no "true" tree that reflects an indisputable evolutionary history. In other words, considerations of horizontal transmission are necessarily relative to some other history.

Several archaeologists and anthropologists have used comparative approaches to test hypotheses related to ethnogenesis and phylogenesis. In what follows, I review some of these methods and discuss their strengths and weaknesses.

The consistency index (CI) and retention index (RI). Chapter 2 provided background on the CI and RI, which are standard statistics calculated in parsimony analyses (see AnthroTree 2.3). Briefly, these statistics reflect the degree to which a character or set of characters are consistent with a tree-like structure, with higher values reflecting greater "tree-like" patterns in the data (i.e., less homoplasy). For this reason, the CI and RI are commonly used in anthropology to infer the degree to which traits are transmitted vertically from parent to daughter populations. The logic underlying their use is that horizontal transmission will often result in homoplasy. Thus, a low CI or RI is often taken as consistent with some degree of horizontal transmission. As we will see, however, there are reasons to be cautious when drawing conclusions from these two statistics.

In one study of cultural evolution that used the CI, Tehrani and Collard (2002) tested whether horizontal or vertical descent best described characteristics of Iranian textiles. Their primary goal was to assess the relative importance of phylogenesis and ethnogenesis in cultural evolution. Using design characters from the period before Russian domination of the Turkmen, the CI was calculated to be 0.68. The authors therefore concluded that most of the characters are shared through common descent. Relatively similar patterns (with perhaps slightly more ethnogenesis) were found in the period following conquest of central Asia by Czarist Russians.

In another example, Jordan and Shennan (2003, 2005) investigated the role of vertical and horizontal transmission in basketry traditions among Californian American Indians. They used three different methods, including the CI. From their analyses, they concluded that horizontal transmission exerts an influence on basket design, but that vertical transmission maintains distinct patterns within American Indian groups.

In the context of estimating horizontal transmission in cultural data, the CI and RI are likely to suffer from several weaknesses that are worth considering (Borgerhoff Mulder et al. 2006; Nunn et al. 2010). First, it is easy to imagine scenarios where a low CI or RI occurs in the absence of horizontal transmission, and other scenarios where a high CI or RI occurs with high levels of horizontal transmission. Thus, a low index could simply reflect homoplasy, that is, the independent evolutionary origins of cultural traits, rather than horizontal transmission; such an effect might happen when rates of evolution are high yet the traits are vertically transmitted. By comparison, a high index could occur in the context of high rates of borrowing if the traits of interest are borrowed as a "package" or unit (e.g., Boyd et al. 1997); in such a case, the phylogenetic signal will remain for those characters, even if other sources of data (genes, language, or other cultural traits) produce a different tree. Second, the CI and RI lack a solid statistical framework to assess the "significance" of the evidence for vertical transmission. For example, does a CI of 0.60 indicate support for vertical or horizontal transmission? Or should one even conclude anything about horizontal transmission, focusing instead on what can be inferred about vertical transmission of traits? All that can be done is to compare values of the CI or RI to values obtained from other traits or systems (Collard et al. 2006b). Last, the CI is affected by sample size (Sanderson and Donoghue 1989), making it difficult to compare different datasets (see chapter 2). This issue is of lesser concern for the RI than the CI (Hauser and Boyajian 1997).

A recent simulation study assessed whether the CI and RI can detect horizontal transmission of individual characters between societies (Nunn et al. 2010). In this study, we simulated a large number of datasets in which we systematically varied rates of horizontal transmission and evolutionary change (as well as other variables, such as rate of extinction and the number of traits). By analyzing data generated with known degrees of horizontal transmission, we tested whether measures of tree consistency detected non–tree-like evolution (i.e., horizontal transmission). We found that variation in the rate of evolution has a much greater impact on the RI than does variation in horizontal transmission, thus calling into question whether the RI is effective in discerning horizontal transmission; it could be, for example, that a low RI simply reflects a higher rate of evolutionary change in a vertically inherited trait. We did find, however, that the largest RIs (the top 1 percent) generally were associated with low rates of horizontal transmission, suggesting that an RI greater than 0.60 probably indicates vertical signal in the data (see also Collard et al. 2006b).

Phylogenetic signal. As discussed in chapter 5, a number of authors have investigated the degree to which cultural traits show evidence for phylogenetic signal. Often this is taken to indicate vertical transmission, meaning the transmission of traits from ancestral to daughter populations (Cavalli-Sforza and Feldman 1981; Borgerhoff Mulder et al. 2006). It is worth emphasizing that the measure of signal is relative to a particular class of other data, that is, the linguistic, genetic or other types of data that were used to infer the tree.

Several authors have investigated phylogenetic signal in African populations. For example, Guglielmino et al. (1995) examined the degree to which
cultural traits in 277 African societies covaried with linguistic, geographic,
and ecological variables. The authors interpreted their findings to show that
most traits exhibit linguistic signal, especially those having to do with family
structure and kinship, but their conclusions were based on overly simplistic
statistical model selection procedures; hence, the analyses should be repeated
with more rigorous methods.

Similarly, Hewlett et al. (2002) found that for a range of traits in African
societies with clear signal, 80 percent showed evidence of inheritance from a
common ancestor. Moylan et al. (2006) also studied African societies. They
used a method from biology (Abouheif 1999, see chapter 5) to measure the
linguistic signal in cross-cultural data and found a mixture of significant and
nonsignificant measures of linguistic signal. In another study, Holden and
Mace (1999) used multiple regression to assess the degree of similarity in
traits related to polygyny, sexual division of labor, stature, and subsistence activities in a wide range of societies. For most traits, they found evidence for
both phylogenetic and geographic signal.

All of the diagnostics related to phylogenetic signal and evolutionary models in chapter 5 can be applied to study cultural traits (see AnthroTree 5.2, 5.3,
and 5.5). Importantly, these methods typically examine the distribution of a
trait on trees derived from data that are independent from the trait of interest.
If the independently derived tree represents the true phylogenetic history of
the traditions, this should provide better insights to phylogenetic signal, compared to calculating the CI and RI on the tree generated from the characters
of interest (Nunn et al. 2010). However, currently available methods for detecting phylogenetic signal in discrete traits also are generally unable to distinguish whether low levels of signal are caused by horizontal transmission or
by high rates of evolution.

**Comparison of trait distributions to expectations under different transmission
models.** Another approach compares observed trait values to distributions
that are expected under different models of transmission, such as horizontal or vertical transmission. It is also possible to generate expected matrices
of differences based on ecological factors, for example, to test whether individuals flexibly tailor their behaviors to particular ecological conditions rather
than assuming vertical inheritance as the null hypothesis (e.g., Guglielmino
et al. 1995). Many of these methods derive from phylogenetic methods and

from approaches based on geographical analyses, such as the Mantel test (Mantel 1967; Smouse et al. 1986; Smouse and Long 1992).

Several authors have used the Kishino-Hasegawa (K-H) test (Kishino and Hasegawa 1989) to assess the correspondence between linguistic relationships, geographic distances, and ecological distances (Jordan and Shennan 2003; Jordan and Mace 2006). This test was devised in biology to investigate whether two phylogenetic topologies are significantly different (see Goldman et al. 2000 for a review and critical evaluation of the K-H test). In anthropological studies, the trees are generated from data on language, geography, or ecology. Jordan and Shennan (2003) applied this method to investigate the role of horizontal and vertical transmission in basketry of American Indians from California. The method revealed evidence for horizontal transmission, particularly for analyses of coiled baskets, yet vertical transmission (measured as language relatedness) retained an effect. In another application of the K-H test, Jordan and Mace (2006) applied the method to sets of cultural traits from American Indians from the Pacific Northwest. They found no clear associations between cultural traits and either geographic distance or linguistic relatedness.

Mantel tests have been used in cross-cultural studies to test whether two distance matrices are correlated. Thus, with matrices for geographic distance, phylogenetic distance, and trait distance, one might predict that geographic distance provides a better estimate of cultural trait variation as the rate of horizontal transmission increases; conversely, the phylogenetic distance matrix should be more highly correlated with trait data as horizontal transmission declines (Nunn et al. 2006). Mantel tests compare two distance matrices; partial Mantel tests compare two matrices while controlling for a third one. Thus, one could compare trait distances with geographic distances while controlling for linguistic distances. It is also possible to include ecological distances, for example to assess whether the characters covary with ecological factors. Statistical significance is assessed by permutation procedures to generate null distributions of the test statistic (Mantel 1967; Smouse et al. 1986). A recent study demonstrated poor performance of Mantel tests in phylogenetic studies (Harmon and Glor 2010). The authors suggested, "matrix-based analyses such as the Mantel test should only be used when the data can only be expressed in the form of pairwise distances" (2,176).

Jordan and Shennan (2003) applied Mantel tests to distinguish horizontal and vertical transmission in their American Indian basketry dataset. They found that geographic distance accounted for 38 percent of the variation in

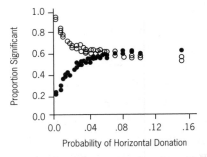

Figure 10.5. Ability of Mantel tests to detect horizontal transmission. Using data simulated under different degrees of horizontal transmission, we tested whether continuously varying trait values correlate significantly with a phylogenetic distance matrix (open circles) and a geographic distance matrix (filled circles). Horizontal transmission results in greater numbers of significant results in analyses of geographic distance, but phylogenetic distances picked up a similar number of significant results even at high levels of horizontal transmission. This probably reflects the high correlation between geographic and phylogenetic distance in the simulations, which also may occur commonly in real data. (Redrawn from Nunn et al. 2006.)

basketry characteristics. Moreover, "adjacency" of cultures in terms of spatial neighbors and geographic distance had a bigger effect than language, which was used as a measure of historical relations. For an example of the Mantel test, see AnthroTree 10.1.

In a simulation test (Nunn et al. 2006), we investigated whether geographic distance provides a better estimate of cultural trait variation as the rate of horizontal transmission increases (and conversely, whether phylogenetic distances covary less strongly with trait distances as horizontal transmission increases). We found general support for these predictions (figure 10.5) with increasing horizontal transmission rendering geographic distance a better predictor of trait distance and phylogenetic distance a poorer predictor of trait distance. Even at high horizontal transmission, however, phylogenetic distances tended to exhibit significant associations with trait distances. This likely reflects that geography and phylogeny themselves are highly correlated, making it difficult to disentangle their effects and again highlights the challenges of inferring process from pattern.

Tests of conflicting signal. Split decomposition (Bandelt and Dress 1992; Huson 1998; Huson and Bryant 2006) is a method that has been used to detect horizontal transmission events in cultural and linguistic data (Gray and Atkinson 2003; Hurles et al. 2003; Bryant et al. 2005; Holden and Gray 2006). Split decomposition methods do not force the data to fit a tree but rather check for conflicting signals that, in the context of cultural traits, might reflect instances of horizontal transmission. The method results in a "splits graph," which can represent conflicts in the data (AnthroTree 10.2). In an extension of split decomposition, one can obtain a delta score, which provides a quantitative estimate of the degree of conflict (Gray et al. 2010). A variety of methods are being developed to deal with non–tree-like or "reticulate" data in general (Posada and Crandall 2001a; Bryant and Moulton 2004; Makarenkov and Legendre 2004; Huson and Bryant 2006; Lipo 2006; Bloomquist and Suchard 2010).

In studying the evolution of Austronesian languages in the Pacific, Hurles and coworkers (2003) used split decomposition to investigate linguistic borrowing. They found that horizontal transmission of words occurred within western and eastern clades of Polynesian languages, but not between them (see figure 2.10). In another application of split decomposition methods to linguistic data, Bryant et al. (2005) were able to identify horizontal borrowing among some Indo-European languages using data on basic vocabulary terms. Thus, even among languages that tend on the whole to show strong tree-like signal, horizontal transmission can be detected in linguistic data (see also Holden and Gray 2006).

From a different intellectual arena, Barbrook et al. (1998) applied split decomposition methods to study forty-eight copies of *The Canterbury Tales*, written by Geoffrey Chaucer in the late fourteenth century. From their analysis, the authors were able to identify the manuscripts that probably were most similar to Chaucer's original version (for another historical example, this one involving musical instruments, see Tëmkin and Eldredge 2007).

Methods of studying tree congruence are also important in analyses of host-parasite cospeciation (Page 2003), and these methods may be usefully applied to investigate cultural traits (Gray et al. 2007; Riede 2009; Tehrani et al. 2010). In their study of American Indians in the Pacific Northwest, Jordan and Mace (2006) applied methods based on host-parasite biology (see figure 1.4; Page 2003) to assess whether trees inferred from different sets of cultural traits are congruent. They also investigated congruence among language trees and sets of traits using these methods. By comparing statistical measures of similarity among different trees, they found that traits were largely independent of one another, and they were independent of both geography and language. In a few cases, however, sets of traits showed associations with one another, such as those related to dress/adornment and marriage practices. This finding suggests that these sets of traits show similar transmission histories through space and time.

Summary and prospects. Due to variation in which traits are borrowed and which are vertically transmitted, cultural histories will weave complex historical trails. We need methods to uncover these trails to place cultural traits in broader perspective (Borgerhoff Mulder et al. 2006; Lipo et al. 2006a; Gray et al. 2010). For example, it will be important to develop methods to identify the cultural "packages" that move together versus alternative models in which traits move individually or as an integrated set (Boyd et al. 1997; Jordan and Shennan 2003; Jordan and Mace 2006; Gray et al. 2007). Recent advances in Bayesian phylogenetics provide a way to partition data to assess whether different types of data produce different trees (Matthews et al. forthcoming). Various other approaches have been or are being developed for discerning trait histories (e.g., McElreath 1997; McMahon and McMahon 2003; Borgerhoff Mulder et al. 2006; Pocklington 2006). Until we have a clearer picture of the strengths and weaknesses of different methods, it makes sense to use a diversity of measures that capture different aspects of the transmission process (Jordan and Shennan 2003; Jordan and Mace 2006; Buchanan and Collard 2007), and this is sure to be a rich area of methodological growth in the future (e.g., Gray et al. 2007, 2010; Tehrani et al. 2010; Matthews et al. forthcoming; M. Franz and C.L. Nunn, unpublished).

What are the Consequences of Borrowing for Comparative Methods?

When using a phylogeny to control for the nonindependence of cultural or biological data, we are assuming that the tree provides an accurate representation of trait history. This is, after all, the basis for reconstructing ancestral states (chapters 3 and 4), inferring models of evolutionary change (chapter 5), and controlling for the nonindependence of comparative data in studies of trait correlations (chapters 6 and 7). When horizontal transmission of one or more cultural traits occurs *across* branches of a tree, however, that tree is no longer an accurate representation of statistical dependencies among the traits—in such a case, the trait moved between different lineages rather than along a branch on the tree. How does horizontal transmission impact the use of phylogenetic methods to study correlated evolution of cultural traits? If horizontal transmission invalidates the use of phylogenetic methods for investigating a cross-cultural dataset, what new methods are needed to incorporate the potential for horizontal transmission more directly?

As we have seen throughout this book, computer simulations provide a compelling way to investigate the statistical performance of a method under different conditions (Martins and Garland 1991; Nunn 1995; Diaz-Uriarte and Garland 1996; Harvey and Rambaut 1998, 2000). Most simulation approaches focus on two aspects of statistical performance. One of these involves type I error rate, or the probability of incorrectly rejecting a null hypothesis of no association between characters; in this case, the investigator simulates two traits that are uncorrelated and calculates the proportion of time that the statistical test detects a significant pattern. The other aspect of statistical performance involves power, where power refers to the probability of correctly detecting an association when one exists (i.e., correctly rejecting the null hypothesis). In this case, we simulate two traits that are evolving in a correlated way and then calculate the proportion of times that the method correctly identifies the traits as being significantly correlated.

In one study, Nunn et al. (2006) designed a simulation approach to investigate the effects of horizontal cultural transmission on the performance of independent contrasts. The simulation begins with a single population that inhabits one cell in a matrix. Populations can expand to one or more adjacent empty cells, or populations can go extinct, with user-defined probabilities of expansion or extinction (figure 10.6). Populations that diversify to new cells are treated as distinct societies in the next round of simulation. Continuous traits evolve under a Brownian motion model of evolution with a user-defined degree of correlation with other continuous traits. Vertical transmission occurs through the inheritance of traits from one generation to the next and through diversification of populations into empty niches (i.e., the traits of the parent generation are carried to the new niche). Horizontal transmission occurs stochastically, with the user defining the probability that borrowing occurs.

With this model, it is easy to generate simulated datasets with known degrees of horizontal transmission. Thus, one can test the performance of different comparative methods under varying degrees of horizontal transmission. In our case (Nunn et al. 2006), we focused on independent contrasts (Felsenstein 1985b). We found that the performance of this phylogenetic method and the nonphylogenetic approach are sensitive to even low levels of horizontal transmission (figure 10.7). More specifically, increased horizontal transmission increased the type I error rate, suggesting that many apparently "significant" results would in fact be nonsignificant. Statistical power was also affected, but this depended on the strength of the correlation and other factors. In some cases, independent contrasts performed slightly better than simply examining correlations among the raw data, but neither approach per-

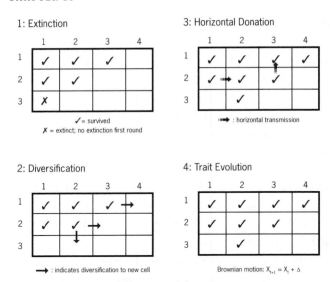

Figure 10.6. Simulating cultural macroevolution: horizontal transmission. A schematic of the simulation procedure used by Nunn et al. (2006) on a 3-row × 4-column spatial matrix shown partway through a simulation run. Each cell on this matrix is potentially filled by a unique society, indicated with a ✓, or empty, indicated by blanks. The following stochastic processes occur in sequence for each generation in the simulation based on user-defined probabilities: (1) extinctions of societies; (2) diversification of societies (and their traits) to empty cells; (3) horizontal transmission among neighboring societies; and (4) Brownian motion trait evolution. (Figure redrawn from Nunn et al. 2006.)

formed well when horizontal transmission occurred. Attempts to use Mantel tests to account for horizontal and vertical transmission also generally failed to improve the statistical performance of tests of correlated evolution.

Importantly, we assumed that horizontal transfer involves the paired transmission of both traits (Nunn et al. 2006). While this might be a reasonable assumption for some trait histories, it is also possible that only a single trait transfers horizontally; in other words, we assumed that the traits were "yoked" together, whereas it could be that some traits are relatively more independent (and the horizontally transmitted trait could subsequently influence the evolution of the other trait in its new lineage; Mace and Pagel 1994). We discussed the importance of the transfer process and alterations on it, such as having the probability of paired transmission set as a function of the correlation among the traits, but took paired transfer as a reasonable starting point for investigating trait transmission (see Nunn et al. 2006, 2010). Using our simulation program, a more recent study investigated independent trait transfer and found improved performance of phylogenetic methods for studying correlated evolution (Currie et al. 2010). Thus, there is a clear need to bet-

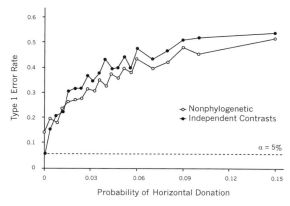

Figure 10.7. Effect of horizontal transmission on the statistical performance of independent contrasts. As horizontal transmission increases, Type I error rates increase, and this is true for analyses of both independent contrasts and the raw data. Moreover, this effect is consistent across different configurations of societies into, for example, a 6 × 6 matrix or a 1 × 36 matrix, and at different levels of societal extinction. (Figure from Nunn et al. 2006.)

ter assess the degree to which two or more traits move horizontally in a correlated way (e.g., Matthews et al. forthcoming) and to develop better statistical models for this process in comparative studies. In addition, future simulations could investigate whether similar results obtain for discretely varying traits using, for example, Pagel's (1994a) discrete method (see chapter 7). This is especially important because most cross-cultural datasets are composed of discretely varying traits (see also Fortunato and Mace 2009).

We should also consider the influence of horizontal transmission on tree inference. In linguistic phylogenetics, a recent computer simulation found that Bayesian approaches to tree inference are relatively robust to horizontal transmission, depending on the rate of borrowing, constraints on borrowing among lineages, and the tree topology (Greenhill et al. 2009). Importantly, even in the worst performance at high rates of borrowing, the Bayesian methods still recovered some of the true phylogenetic signal. Horizontal transmission tended to result in a bias toward underestimating the age of the root. In this simulation study, the authors limited borrowing by the time since two lineages last shared a common ancestor, rather than through explicit incorporation of the potential for spatial contact among societies (as in Nunn et al. 2006; for another simulation study of linguistic borrowing, see McMahon and McMahon 2005).

Returning to the issues of comparative methods, given that phylogeny-based comparative methods may perform poorly under conditions of horizontal transfer, what should someone interested in cross-cultural research do? First, it is essential to diagnose the degree to which vertical and horizontal trait transmission occur (Borgerhoff Mulder et al. 2006), for example, by using methods described earlier in this chapter. One should keep in mind that convergence due to similar ecological conditions may be difficult to distinguish from horizontal transmission; both processes could produce similar patterns of trait evolution on the tree. Second, it is important to ascertain whether the traits are transmitted as a pair, individually or as some function of their degree of correlation (Nunn et al. 2006, 2010; Currie et al. 2010). If traits are transmitted vertically, then phylogeny-based methods may be appropriate. If horizontal transmission occurs, however, other options for studying correlated trait change should be considered, especially if traits are transmitted as a pair. One possibility is to use spatial autocorrelation approaches that control for both phylogenetic and geographic distances among populations (Dow 1984) where geographic distance is used as a proxy for the possibility of horizontal trait transmission. Another more recently developed option is to use two-stages least squares (Dow 2007). It seems likely that new tools will build on generalized least squares approaches discussed in chapter 7, possibly by incorporating multiple sources of nonindependence into a single GLS model.

Similar effort is needed to develop valid proxies for the probability of horizontal transmission. Some authors have investigated the effects of both geographic distance and the adjacency of populations (Jordan and Shennan 2003), or have devised measures that are a declining function of actual geographical distance to account for the possibility that diffusion mainly operates over very short distances (White et al. 1981). Another approach for the future is to make use of loanwords as a proxy for the probability that cultural traits transmit horizontally (Borgerhoff Mulder et al. 2006; Nunn et al. 2006). Loanwords are particularly promising because they may be able to detect unidirectional patterns of transmission rather than assuming that all societies are likely to borrow from other societies based only on relative proximity. This of course assumes that horizontal transmission of cultural behaviors covaries with rates of horizontal transmission of words (for which at least some evidence exists, e.g., Holden and Gray 2006). Last, methods based on independent contrasts and developed for biological data could be useful. One recent method, for example, assesses the degree to which geography and phylogeny account for variation in biological data (Freckleton and Jetz 2008). A similar approach could be used to diagnose vertical versus horizontal transmission signal and possibly even control for these effects.

Testing Hypotheses with Linguistic and Other Data

In trying to decipher the distributions of human populations, a variety of fascinating questions arise. For example, did Indo-European populations spread through the domestication of horses, or through demic expansion associated with farming (Renfrew 1987; Gray and Atkinson 2003)? What factors lead to the Bantu expansion in Africa? At what point did American Indians enter the New World, and how many migrations took place? While data on cultural behaviors and artifacts can help to disentangle some of these questions (e.g., Buchanan and Collard 2007; Anthony 2008), linguistic traits often provide even more fine-grained resolution, specifically, by providing a larger number of characters and a wider array of character types (e.g., grammar, dialect, and cognates). Phylogenetic methods—and extensions of these methods designed specifically for linguistic data (e.g., McMahon and McMahon 2005; Nakhleh et al. 2005; Nicholls and Gray 2006)—are providing new tools to investigate human prehistory.

Linguistic trees are likely to be a familiar sight for most readers of this volume, for example, in the phylogeny of Indo-European languages shown in figure 2.9. The branch of historical linguistics known as comparative linguistics (or comparative philology) is the field generally responsible for producing language trees. Chapter 2 considered issues involved in reconstructing linguistic trees, while chapter 4 gave an overview of methods for reconstructing protolanguages. In this section, we are concerned with how phylogenies and comparative methods are providing new insights to human population movements and linguistic diversity more generally.

Historical migration patterns. Many studies have used genetic data to test hypotheses about human dispersal patterns. Perhaps the best known of these involve path-breaking work by Cavalli-Sforza on the distribution of genes in European populations (Cavalli-Sforza et al. 1994), and the "mitochondrial Eve" hypothesis, which revealed support for an African origin of human populations (Cann et al. 1987; Vigilant et al. 1991). Researchers also have considered linguistic evidence for patterns of human migration and diversification (Nichols 1997; Bellwood 2001), with recent research bringing more rigorous phylogenetic approaches to uncover human population movements (e.g., Gray and Jordan 2000; Holden 2002; Gray and Atkinson 2003; Gray et al. 2007). In what follows, I briefly sample some examples.

Gray and Jordan (2000) applied phylogenetic techniques to assess two hy-

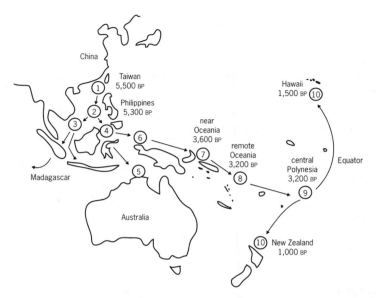

Figure 10.8. The spread of Austronesian languages. Gray and Jordan (2000) developed a scenario for the expected colonization of Polynesia under the "express train" hypothesis (Diamond 1988) and then mapped this sequential ordering onto a phylogenetic topology derived from languages from these people (see also Hurles et al. 2003). Dates indicate approximate ages of settlement based on archeological evidence. (Redrawn from Gray and Jordan 2000.)

potheses for the peopling of the Pacific Ocean using linguistic data. The "express train" hypothesis proposes that the Polynesian islands were populated in a rapid sequence arising from an initial population in Taiwan. In contrast, the "entangled bank" hypothesis proposes that the Polynesian islanders experienced longer-term prehistoric contact and interaction. The express-train model predicts a rapid series of stepwise colonization events with clear phylogenetic signal in the data, while the entangled bank hypothesis predicts less clear signal in the branching of linguistic groups relative to distance into the far reaches of the eastern Pacific Ocean (see also Hurles et al. 2003; Gray et al. 2010). To test these two hypotheses, Gray and Jordan used parsimony to infer a phylogeny for Austronesian languages and then mapped key events in the peopling of the Pacific onto the resulting tree (figure 10.8). The analysis revealed strong support for the express train hypothesis, with the character mapping requiring only 9 steps in a parsimony analysis (compared to 43–53 steps under a null model). Some linguistic contact was likely to occur, however, based on a low CI (0.25) and, more convincingly, from a later analysis that used split decomposition (Hurles et al. 2003; figure 2.10; see also Gray et al. 2010).

Two more recent studies, both published in the same issue of *Science*, deserve mention in this context. In the first of these, Gray et al. (2009) reexamined the settlement patterns of Polynesia using Bayesian phylogenetic analyses and an expanded linguistic dataset. With better resolution of the timing of diversification events, the authors were able to detect "pulses" and "pauses" in the movement of Austronesian speakers in the Pacific, with these patterns linked to technological innovations. In the other article, Moodley et al. (2009) investigated the geographical distribution of strains of a bacterial parasite, *Helicobacter pylori* (using genetic rather than linguistic characters!), again with the goal of inferring human population movements. They found evidence that Polynesians carry a strain with an origin in Taiwan that is distinct from the strain found in people who colonized Australia. Thus, the bacteria hitched a ride on the dispersing humans, leading to congruence in parasite genomes and human languages.

It is important to keep in mind that in addition to linguistic characters, other cultural traits can be used to infer past population movements. For example, Buchanan and Collard (2007) examined a phylogeny of Paleo-Indian projectile points to test among competing hypotheses for the colonization of the New World (see figure 10.4). Their phylogenetic analysis of projectile points from across the continental United States revealed that populations using Clovis and similar technologies spread rapidly in North America. Moreover, the authors found that users of this technology entered the New World from the northwest rather than under alternative hypotheses that propose colonization of North America from South America or from the mid-Atlantic region. Ultimately, multiple lines of evidence are needed to uncover past population movements, including linguistics, genetics, archeology and human cultural behavior (Diamond and Bellwood 2003; Hurles et al. 2003).

Linguistic diversity. As touched upon in chapter 1, languages are spread unevenly around the world (Nichols 1992; Nettle 1999b; Currie and Mace 2009). Some areas, such as New Guinea, harbor a tremendous diversity of languages, possibly up to 15 percent of the global total recognized extant languages (figure 10.9a). Other areas of high language diversity include the Amazon, parts of Southeast Asia, and central Africa. In contrast, areas such as southern and northern Africa have relatively few languages per unit area. Many of these

(A)

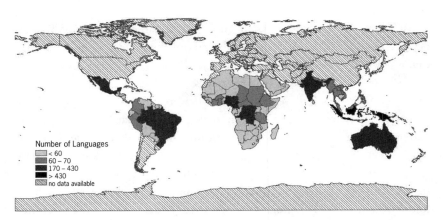

(B)

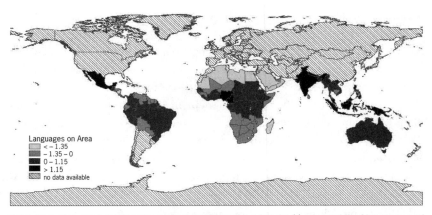

Figure 10.9. The global distribution of languages. Plots show data for (a) raw number of languages and (b) number of languages regressed on area. Data were not available for all countries. In all cases there is a tendency for languages to concentrate near the equator (Mace and Pagel 1995; Nettle 1998). (Data from Nettle 1998.)

patterns generally hold when controlling for population size or geographic area (figure 10.9b).

A fundamental question concerns the factors that account for this variation in linguistic diversity. Mace and Pagel (1995) describe the search for factors underlying language diversity as "linguistic ecology," and Pagel's (2000b) research shows how methods used to study the diversification of species (chapter 8) can be used to study the diversification of languages. For example, using a simple birth-death model from phylogenetics and assuming exponential growth, Pagel (2000b) considered how the number of languages at a given time (n_t) is a function of the number of founder languages (n_0), the birth rate (b), the death rate (d), and the time elapsed (t), as captured in the following equation:

$$n_t = n_0 \, e^{[(b-d)t]}$$

With this equation and the assumptions embedded in it, one can examine patterns of global linguistic diversity through time based on specified dates for the origins of a group of languages, or one can fit the parameters of the model to estimate rates of language accumulation needed to reach the number of languages spoken today. In addition, this formula can be used to infer the total number of languages that ever have been spoken (i.e., including those that went extinct). From this, Pagel (2000b) proposed that 80 percent to 99 percent of the languages that ever existed have gone extinct.

Several researchers have identified geographical and environmental drivers of language diversity, with the number of languages increasing in coastal areas, at low altitude, with greater rainfall, and with increasing climatic variability and number of habitats (see chapter 1 and Nichols 1994, 1997; Mace and Pagel 1995; Nettle 1998; Collard and Foley 2002). Some studies have even proposed that parasites are a driver of linguistic diversity, perhaps by acting as a "wedge" when behavioral avoidance of other humans results in the subdivision of populations (Fincher and Thornhill 2008). Latitude turns out to be a major determinant of the distribution of languages, with an increase in language diversity as one moves towards the equator, in both the New World (Mace and Pagel 1995) and the Old World (Collard and Foley 2002).

In a global study of linguistic diversity and its drivers, Sutherland (2003) found that the number of languages increases with area and, more weakly, with altitude. In addition, he showed that language diversity declines with latitude at a global scale, thus matching patterns found by Mace and Pagel (1995) at more regional scales (see figure 1.3). Areas with relatively high linguistic diversity also possessed relatively high biological diversity. Smith (2001) also found an association between some measures of linguistic and biological diversity in native North Americans. Similarly, Moore et al. (2002) found that language richness covaries positively with biological diversity of other vertebrates (measured as species richness), and that environmental variables involving precipitation, climatic variability, and the length of the growing season explained the most variation in language richness. The environmental variables examined by Moore et al. (2002) accounted for more variation in language diversity than did vertebrate species richness. Some differences were found in how environmental factors influence species diversity and language diversity, which are likely to reflect fundamental differences in the drivers of these two factors, including the possibility that historical contingency is more important for linguistic than species diversity.

Researchers also have incorporated human behavioral ecology more directly into models of linguistic variation. Nettle (1999b) proposed ways to investigate whether human ecology, social network patterns and linguistic diversity are associated. If ecological risk is viewed as the probability that a household will face a shortfall, Nettle suggested that increased risk would be associated with a larger social network, specifically as a means to reduce this risk (e.g., through exchange). Because social networks require some form of communication and should therefore favor a wider distribution of a given language, we might expect that areas with more ecological risk are associated with fewer languages per area (or per population size). In a comparative study across human populations, Nettle found support for this predicted link between ecological risk and language area. Specifically, the number of languages in 74 countries correlated positively with the mean growing season, where a longer growing season was assumed to represent less ecological risk (figure 10.10). Exceptions to this global pattern can be further examined to understand differences among regions of the world, including the fewer than expected numbers of languages in areas of the New World (e.g., Cuba in figure 10.10). As noted by Nettle, one explanation could involve the disappearance of large numbers of populations due to disease, in many cases before their languages could be documented.

A major future direction involves more explicit incorporation of geography into analyses of linguistic diversity (Moore et al. 2002). In one promising example of this approach, Currie and Mace (2009) examined the geographical distribution of languages relative to environmental, ecological and ethnographic data. Using a geographic information system (GIS) and data on the geographical range of languages at a global scale, they discovered that increased political complexity and subsistence strategy are among the best predictors of the area that a language covers. Political complexity reflects the division of labor within societies, hereditary inequalities, and the coordination of greater numbers of individuals in the economic system; this is consistent with a greater capacity for more complex societies to expand at the

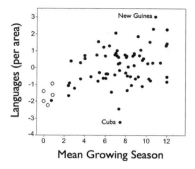

Figure 10.10. Languages and mean growing season. Number of languages is represented as a residual from the linear regression of languages on area, where languages and area were logarithmically transformed. For illustrative purposes, two countries are labeled: New Guinea, which has the highest language diversity per area, and Cuba, with the lowest diversity (possibly related to the extermination of native peoples during the conquest of the Americas). Note also that five of the data points to the far left of the plot (open circles) involve countries spatially located in the Middle East (United Arab Emirates, Oman, Yemen, Saudia Arabia, and Egypt). This illustrates the effects of geographic proximity as a factor influencing the nonindependence of data in comparative studies; similar effects are found in other regions of the world using this data set. (Data from Nettle 1998.)

expense of neighbors. In terms of subsistence strategy, languages spoken by agriculturalists tended to have smaller areas than languages spoken by pastoralists, which was expected, because agriculture requires less mobility and can support higher population densities. Consistent with previous research cited above, Currie and Mace also found support for effects of latitude, mean length of growing season, and species diversity.

Summary and Synthesis

As in biology, comparison has long played a central role in cross-cultural research. Also as in biology, phylogeny-based methods recently have fueled a number of exciting discoveries in cross-cultural research. This chapter provided an overview of the questions that are being addressed in cross-cultural research, and it highlighted some of the methods that have been used.

In the future, a need exists to examine the degree of phylogenetic signal for more datasets, as many different studies at different geographic locations and scales have found variable support for the degree of vertical and horizontal transmission. We also need a common set of methodological approaches to investigate these questions, and to do so rigorously and in a way that allows studies to be compared. Coupled with this, greater attention to the geographical context and explicit analysis of georeferenced data are needed (Moore et al. 2002; Currie and Mace 2009). Importantly, we also simply need more high-quality cross-cultural data to empirically investigate hypotheses (Gray et al. 2010). Last, computer simulation approaches are likely to play an increasingly important role in the future development of statistical tools for cross-cultural comparison (Borgerhoff Mulder et al. 2006; Nunn et al. 2006; 2010).

Pointers to related topics in future chapters:

- Conservation of linguistic and cultural diversity: chapter 11
- Applying phylogenetic community ecology to cultural diversity: chapter 11
- The consistency index: chapter 12

11 Behavior, Ecology, and Conservation of Biological and Cultural Diversity

Primates exhibit striking variation in behavioral, ecological, and cognitive features, and comparison has played a major role in elucidating the drivers of this variation. Milton and May (1976) and Clutton-Brock and Harvey (1977) were among the first to take a quantitative approach to comparatively investigate behavioral and ecological variation in primates. Their research built on previous comparative perspectives in primatology (DeVore 1963; Crook and Gartlan 1966; Crook 1972; Goss-Custard et al. 1972) and allometry and comparative physiology (Kleiber 1961; McNab 1963; Schmidt-Nielsen 1975).

The early studies on primates revealed patterns that we now consider to be fundamental and commonsense, including that home range size increases with body mass, that frugivores have larger home ranges than folivores for a given body mass, and that sexual dimorphism in body mass is related to the sex ratio (with more dimorphism associated with more female-biased sex ratios; Milton and May 1976; Clutton-Brock and Harvey 1977; Clutton-Brock et al. 1977). Clutton-Brock and Harvey's research is notable for being among the first to address issues related to phylogenetic nonindependence. Their solution was to average species values to higher taxonomic levels. Although the best method available at the time, this "higher nodes" approach is now known to be insufficient, as higher taxonomic groups are differentially related to one another and thus also not statistically independent units for analysis (see figure 7.12). More specifically, one of their associations—a positive correlation between body mass and group size—was, at best, weakly supported and is probably driven by patterns of activity period and substrate use that are shared through common descent (Nunn and Barton 2001).

These first papers were groundbreaking, and comparative research flourished and continues to play an important role in studies of primate behavior and ecology, often by leaders in primatology and behavioral ecology (e.g., Harcourt et al. 1981; Cheverud et al. 1985; Martin 1990; Davies et al. 1991; Dunbar 1992; Dunbar and Cowlishaw 1992; Grant et al. 1992; Plavcan and Van Schaik

1992; Hauser 1993; Sillén-Tullberg and Møller 1993; Garber 1994; Mitani et al. 1996a; Kappeler 1997; Mitani and Watts 1997; Van Schaik and Kappeler 1997; Lindenfors and Tullberg 1998; Lee 1999; Dixson and Anderson 2004; Ossi and Kamilar 2006; Pagel and Meade 2006a; Ostner et al. 2008). Given that behavior and ecology do not generally leave a strong imprint in the fossil record, methods of "statistical paleontology" (Pagel 1997) described in previous chapters will continue to play an essential role in studies of behavioral evolution. Cross-species comparison is also increasingly important for conservation of biodiversity, while phylogenetic approaches are becoming de rigueur for studies of community ecology.

This chapter covers three main topics involving the application of phylogenetics to ecological and behavioral research. First, I briefly review new phylogenetic approaches to study community ecology and biogeography (Webb 2000; Webb et al. 2002) and how these approaches might apply to questions in evolutionary anthropology. Second, I review the application of phylogenetic approaches and perspectives to the conservation of biodiversity (Faith 1994; Fisher and Owens 2004; Purvis et al. 2005b). Many of these issues also apply to cultural diversity, including the diversity and conservation of human languages, which I review briefly. Last, I consider several special issues that commonly arise in the study of primate behavior, including behavioral homology and new ways to control for intraspecific variation. Although intraspecific variation is thought to be most relevant in studies of behavioral and ecological traits, it is worth noting that these methods also should be useful for other types of data, including morphological and life history traits.

To help set the stage for what follows, let us first consider one example of how comparative research has enhanced our understanding of primate sociality and mating systems by reviewing comparative research on a fundamental question: What factors influence the number of males in primate social groups?

The number of males in primates groups. Early cross-species studies revealed that the number of males increases with the number of females in primate social groups (figure 11.1), a pattern that also holds when using independent contrasts (Nunn 1999b; Lindenfors et al. 2004). Looking more closely at the raw data revealed, however, that some species are characterized by many females in the group, yet they have very few males—in some cases only one (open circles in figure 11.1; see also Andelman 1986). In other words, exceptions exist to the general association between the number of males and num-

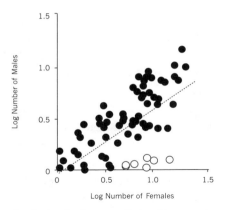

Figure 11.1. The number of males in primate groups. Each point represents a different species of non-human primate, using data from Nunn and Barton (2000). Open circles represent a variety of species, all characterized by having a small number of males relative to female group size.

ber of females, presumably because additional factors influence the ability of a single male to monopolize access to a group of females.

Why do some groups have single males, while other groups are multi-male? This is an important issue, because the factors that influence group composition are critical for many aspects of primate social life. For example, the risk of infant death is often higher in single-male groups, as turnover of males often results in new males killing the infants sired by the previous male, that is, sexually selected infanticide (Hrdy 1974; Van Schaik and Janson 2000). The number of males also impacts patterns of reproductive skew within groups, with a recent comparative study showing that an increase in the number of males reduces the dominant male's mating success (Kutsukake and Nunn 2006).When a group contains more males, this increases the resources that are needed for the group as a whole, resulting in a larger home range (Clutton-Brock and Harvey 1977; Martin 1981; Nunn and Barton 2000). Increased demands on foraging time mean that less time is available for other fitness-enhancing activities, such as defending the range, socializing, and sleeping (Altmann 1980; Dunbar and Dunbar 1988; Elgar et al. 1988; Lesku et al. 2006; Acerbi et al. 2008; Capellini et al. 2008).

Primatologists have focused on two main sets of factors that influence the number of males in a primate group. First, as already discussed, the number of females is likely to have an effect on the number of males in a group (Andelman 1986; Altmann 1990; Mitani et al. 1996b). This effect arises from competition and efforts by males to monopolize females; when there are more females, competition for sexual access from males outside the group increases, and a single male is unable to maintain exclusive access to the group of females (resulting in multi-male groups). Second, researchers have considered the role of female synchrony in the timing of estrus, which again relates to competition over mating access (Emlen and Oring 1977; Ridley 1983; Dunbar 1988). When females overlap more in their timing of estrus, this reduces the ability of a single male to maintain complete control over a group of females, resulting in multi-male groups. These hypotheses are not mutually exclusive, however, and are possibly even related. For example, a larger number of females should result in greater estrous overlap simply because there are more opportunities for two or more females to come into estrus simultaneously.

In a piece of comparative work that brought new clarity and attention to this fundamental question, Mitani et al. (1996b) investigated whether the number of females or female synchrony accounts for the number of males in primate groups. Building on a previous study (Ridley 1986), they reasoned that females are more likely to exhibit estrous overlap in species characterized by seasonal breeding, with a shorter breeding season resulting in less time for females to come into estrus and thus more overlap. Mitani et al. (1996b) obtained data on the number of males and females in primate groups, along with information on seasonality, and coded species as single-male versus multi-male and as seasonal versus nonseasonal breeders. Using phylogenetic methods and treating the hypotheses as mutually exclusive, they found that the number of females—but not breeding seasonality—accounted for variation in the number of males in primate groups.

This research raised new questions about the mechanisms that underlie the association between the number of males and females. Thus, I extended these results by noting that continuous values of traits provide more statistical power than discrete codes in comparative tests (Nunn 1999b; see also chapter 9; Garland et al. 1993). Moreover, the hypotheses are not mutually exclusive, and measures of breeding seasonality can only serve as indirect measures of female overlap. Even in species that have the shortest breeding seasons—which is thought to predict greater estrous synchrony—females are known to come into estrus *asynchronously*, a situation that is illustrated vividly by research on ring-tailed lemurs. While this species was correctly coded as having the shortest breeding season in Mitani et al.'s (1996b) dataset, Pereira (1991) found that the very short estrous periods of ring-tailed lemurs actually tend to be significantly overdispersed, with females rarely mating on the same day.

Thus, to more powerfully test the hypotheses, information is needed on continuously varying traits, better measures of estrous overlap, and a framework that considers the number of females and estrous overlap in a multivariate model (and thus addresses the fact that the two hypotheses are not mutually exclusive). Using this approach, I found support for both hypotheses (Nunn 1999b), with a significant effect of female synchrony even after controlling for the fact that synchrony should increase with the number of females in the group. Consistent with the results of Mitani et al. (1996b), the number of females tended to explain more variation in the number of males than did three different measures of female estrous overlap.

For fundamental questions about behavioral variation, cross-species comparisons often provide some of the most convincing tests. Rarely are within-species data on behavioral traits sufficiently available for testing these hypotheses (e.g., Altmann 2000)—for example, across multiple groups of the same species—and when they are, such data would fail to address questions about broader evolutionary patterns. Thus, while intraspecific variation often provides more fine-grained, higher-quality data, a phylogenetic comparative perspective is essential for understanding the generality of patterns in broader evolutionary context. In many cases, comparative research also identifies new questions to test. In studies of the number of males, for example, the spatial dispersion of females in a group could also influence a male's ability to monopolize the group of females. In other words, when the average distance among females in a group increases, this could make it more difficult for males to defend access to those females. Such a hypothesis could be tested within the comparative framework developed thus far by adding information on the diameter of primate groups, that is, measures of "group spread."

Phylogenetic Community Ecology

Community ecologists study the distribution, abundance, and interactions among coexisting populations of organisms. In studies of primates, community ecology includes studies of primates that overlap spatially in a forest reserve, the position that a primate species occupies in a food web, or heterogeneity in plant communities in the home range of a group of primates. Similarly, studies of parasite species richness in primates involve basic principles from community ecology, such as island biogeography theory as a framework for understanding the number of parasites that infect a given host (Kuris et al. 1980). Community ecology also involves studying niche space and how it is filled over "replicate" runs of evolution, such as on multiple islands

(Losos 1992, 1995). All of these questions can be applied to both living and extinct populations (Valkenburgh 1988; Fleagle and Reed 1999).

Why is phylogeny integral to understanding community ecology? First, community ecology addresses fundamental questions involving niche space, and it is widely believed that closely related species share similar niches. Thus, we might expect that closer phylogenetic relationships lead to exclusion in niche space: when close relatives coexist, we expect them to inhabit different niches; conversely, when close relatives have very similar niches, we expect them to have distinct geographic ranges (i.e., to be in different communities). Second, measures of species richness can be informed by phylogenetic relatedness. Thus, it might be more informative to quantify diversity in relation to phylogenetic dispersion among taxa in a community rather than simply as counts of the number of species. Last, the age of an adaptive radiation and ecological characteristics of its first lineages may influence the subsequent patterns of speciation and community formation (Price et al. 2000). In other words, historical contingency can play a role in the structuring of communities, and phylogenetic perspectives can help to study this possibility.

In the 1990s, a number of biologists began to investigate the role of phylogeny in community ecology using phylogenetic approaches (Wanntorp et al. 1990; Brooks and McLennan 1991; Losos 1996; McPeek and Miller 1996). Specific applications included studies of *Anolis* lizards (Losos 1995), rainforest trees (Webb 2000), and primates (Fleagle and Reed 1999). Biologists have developed new phylogenetic tools to probe community ecology (Webb et al. 2002), resulting in a variety of toolboxes and programs that are now available to study patterns of community diversity in a phylogenetic context (AnthroTree 11.1). Vamosi et al. (2009) and Cavender-Bares et al. (2009) provide overviews of recent studies, including general findings, the strengths and weaknesses of different approaches, and the importance of spatial and temporal scale in studies of phylogenetic community ecology.

Some of these methods are focused on estimating the phylogenetic relatedness of a community and comparing the observed level of relatedness to that expected under null models of community formation. For example, Webb (2000) examined tree species composition of twenty-eight forest plots in Borneo. After obtaining a measure of phylogenetic relatedness for plants in each of the twenty-eight plots, he found that plots were composed of more closely related species than expected under a null model. This may reflect that plant species share adaptations for different ecological conditions through descent, with ecologically more similar species tending to co-occur in the same plot when habitat conditions vary among the plots.

Several studies have investigated phylogenetic aspects of primate communities. Houle (1997) investigated the roles of phylogeny and behavioral competition on the coexistence of different primate species. For this, he collected data on forty-one primate communities from all major primate radiations, and he assumed that species in the same subgenus or species group are more closely related than those who group in a higher taxonomic rank. According to this taxonomic definition of phylogenetic propinquity, some very close relatives do coexist, such as *Cercopithecus pogonias* and *C. mona* in Bakundu Forest in Africa, or *Macaca mulatta* and *M. fascicularis* in Huay Kha Khaeng Forest in Southeast Asia. However, it is more common for primate communities to be composed of more distantly related species. Houle also found that more closely related species exhibit higher rates of interspecific aggression and spend less time grooming. These results indicate that competition for niche space plays an important role in the assembly of primate communities.

A more recent study applied phylogenetic methods to the question of competition within mammalian communities. Cooper et al. (2008) examined patterns of phylogenetic dispersion in New World primates, North American ground squirrels, and Australian possums. A schematic of possible patterns of dispersion are illustrated in figure 11.2. A clumped distribution of taxa among community samples suggests that habitat use is a conserved trait (as described in the example of tropical trees above; Webb 2000). In contrast, an overdispersed distribution of species suggests exclusion of closely related species, which are likely to have similar niches; in other words, overdispersion is consistent with interspecific competition for niche space. Cooper et al. (2008) found that primates and ground squirrels exhibited significant community overdispersion on average (although most individual communities were judged to be phylogenetically random). In a study from around the same time, Cardillo et al. (2008) found that most mammalian island assemblages were indistinguishable from random compositions.

Another study combined trait data and phylogeny to investigate primate community ecology. Fleagle and Reed (1999) examined primate communities in phylogenetic perspective in four major biogeographical regions: South America, Africa, Madagascar, and Asia. Among their analyses, they examined the relatedness of species within primate communities. They found that African and Asian communities are composed of lineages that split over a

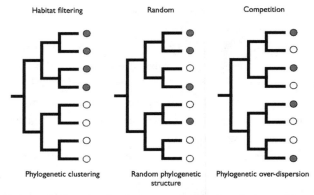

Figure 11.2. Phylogenetic dispersion. Panels show membership of species in two different communities in relation to phylogeny. Communities are indicated by the color of the circle (open or filled). On the left, habitat filtering results from the evolution of a character that enables species to persist in a community. It results in phylogenetic clustering in communities. On the right, competition results in closely related species being unable to coexist, resulting in phylogenetic overdispersion. These two extremes can be compared to random phylogenetic assemblages. (Adapted from Cooper et al. 2008; see also Cavender-Bares et al. 2009.)

longer period of time, compared to the communities of South America and Madagascar, which have more balanced phylogenies (for two examples, see figure 11.3). These differences may reflect that South American and Malagasy primates underwent explosive radiations shortly after colonizing these areas and possibly also that many African and Asian communities contain both strepsirrhines (lemurs and lorisids) and haplorrhines (monkeys, apes, and the tarsier). Fleagle and Reed (1999) went on to investigate the links between phylogenetic divergence and ecological diversity (using data from Fleagle and Reed 1996). They found a positive association between phylogenetic distance and ecological distance using a Mantel test. This pattern was retained within the African and Asian communities, but less so for those in South America and Madagascar.

In another study, Heard and Cox (2007) investigated patterns of *diversity skewness* in primate communities. High skewness refers to a situation in which some lineages have diversified more rapidly than others, while low skewness indicates that all lineages have experienced similar diversification rates (see also discussion of "tree shape" in chapter 8; Mooers and Heard 1997). After developing an approach that provides a null hypothesis based on the availability of species at regional and ecological levels, they found that African communities showed significantly higher diversity skewness than expected, while the South American communities had lower than expected diversity skewness.

(A)

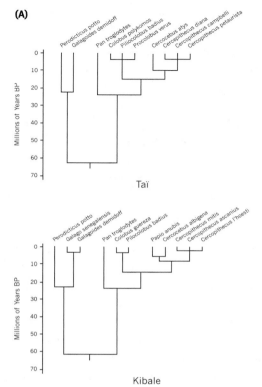

Taï

Kibale

Figure 11.3. Primate communities in phylogenetic context. The phylogenetic relationships among primates in (a) African communities and (b) South American communities. (Redrawn from Fleagle and Reed 1999.)

(B)

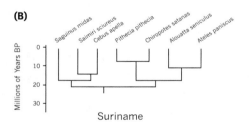

Suriname

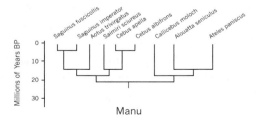

Manu

Just as we can examine the community diversity of primates, we can also examine the diversity of parasites and pathogens on primate hosts (Combes 2001; Nunn and Altizer 2006). One recent study (Davies and Pedersen 2008) investigated a fundamental question involving the factors that influence similarity in the parasite communities of different primate hosts where parasites include microparasites, such as viruses and protozoa, and macroparasites, such as helminths. Do hosts harbor the same parasites through common descent (i.e., phylogeny), or through sharing of parasites among other hosts that they overlap with geographically? The authors reported that the mean parasite community was 6.9 parasites per primate host, 28 percent of all possible combinations of primate hosts shared at least one parasite, and the gorilla and chimpanzee shared the most parasites (17 in total were shared). In analyses that examined the drivers of parasite sharing among hosts, they found that both geographic overlap and phylogenetic history accounted for parasite sharing. For viruses, geography was a better predictor of sharing than was phylogeny, possibly because high mutation rates and short generation times of viruses facilitated host shifting (see also Pedersen et al. 2005). By contrast, phylogeny was a better explanation than geography for sharing among protozoan parasites and in some analyses of helminths. Davies and Pedersen (2008) also examined the degree to which humans share parasites with nonhuman primates. They found that humans are indeed more likely to share pathogens with more closely related primates, and in a later study, they identified geographic areas where spillover from nonhuman primates to humans is most likely to occur (Pedersen and Davies 2009).

In the future, it may prove valuable to apply phylogenetic community ecology approaches to study cultural traits. For example, Kroeber proposed that cultures are composites of "more or less fused aggregates of elements of various origin, ancient and recent, native and foreign" (1931, 149). In this view, cultures are more like communities of species than like an organism itself, and this raises the possibility that methods from phylogenetic community ecology could be useful for studying cultural assemblages. For example, we might predict that areas of geographic overlap among human populations are limited to populations coming from more divergent linguistic and cultural histories, which are thus more likely to have different technologies for exploiting different niches. Similarly, controlling for geographic proximity, is warfare more common among phylogenetically close groups, who might compete for similar resources, than those who are more distantly related? The methods just discussed, as well as others from previous chapters, could play a role in investigating these possibilities.

Conservation of Biological (and Cultural) Diversity

Conservation is about preserving biodiversity for the future, and phylogenetic comparative perspectives play several key roles in this effort. In what follows, I consider how phylogenetic approaches are being used to quantify biodiversity and to target clades that require greater conservation effort, and I review how comparative research can provide new insights to conservation questions, especially when phylogeny is taken into account. Indeed, studies of this type have yielded several surprises, and as we will see, similar approaches can be applied in efforts to preserve cultural diversity.

Quantifying biodiversity. For those of us who value biodiversity for its own sake, it is a tragedy whenever a species goes extinct. Given that resources are finite and the pressures on wildlife are so great, however, we cannot hope to preserve all of the world's biodiversity. We have to prioritize our efforts in the face of these constraints, and prioritization requires that we quantify biodiversity in a meaningful and comparative way. Phylogeny has impacted several aspects of this important effort.

First, the *phylogenetic species concept* (PSC) has been used to identify species and conservation units, including in primates (Goodman et al. 1998; Groves 2001a, 2004). While definitions of PSC vary, the essence of this view is that species can be recognized based on phylogenetic relatedness. PSC definitions focus on identifying species as the least inclusive groups on a phylogeny (e.g., Agapow et al. 2004), or, in the words of Joel Cracraft, "a species is the smallest diagnosable cluster of individual organisms within which there is a parental pattern of ancestry and descent" (1992, 103). This is obviously a useful definition in the context of phylogeny-based comparative studies, which make use of the tips of the tree as the units of analysis.

In addition to the PSC, dozens of "species definitions" have been proposed, and many of these definitions have overlapping components (see Kimbel and Martin 1993; Mayden 1997). Probably the best known is the *biological species concept* (BSC), which identifies species as groups of breeding (or potentially breeding) populations of individuals that are reproductively isolated from other such groups (Mayr 1963). The BSC is thus a definition based on a process—reproductive isolation—rather than a pattern, as in the PSC (Groves 2004). We have seen throughout this book that getting a handle on process, while desirable, is often much more challenging than quantifying pattern (see chapter 5). Thus, the BSC has several major weaknesses, particularly involving difficulties inherent to testing whether two taxa exhibit successful reproduction and in its application to asexual organisms. Partly as a consequence, the past twenty years have seen a surge of interest in developing more operational and evolutionary based approaches to demarcating species, such as the PSC.

How does using a PSC impact studies of conservation biology? Agapow et al. (2004) investigated this question by compiling data on a wide variety of organisms that have been classified under both PSC and BSC, including two primate taxa (titi monkeys and brown lemurs). They found that using a PSC increased the number of species by at least 49 percent across their entire sample, compared to BSC (see also Groves 2001a; Isaac et al. 2004). For mammals, the increase was 87 percent. This "splitting" of taxa will tend to reduce the geographic ranges and population sizes of the species, resulting in identification of more endangered species (because geographic range and population size are often used to quantify threat status). In addition, changes to threat categories under different species concepts tend to cast doubt on the scientific basis of conservation planning, thus potentially weakening efforts to conserve biodiversity. Agapow et al. (2004) considered several possible remedies to this dilemma, including quantifying conservation value on the basis of economic significance, trait diversity, and other measures.

Another way that phylogeny has impacted conservation effort is by assigning priority to evolutionarily distinct lineages. For example, Faith (1992) proposed that we can use phylogenetic information to quantify biodiversity in the context of conservation, specifically by using a measure of *phylogenetic diversity* (PD; see also Faith 1994, 2002; Crozier 1997; Purvis et al. 2005b). The essence of this approach lies in conserving the diversity of organismal features, or, *feature diversity*. Because it is impractical to quantify the total feature diversity for all species (or even for one species!), we need a surrogate measure. One such measure of feature diversity is evolutionary time: all else equal, lineages with a most recent common ancestor further in the past should have accumulated more evolutionarily novel features (i.e., their branches connect further back in time). Thus, in attempting to prioritize conservation efforts, PD aims to preserve evolutionary time, measured as the sum of branch lengths that will be lost when particular species are lost. As shown in the dated phylogeny of figure 11.4, if we lose species C, we will lose

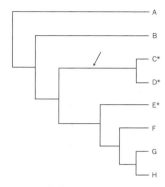

Figure 11.4. Loss of phylogenetic diversity (PD). PD is measured in units of branch length. Thus, priority will be given to lineages that have longer terminal branches. Moreover, loss of two sister species will result in the branch connecting their ancestor to the rest of the tree, as indicated by the arrow if species C and D are both lost. Thus, phylogenetic clumping of extinction risk in living species can result in greater potential for losing PD. Asterisks indicate species with higher threat status, as described in the text. Branch lengths are in units of time (a dated phylogeny).

the short branch connecting C to its last common ancestor with D. If species E is lost, however, a larger chunk of evolutionary history will disappear. In this example, then, species E has higher PD than species C (and thus should also possess greater feature diversity). In general, measures of PD will give greater weight to lineages emanating from more basal nodes, compared to those from nodes closer to the tips of the tree.

PD provides critical insights to the loss of biodiversity, which was nicely illustrated by Purvis et al. (2000a) in a phylogenetic study of primate and carnivore extinction risk. These authors asked a simple yet critically important question: What happens to overall biodiversity when species are lost at random, compared to when they are lost in relation to their actual probabilities of extinction? We might reasonably expect that more threatened primates are more likely to go extinct than less threatened primate species. In the PD framework, if the probability of extinction is clumped phylogenetically—that is, if more closely related species have more similar threat levels—this will result in greater loss of phylogenetic diversity. Consider, for example, that species C, D, and E in figure 11.4 are at greater risk of extinction (indicated by asterisks), as measured for example with data from the International Union for Conservation of Nature Red List of Threatened Species (http://www.iucn redlist.org/; Hilton-Taylor 2002). If species C and D are lost, this results in the loss of the two branches emanating from their common ancestor *and also the branch connecting this ancestor to the other lineages on the tree* (i.e., to the node shared with clade containing species E, F, G, and H, indicated by an

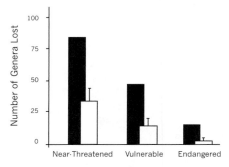

Figure 11.5. Loss of diversity through clumping of extinction risk in mammals. This figure compares the loss of diversity based on actual measures of threat status (black bars) to that expected if species go extinct randomly (white bars). The sets of bars represent loss of species at or above the threat category of near-threatened (left), vulnerable (middle) and endangered (right), with results shown for polytypic genera. (Redrawn from Purvis et al. 2000a.)

arrow on figure 11.4). In contrast, extinction of species E results in the loss of only one branch on the phylogeny.

Thus, phylogenetic clustering of extinction risk predicts that more PD will be lost when threatened species go extinct, compared to random extinctions on the tree; indeed, research on this topic was motivated in response to a "field of bullets" extinction scenario, that is, random extinction across the tips of the tree (Nee and May 1997). The research of Purvis et al. (2000a) revealed that extinction risk is highly clumped in primates and carnivores, with much greater loss of biodiversity than would be predicted if phylogeny were ignored (figure 11.5). For similar perspectives on the loss of more threatened species, see Heard and Mooers (2000), who used a simulation approach, and Jernvall and Wright (1998), who examined future primate ecological and morphological diversity in the context of progressive culling of taxa that are currently threatened.

A different approach to quantifying PD was taken by Vane-Wright et al. (1991) and extended by Posadas et al. (2001). These authors calculated measures of PD based on the number of nodes on a tree. This is what might be expected under a speciational model (see chapter 5), which is perhaps an appropriate framework for quantifying biodiversity if much of the feature diversity accumulates through the speciation process itself. Their index I measures the number of monophyletic groups to which an organism belongs, while their index W measures the proportion that each taxon contributes to the diversity of a group.

Phylogenetic approaches to prioritizing biodiversity have been applied to primate conservation. Cowlishaw and Dunbar (2000), for example, calculated a measure of "independent evolutionary history" for primates, defined as the time since a species last shared an ancestor with its closest relative (based on Purvis's 1995 inference of primate phylogeny). Three of the five highest scoring species were lemurs (*Lepilemur mustilinus*, *Varecia variegata*, and *Daubentonia madagascariensis*), with *Cebus apella* and *Pongo pygmaeus* rounding out the top five. In another study, Sechrest et al. (2002) found that biodiversity hotspots contain 71 percent of phylogenetic history for primates and carnivores, which is significantly higher than predicted under a null model (a reanalysis of the primate data found that this association was driven by Malagasy primates; see Spathelf and Waite 2007). Lehman (2006) applied phylogenetic approaches to conservation of Malagasy lemurs in both taxonomic and geographic context, while McGoogan et al. (2007) calculated PD scores for African primates and for countries and regions of mainland Africa (excluding Madagascar). The Democratic Republic of the Congo, Cameroon, and the Republic of Congo were identified as the countries with the highest PD primate species.

Comparative studies of extinction risk. It is also possible to use comparative methods to investigate the factors that influence extinction risk (Purvis et al. 2000b, 2005a; Fisher and Owens 2004). A critical first step is to quantify extinction risk in a comparable way across the species of interest. Many of the studies that have conducted comparative tests of extinction risk have relied on measures of threat status, particularly the IUCN Red List. It is important be aware of potential circularity when using IUCN threat status level. In particular, the measures used to quantify threat status in the Red List—such as having a small geographic range—will also, by definition, predict extinction risk if entered as a predictor variable in a comparative test. Despite this caveat, many interesting questions concerning extinction risk have been addressed comparatively, as illustrated in the examples that follow.

What are some of the traits that might increase the risk of extinction? Several factors have been proposed. A slow life history might make it more difficult for a species that is hit by demographic changes or natural disasters to recover; thus, extinction risk levels might be higher among species in which individuals have smaller litter sizes or more widely spaced births (i.e., a long interbirth interval). Similarly, a species in which individuals live at low population density might be unable to maintain group sizes that are sufficiently large and stable for reproduction. Geographic range size might be important because a small geographic range predisposes a species to more threats, either from humans or natural events (although as just noted, it is also a primary variable used to quantify extinction threat by the ICUN, creating potential circularity when using it as a predictor of extinction threat). Island endemicity may also put a species at risk, for example, if such species evolve in isolation of predators or infectious diseases that can decimate them when they are introduced (e.g., in the case of malaria and native Hawaiian birds; Van Riper et al. 1986). A species that exists at a higher trophic level will experience more risk when other species lower in the food chain are lost; this puts carnivorous species at greater risk than herbivorous ones. Lastly, body mass might be an important predictor of threat level, as this variable is correlated with some of the other variables that influence extinction risk (particularly life history traits and population density). Anthropogenic factors are also important, such as human population density (Harcourt and Parks 2003).

Purvis et al. (2000b) investigated extinction risk factors in primates and carnivores using data on threat status from the IUCN Red List. To perform the analyses, they created a numerical index with six categories—"lower risk least threatened" species were scored lowest, while "extinct" species were scored highest. "Data deficient" species were excluded. Using multivariate models, they found that primate threat levels increase in species with lower population density and larger body mass (see also Matthews et al. 2010). In addition, primates at a higher trophic level were more likely to experience higher threat status. Interestingly, life history traits were not significant predictors of threat status for primates, but were significant for carnivores. Last, a smaller geographic range size predicted greater extinction risk, which was expected, given that the IUCN threat categories are based in part on geographic range size.

Harcourt et al. (2002) took a somewhat different approach (see also Cowlishaw and Dunbar 2000). Harcourt et al. focused on rarity, which was defined by attributes including a small geographic range, low density, and narrow habitat specificity. In comparative tests, they found that rarity covaries with measures of specialization, but not with measures of resource use or recovery rate. Collectively, these results show that it is possible to predict a species' vulnerability to extinction from knowledge of the characteristics of that species and its close relatives. Phylogeny plays a key role in this endeavor by controlling for the factors that result in the clumping of extinction risk, that is, the nonindependence of data points (see chapter 7).

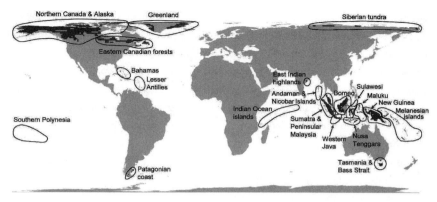

Figure 11.6. Areas of highest latent extinction risk for mammals. The map shows the areas of the world with the biggest differences between actual risk levels and expected risk levels based on the biology of the organisms (Cardillo et al. 2006). (Figure provided by M. Cardillo.)

This general approach has been extended in more recent analyses (Purvis et al. 2005b; Kamilar and Paciulli 2008). In one study, Cardillo et al. (2006) incorporated geographic information to map the distribution of mammalian extinction risk at a global scale. They also used biological predictors of extinction risk to map predicted areas of extinction risk. By comparing these two sources of information for the species involved, the authors quantified *latent extinction risk*, which is defined as the discrepancy between a species' current threat status and its inherent risk based on biological traits. Of interest for conservation action are areas where biological risk is high based on biological traits of the species, but threat status is currently relatively low (i.e., "positive" latent extinction risk). Areas of positive latent risk may reveal new insights into the factors that allow species to persist. At a practical level, they pinpoint locations that require close monitoring to ensure that species remain immune from anthropogenic and other factors that might tilt these populations towards extinction in the future.

The analyses of Cardillo et al. (2006) revealed that both tropical and temperate areas hold species at high risk (figure 11.6) and that hotspots of latent extinction risk do not match the hotspots of mammalian species richness. For example, large swaths of northern North American mammals show high latent extinction risk. While many of the large-bodied mammals in these areas are not currently threatened, their slow life histories make them vulnerable to extinction in the future (and this is likely to be further affected by climate change).

Comparative methods can often be used to address unexpected aspects of conservation biology. For example, crowding animals into smaller patches of habitat elevates stress levels and thereby can negatively impact immunocompetence, lowering resistance to disease and potentially intensifying the severity of infections (Coe 1993; Lyles and Dobson 1993; Lloyd 1995; Capitanio and Lerche 1998; Friedman and Lawrence 2002). Which species are most vulnerable to such effects? Clubb and Mason (2003) addressed this question comparatively with data on stereotypic pacing in captive carnivores, which they used as a measure of stress-related behaviors. They found that species with larger home ranges in the wild exhibited more stress-related behaviors and experienced higher infant mortality in captivity, thus predicting that these species will experience more stress as habitats decline in size.

Another important direction is to identify geographical areas and clades where biodiversity remains undiscovered. In one study, for example, Collen and colleagues used a comparative approach to investigate the factors that influence the discovery of new species of primates and carnivores (see also Purvis et al. 2003). Hence, they assumed that "species remaining to be discovered are more similar to recently described species than to species named in the past" (Collen et al. 2004, 459). Using both phylogenetically independent contrasts and species data, they regressed the year in which a species was discovered on the following traits: body mass, geographic range size, population density, home range size, day range, group size, and diurnality. They reasoned that all of these factors tend to increase the rate at which species are discovered by field biologists. In addition, they investigated whether having fewer other species in the genus increased the rate of discovery, on the assumption that fewer congenerics makes species within the genus easier to identify based on morphological or other grounds. Collen et al. found that primate species were discovered earlier when they have a larger home range, when they had fewer congenerics, and when they were diurnal rather than nocturnal.

In summary, a wide array of approaches has been taken to investigate questions of conservation importance in primates. This should not be surprising, as the IUCN Red List is inherently a comparative measure, and understanding the factors that influence risk has obvious conservation value. For the future, the most exciting developments are likely to link phylogenetics with geography, as illustrated above in figure 11.6 and, less directly, as shown by studies of community composition at particular research sites. It will be useful to conduct comparative studies at a more local level and to incorporate both intrinsic features of the organisms and anthropogenic drivers (Brashares 2003; Harcourt and Parks 2003; Isaac and Cowlishaw 2004; Kamilar and Paciulli 2008).

Cultural and linguistic extinction. A number of researchers are working at the interface of biology and anthropology to understand patterns of cultural variation at regional and even global scales (Mace and Pagel 1995; Collard and Foley 2002; Maffi 2005). One particularly interesting set of analyses examined the extinction risk of different languages—just as different biological species vary in their risk of extinction, languages can go extinct, and this risk can be quantified and compared using linguistic "threat status" categories. Thus, similar to the IUCN Red List for biological species, endangered languages have a Red Book through UNESCO, and various other print and online resources are devoted to documenting and preserving endangered languages (Wurm et al. 2001; Gordon 2005).

Sutherland (2003) used existing data on speakers of different languages to classify threat levels in nearly 7,000 languages. Importantly, he followed criteria established for biological species by the IUCN, which allowed him to compare his results to those obtained for biological diversity. Thus, languages ranged from nonthreatened (1,698 languages, or about 25 percent of the total) to critically endangered, indicated by having fewer than 250 speakers and/or declining populations of speakers (438 languages, or 6.4 percent of the total when "data deficient" languages were not assumed to be critically endangered). By comparing patterns of extinction risk in birds, mammals, and languages at a global scale, Sutherland discovered that languages are more threatened than these two groups of vertebrates, based on the proportions of each set that fall into given threat categories. Thus, in comparison to the 6.4 percent of threatened languages, only 1.9 percent of 9,797 birds and 4.1 percent of 4,630 mammals are "critically endangered." While similar factors covary with the diversity of biological and cultural diversity (see chapter 10), it is important to remember that the specific processes of extinction are likely to be very different in biological and cultural systems. In particular, extinction of languages does not require the actual death of speakers of that language; it can happen more simply as younger generations switch to an economically more relevant language.

Extinction is a process that happens naturally in biological systems, and the same is likely to be true for languages. Thus, statistical methods and tools reviewed in chapter 8, such as lineages through time plots, could also be applied to study rates of linguistic extinction. A cursory glance at a dated language tree suggests that the rate of background extinction could be very high. Referring back to figure 2.9, for example, we see what appears to be a higher rate of cladogenesis as we approach the tips of the Indo-European language tree; as discussed in chapter 8, this is consistent with a strong "pull of the present" in lineage through time plots (see figure 8.4), which indicates a high rate of extinction. We currently lack estimates of background extinction rates for language trees. Regardless of what that extinction rate is, however, as the world's populations become more economically linked, more languages are likely to go extinct, thus drastically pruning the tree of language history to a handful of dominant tongues.

Common Concerns in Comparative Studies of Primate Behavior

Although many researchers interested in primate behavior and ecology use comparative methods, some field primatologists have raised questions about phylogenetic methods and comparative studies of different species (Struhsaker 2000; Strier 2003) or have suggested that more attention needs to be focused on carefully controlled comparisons that make use of intraspecific variation (Chapman and Rothman 2009). In what follows, I cover two main areas of concern when applying comparative methods to study behavior: intraspecific variation and identifying homology. These issues are not limited to studies of behavior and ecology but are commonly raised for these types of data.

Intraspecific variation. Intraspecific variation is a major issue in studies of primate behavior, because many traits, such as group size, exhibit marked variation among social units or populations (see Chapman and Rothman 2009). In one study of rhesus macaques, for example, six social groups ranged in size from six to ninety individuals (Makwana 1978), while another study of five maroon leaf monkey groups yielded variation of three to eight individuals (Supriatna et al. 1986). Thus, one might reasonably ask of a comparative biologist: How do you decide what is a *mean* group size for a baboon, a capuchin, and a gorilla, given the wide variation observed across groups and among populations?

Several issues arise in the context of intraspecific variation: (1) Does intraspecific variation invalidate the cross-species comparative approach? (2) Can intraspecific variation be used to test hypotheses, and more specifically, can it be combined with cross-species data in such tests? (3) What methods are available to incorporate intraspecific variation into statistical models for cross-species studies? I briefly address each of these questions in turn.

(1) Intraspecific variation does not, in general, invalidate cross-species

comparative studies. As someone who has spent many hours observing lemur behavior, I can sympathize with the concerns about intraspecific variation (as do other comparative biologists; Gittleman 1989). Yet the evidence suggests that for most traits of interest to primatologists, differences among species are much greater than intraspecific variation. Thus, a wild ring-tailed lemur very consistently finds herself in closer proximity to a greater number of individual conspecifics than does a sportive lemur, and the mating system of a typical gibbon (i.e., usually monogamous) is strikingly different from that of a typical Asian colobine monkey (i.e., often single male), despite observations that some gibbons are in fact multi-male (Barelli et al. 2008) and some colobines live in pairs, that is, monogamous groups (Tilson and Tenaza 1976).

(2) It is possible to use intraspecific variation to test adaptive hypotheses, and even to include this variation in cross-species comparative tests. Some studies have used intraspecific variation when building comparative datasets or testing hypotheses, including tests of variation itself (Ossi and Kamilar 2006). I noted above, for example, that rhesus monkeys and maroon langurs exhibit variation in group size (Makwana 1978; Supriatna et al. 1986). In both cases, positive associations can be demonstrated between group metabolic needs and home range size intraspecifically, which is generally congruent with patterns found across species (e.g., Nunn and Barton 2000).

As another example of intraspecific comparisons, Srivastava and Dunbar (1996) collated data on twenty-three populations of Hanuman langurs to investigate why some populations are single male and others are multi-male. They found that populations in which females exhibit a longer birth season are more likely to have single-male groups, consistent with the ease with which males can monopolize females when their estrous periods are asynchronous (Dunbar 1988; Nunn 1999b). Other factors, such as the distance between groups, were also important predictors of Hanuman langur mating systems (Srivastava and Dunbar 1996). However, we should be aware that when we run a nonhistorical test across different populations, we are assuming that the population level units are statistically independent. As we saw in studies of cultural variation in different human populations, this need not be the case, and the same is true in biological systems (Edwards and Kot 1995).

Other researchers have added population-level data points to a species level phylogeny to increase the sample size of comparative tests. For example, Ostner et al. (2008) included several populations in their comparative analysis of reproductive skew. This was achieved by creating diversification events at the tips of the tree to represent different populations of the same species, which increased their sample size from nineteen to twenty-seven. They also conducted the analysis at the level of social groups (n = 43) by adding groups to the tips of the population-level tree. All three sets of tests produced congruent results (Ostner et al. 2008). An issue with this approach is that it can be difficult to decide on branch lengths for the populations or social groups, especially because gene flow means that these units are not reproductively discrete with distinct evolutionary histories. New methods, described next, take a more statistically rigorous approach to incorporating intraspecific variation.

(3) Methods are now available to incorporate intraspecific variation more explicitly into cross species tests. Several evolutionary biologists have recognized the importance of intraspecific variation in comparative biology (Lynch 1991; Martins and Hansen 1997; Housworth et al. 2004; Harmon and Losos 2005; Ives et al. 2007; Felsenstein 2008). In statistical terms, intraspecific variation can be modeled as "error," which in this case refers to variation that is unrelated to the other variables and results in the true value of X being unknown; it can arise from different measurement techniques, natural variation in the data, measurement error, or sampling variation.

To date, these methods have not yet played a significant role in comparative biology. Two problems have hampered the application of these methods, both of which have been overcome in recent years. First, it has been difficult to implement the methods and to understand their importance. Recently, however, a paper presented methods for dealing with intraspecific variation and provided computer code and worked examples that demonstrated the importance of controlling for measurement error (Ives et al. 2007, AnthroTree 11.2). Second, most comparative datasets provide only mean values for the different traits that were studied. As informatics tools and databases grow in use, it will be increasingly easy to compile measures of variability, thus facilitating the application of methods to incorporate intraspecific variation in comparative studies.

While behavioral data are most commonly suggested to exhibit high levels of intraspecific variation, morphological and other traits also exhibit variation among individuals, groups, or populations. Thus, methods to incorporate intraspecific variation are likely to be useful for many types of data in primate comparative biology. Using methods based on Ives et al. (2007), for example, Lindenfors et al. (2010) performed a study of aerobic function in primates while taking into account intraspecific variation in physiological values (hematocrit and red blood cell counts). In this case, including intraspecific variation had no impact on the results, which involved analyses of sex- and age-based differences using a phylogenetic paired t test.

Behavioral homology. As discussed in chapter 2, homology refers to traits that are shared through common descent, and much of the phylogenetic toolkit relies on identifying homologous traits. Given that behavioral traits often exhibit flexibility in relation to social conditions or the environment, how can we be sure that they are inherited behaviors, rather than individual responses to external conditions? And for continuous characters such as body mass or home range size, how is homology assessed?

Many authors have discussed behavioral homology in the context of comparative studies and phylogeny (Lorenz 1941; Baerends 1958; Lauder 1986; Brooks and McLennan 1991; Wenzel 1992; Gittleman and Decker 1994; Greene 1994; Lockwood and Fleagle 1999; Robson-Brown 1999; Rendall and Di Fiore 2007). Homology is difficult to demonstrate, and possibly easier to show for morphological traits, where structure, development and other aspects of trait expression provide stronger insights to homology. In previous work, ethologists attempted to identify homology through the similarity in how a behavior was implemented at the nervous and muscular level, thus linking behavior more firmly to structure. This is one justification for studying fixed action patterns, which are instinctive behavioral patterns elicited by an external stimulus and are assumed to have an underlying, evolved neurological basis (Alcock 1998); when these behaviors are performed in different species and involve homologous muscular and nervous system mechanisms, they can be used as criteria for behavioral homology. Similarly, studies of behavioral ontogeny can shed light on homology, much as development also can be used to assess morphological homology.

Another approach to the question of behavioral homology is to consider empirically whether behavioral traits can be used to infer phylogeny as effectively as other types of traits, such as morphological characters (Rendall and Di Fiore 2007). In one study, for example, de Queiroz and Wimberger (1993) compared levels of homoplasy in datasets that contained both behavioral and morphological data. Under the assumption that behavioral traits are more labile (and thus homology less easy to evaluate), measures of homoplasy should be higher in phylogenies constructed of behavioral than morphological traits. The authors found, however, that the consistency index (see chapters 2 and 10) is not significantly different in the two types of data. Other studies, however, suggested greater variability in behavioral than morphological traits

(Gittleman et al. 1996; Blomberg et al. 2003). This should not be surprising, as all types of traits exist on a spectrum from less to more responsive to environmental stimuli and with rates of evolution that make them more or less variable at different phylogenetic scales (Rendall and Di Fiore 2007). Indeed, many authors have shown that primate behavioral traits exhibit evidence for phylogenetic signal (Blomberg et al. 2003; Kutsukake and Nunn 2006; Ostner et al. 2008; Thierry et al. 2008).

Thus, the homology of traits is a challenging issue, and if important to a particular comparative study or phylogenetic inference, homology should be considered on a trait-by-trait basis, for example, by investigating the neural or morphological underpinnings of the behavior. On the whole, however, questions raised about the homology of behavior in different organisms have done little to deter others from convincingly addressing behavioral questions using comparative methods (Martins 1996a; Greene 1999; Robson-Brown 1999; Rendall and Di Fiore 2007).

Summary and Synthesis

Quantitative comparison has long played a major role in studies of primate behavior and ecology (Milton and May 1976; Clutton-Brock and Harvey 1977; Harcourt et al. 1981). Phylogenetic methods to study behavior and ecology continue to improve, including methods to examine primate community ecology, incorporate intraspecific variation, and reconstruct the evolution of behavioral traits in relation to morphology and ecology. Similarly, comparative studies of extinction risk have proceeded from tests of the correlates of threat and rarity to more sophisticated analyses that take geography into account more explicitly. A practical direction for future development involves the construction of relational databases from which intraspecific variation can be easily shared with others and extracted for analysis (see Chapter 13).

In addition to the concerns commonly raised above, I have heard behavioral ecologists question whether our knowledge of primates is complete enough to undertake comparative research with the available data. The implication is that research effort would be better placed on acquiring new data rather than studying what we have already collected. In this context, it is important to keep in mind that cross-species comparisons can inform fieldwork, for example, by uncovering patterns that lead to new hypotheses to test in the field. Similarly, new comparative methods can be used to identify the primate species that are most worthwhile to study (Arnold 2008; Arnold and Nunn

2010; see chapter 13)—just as GIS methods have been used to identify the primate populations that represent "gaps" in our knowledge of primate infectious diseases (Hopkins and Nunn 2007, 2010).

With these arguments in mind, however, it is also important for comparative biologists to appreciate that their research depends on the hard work of field biologists who are collecting original data. If we are to see comparative primatology grow in the future, it will be essential to fund field research, including research on the natural history of unstudied primate populations and species; similar arguments apply to field studies of human cultural traits and languages. It is also essential for comparative biologists interested in primate behavior to gain field experience. Observing real animals interacting with one another and their environment can serve as inspiration for developing new comparative research projects, as I have found in my own research. In addition, field experience helps comparative biologists appreciate the potential error involved in estimating some behavioral and ecological measures from field data, for example, in terms of the challenges of estimating home range size from necessarily incomplete field observations.

Last, what about the role of behavior in influencing evolutionary trajectories? While this has generated much discussion in the context of niche construction effects (Laland et al. 2000), relatively few studies have investigated this possibility phylogenetically (e.g., Matthews et al. n.d.). It seems likely that future studies will begin to probe these questions in a phylogenetically rigorous way, which is essential for understanding the role of behavior in shaping evolutionary patterns.

Pointers to related topics in future chapters:

* Species concepts and comparative databases: chapter 13
* Intraspecific variation and comparative databases: chapter 13

12 Investigating Evolutionary Singularities

Studies of human evolution are replete with broad-scale comparative patterns. Many readers will be familiar with maps indicating the likely migration routes of early humans out of Africa and into Asia, Europe, Australia, and across to the New World, or maps showing the movement of humans across the vast expanses of the Pacific during the peopling of Polynesia (see figure 10.8). These maps typically are based on broad-scale comparisons of archeological, linguistic, and genetic data. Similarly, phylogenetic hypotheses for the relationships among fossil hominins are inferred using morphological data from fossils and living primate species (see figure 2.5; Strait et al. 2007). Fossil excavation is a broad-scale comparative enterprise, often involving careful mapping of strata across large spatial scales. Similarly, biological anthropologists have long been interested in adaptive explanations for patterns of physical variation among living and recently extinct human populations (e.g., Howells 1973, 1989; Beals et al. 1984; Relethford 1994; Jablonski and Chaplin 2000).

Many of these questions fit neatly within the standard framework of comparative biology described throughout this book in which the comparative method is used to test adaptive hypotheses by identifying cases of trait convergence (see chapter 6; Pagel 1994b). The convergence approach capitalizes on "evolutionary coincidences," and it uses a phylogeny to make inferences about these coincidences (i.e., independent evolutionary events). Provided that there are enough evolutionary origins of a trait, the convergence approach provides a means to test *statistically* whether trait origins are correlated with particular factors. The convergence approach also gives a sense of the generality of evolutionary patterns and can be used to identify "exceptions to the rule" that might be biologically interesting.

Evolutionary convergence thus offers valuable data for testing adaptive hypotheses, but for many interesting evolutionary questions we have only a single evolutionary origin available to study. Few questions are as interesting in this regard as human origins and the ecological and social factors that made us what we are today. Indeed, humans are strange creatures with many unique or uncommon characteristics, or, as Tooby and Devore put it, "zoologically unprecedented capacities" (1987, 183). We walk on two legs with a striding gait. We use language to describe the world and represent abstract concepts. We have relatively large brains and exhibit long periods of parental care. We lack hair, and we wear clothing. We possess complex cultural traits that build on other cultural traits, and thus exhibit cumulative cultural evolution. While some of these traits can be examined in broad phylogenetic context, such as brain size and parental care, most of these traits cannot, which makes it difficult to quantitatively investigate the factors that influenced their evolution.

The challenges of studying singular events is not restricted to human evolution; other traits have evolved only one time among anthropoid primates, including twinning in callitrichids, foregut fermentation among colobines, and nocturnality in owl monkeys. An important difference in these non-human examples is that the traits are shared by multiple species on the tree, that is, they are *synapomorphies*. Another difference is that while they may be singularities in primate evolution, these traits (or similar ones) are present in other mammals, opening the possibility to expand the comparative context (and increase the number of independent origins) by including nonprimates. Humans are but a twig on the tree of life, and many of the traits discussed above are *autapomorphic*, that is, derived characters that are unique to us. Evolutionary anthropologists are increasingly interested in probing how we differ from our closest relatives (e.g., Kappeler and Silk 2009) and, thus, in identifying and studying these autapomorphies.

In this chapter, I refer to this issue as the *singularity problem*. I focus on human evolution, yet the principles discussed here apply more broadly to any situation in which a trait evolves on only one branch of the phylogeny. We will find that by using phylogenetic comparative methods, singularity problems can be studied more rigorously than they usually are, but that our conclusions must often be more cautious than would be the case when studying traits with multiple evolutionary origins. Perhaps counterintuitively, we will also discover that singularities actually make a strong case for broader phylogenetic study of primate behavior and ecology, with benefits arising from studying monkeys, apes, and even strepsirrhines, rather than focusing only on African great apes as models for human evolution (see also Martin 2002). By expanding the taxonomic scale in this way, for example, we can narrow the confidence intervals on reconstructed ancestral states in apes (see chapter 3), and we can better decide if the amount of evolutionary change for a quantitative trait along the human lineage is remarkable in the broader context of primate evolution. Similarly, we can use the variation across primates to build predictive models for the state of some character in humans, which can then be compared to the observed value in humans. If the observed trait values fall outside the range predicted by the statistical model, we would have strong evidence that humans are unique relative to other primates (Organ et al. n.d.).

The chapter has two sections. I start by briefly reviewing *referential* and *strategic* modeling as two general approaches to studying human evolution. *Referential models* make use of some real phenomenon that serves as a referent for another phenomenon of interest that is less easily studied, while *strategic models* reflect sets of theoretical concepts, evolutionary principles and relationships among traits (Tooby and DeVore 1987). I suggest that two additional types (or actually subtypes) of models are worthwhile to consider in the context of this book: a *phylogenetic referential model*, which involves reconstructing ancestral states based on data from living species (Ghiglieri 1987; Wrangham 1987), and a *comparative model*, which involves assessing correlations among traits to infer the likely suite of characters in an organism and the evolutionary trajectory of these characters (Foley and Lee 1989; Nunn and Van Schaik 2002; Organ et al. n.d.).

The second section describes phylogeny-based methods that can be used to statistically identify and investigate evolutionary singularities. I discuss how four methodological approaches from previous sections of the book can be used to investigate human evolution with greater statistical and phylogenetic rigor: (1) reconstruction of ancestral states using Bayesian methods to control for uncertainty in ancestral state reconstructions; (2) modeling trait change, such as changes in selective regimes, with explicit evolutionary models; (3) testing whether a single branch has experienced a larger than expected change in a quantitative trait; and (4) testing for correlated evolution using more general modeling approaches (rather than relying on multiple independent origins).

In some ways, this chapter—especially the latter part—is a review of material presented in previous chapters. When appropriate, I refer back to earlier chapters to avoid repetition and to keep the focus on how the methods can be used to study evolutionary singularities. In addition, it is worth reemphasizing a cautionary note from above. Although these methods can produce *p* values that are useful for testing hypotheses, we should keep in mind that the most convincing comparative test of an adaptive hypothesis generally comes from the repeated evolution of traits. For example, having only a single transition means that it will be difficult, if not impossible, to rule out an influence of other variables that also changed along the branch in question (see Garland and Adolph 1994). It is therefore important to use multiple approaches when studying evolutionary singularities, and to link up data on extant species, fossils, and phylogeny as much as possible.

Types of Models

Referential models have played a significant role in efforts to reconstruct the behavior of hominins. Referential models generally use a living species as a surrogate for our ancestors, thus assuming that behavior of the reference species provides insights to the behavior of early humans. Two types of approaches have been used to develop referential models of human behavior. In the homology approach, a researcher chooses a close relative that is likely to share many ancestral characters with the species of interest. Thus, one could use chimpanzees (McGrew 1981) or bonobos (Zihlman et al. 1978; Susman 1987) as models for human evolution on the basis that they are the closest living relatives of humans (see also Wrangham and Pilbeam 2001). In the analogy approach, a researcher examines characteristics of species that fill an ecological niche similar to what is thought to have characterized early humans. Baboons and their close relatives often have been used as analogous referential models, specifically, as ecological analogies in terms of their diets, open savanna lifestyles, and potential for hybridization (DeVore and Washburn 1963; Jolly 1970, 2001).

The other major approach involves *strategic* or *conceptual modeling*. In this case, a well-validated theoretical model from living species provides a way to infer behavior in extinct species. When combined with details on the morphology and ecology of particular hominin species, theoretical models are thought to provide valid inferences of behavior for that fossil taxon. Tooby and DeVore (1987) provide an overview of strategic modeling, with further useful perspective given by Ghiglieri (1987) and Stanford and Allen (1991). Comparison plays an important role in strategic models; specifically, it generates insights to the linkages among key traits. Thus, "by following patterns of convergence and divergence among homologous and uniquely derived features, and analogous and non-analogous adaptations, the comparative approach allows some inference to be made about functional and ontogenetic interdependence" (Tooby and DeVore 1987,198–99).

Both of these major types of models have value in reconstructing hominin behavior. In the context of the approaches presented in this book, however, two types of approaches stand out as particularly relevant for further discussion: phylogenetic referential models and comparative models. These models, in fact, are connected methodologically in that both use a hypothesis of evolutionary relationships and comparative data.

Phylogenetic referential models. Wrangham (1987) provided an example of phylogenetic referential modeling that builds on the homology approach discussed above. In this approach, data on closely related species are used to infer behavior for some common ancestor of those species. In his application of this concept, Wrangham reconstructed the last common ancestor of humans and all other African apes (i.e., the gorilla, chimpanzee, and bonobo). However, phylogenetic referential modeling can be used to reconstruct behavior in any group of organisms, or even among human populations (e.g., linguistic groups). Wrangham did not use formal phylogenetic methods to reconstruct ancestral states, which were relatively undeveloped at the time. Instead, he reasoned that if a trait "occurs in all four species [of African apes], it is likely (though not certain) to have occurred in the common ancestor because otherwise it must have evolved independently at least twice" (52–53). The decision to use all species of African apes was based on uncertainty at the time concerning whether the chimpanzee/bonobo or the gorilla is more closely related to modern humans. Thus, Wrangham's analysis essentially assumed the phylogeny shown in figure 12.1a.

In analyses of grouping patterns, intrasexual relationships, and sexual behavior, Wrangham (1987) found that eight of fourteen traits had similar character states across the African apes. Based on this, he proposed that the common ancestor had closed social networks, hostile intergroup relationships maintained by males (but not necessarily characterized by territoriality), female dispersal, weak social bonds among females, and nonmonogamous sexual relationships. Wrangham was suitably cautious in the conclusions that can be drawn from comparisons of this type, noting for example that phylogenetic referential modeling fails to make use of paleontological reconstructions, it ignores intra-specific variation, and it can only reconstruct the traits shared by all descendants from a node (rather than unique characteristics along a branch of interest, such as bipedality in humans).

Ghiglieri (1987) took a similar approach to building a phylogenetic referential model. Importantly, he used updated phylogenetic information based on

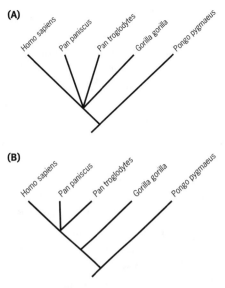

Figure 12.1. Phylogenetic referential model. (a) Based on evidence available at the time, Wrangham (1987) assumed that *Pan*, *Gorilla*, and *Homo* were equally related. Thus, only character states that were found in all African apes were assumed to exist in the last common ancestor of *Homo* and the nonhuman primates. (b) In contrast, Ghiglieri (1987) made use of evidence for a closer association between *Pan* and *Homo*.

molecular evidence, which indicated a closer relationship between *Pan* and *Homo* than between *Gorilla* and *Homo* (figure 12.1b). Better phylogenetic information should produce stronger insights to the array of character states present in the most recent common ancestor of humans and other great apes. Ghiglieri also considered ecological variables, whereas Wrangham (1987) included only social and mating system variables in his analysis. Inclusion of ecology provides a way to incorporate a strategic modeling component into the phylogenetic model, at least in terms of determining the ecological influences that likely impinge upon the social factors. From this analysis, Ghiglieri identified the following traits in the common ancestor: female dispersal, closed social groups, stable multi-male/multi-female communities, male territorial behavior, male intrasexual competition, and fission-fusion sociality with individuals of both sexes at least occasionally traveling alone (and females doing so more than males).

A phylogenetic referential model provides a useful puzzle piece in the larger goal of reconstructing hominin evolution (see also Di Fiore and Rendall 1994). By using modern phylogenetic methods, a phylogenetic referential model can be constructed with much greater statistical rigor; even so, we should keep in mind the limitations of this approach. Importantly, such an approach fails to capture the many unique characteristics of humans that could have played a central role in our evolutionary success, including bipedality, language, pair bonding, and paternal care. These and other important reconstructions of human evolution will require a combination of referential and conceptual models, coupled with better incorporation of the fossil record into comparative studies (e.g., Organ et al. n.d.).

Comparative models. A second major approach that is relevant to this book is to investigate correlations among suites of traits that were likely to have played key roles in human evolution, or, a *comparative model* of human evolution. When these traits are grounded in theory, such a model falls on the conceptual end of the spectrum, and when it is based more phenomenologically on patterns of trait correlation with little concern for the functional relationships among traits, it falls on the referential end of the spectrum. Tooby and DeVore (1987) provided an overview of comparative approaches in the context of conceptual models. They note that caution should be used because such inferences are probabilistic: many exceptions to the rule can occur, especially when functional linkages among the traits are unknown or the association is weak. Phylogenetic comparative methods could be brought to bear on this issue by providing phylogenetically rigorous tests of association among the traits (Nunn and Van Schaik 2002).

An example of a comparative model is provided by Foley and Lee (1989; see also Rodseth et al. 1991). These authors modeled the evolutionary pathways of primate behavior based on patterns of variation found in living primates, and then applied this approach to infer hominin behavior. Their approach avoided using a living species, such as chimpanzees, as an exemplar of early hominins and instead focused on the possible traits and their probable transitions. They considered just three traits: the female distribution, the male distribution, and the stability of male-female associations. Given the number of character states for each trait (four, four, and two, respectively), this produced thirty-two possible combinations of character states (figure 12.2). As with the step matrices used in chapter 3 to reconstruct ancestral states using parsimony (see AnthroTree 3.3), Foley and Lee assumed that transitions among adjacent positions in this matrix are more likely than longer "jumps" across the matrix.

From this parsimony assumption, Foley and Lee (1989) generated a cost

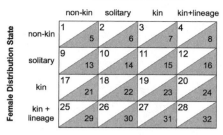

Figure 12.2. Matrix of ordered socioecological character states. The matrix shows three characters: the distribution of males (in columns), the distribution of females (in rows), and intersexual relationships dividing each cell (open triangle indicates stable male-female relationships, filled triangle indicates transitory male-female relationships). The category "kin + lineage" refers to kinship recognized across groups, which the authors assumed only applied to hominins (including modern humans). All primate species can be assigned to a given triangle in this matrix, which are indicated by numbers 1 to 32. Thus, Foley and Lee (1989) placed the callitrichids in cell 11, while langurs were included in cell 18. A single step is achieved by moving from one triangle to another within a cell (e.g., cell 10 to 14), or by moving on the flat sides of the cells but remaining in the same triangle shading (e.g., cell 17 to 9). Moving from cell 17 to 22 would require a minimum of two steps (through state 18 or 21). (Redrawn from Foley and Lee 1989.)

matrix for all possible transitions, and they predicted the pathways through which social systems would evolve most parsimoniously (figure 12.3). Each line in this figure indicates change in one of the characters (similar to an evolutionary step in a parsimony analysis), and multiple paths can be taken to reach the same state (i.e., notice that states around the edge of the network appear more than one time). Starting at the center of figure 12.3 with, for example, solitary males and females that lack stable associations (state 14), it is possible to follow lines to move to other states. Importantly, many of these states do not occur in primates, such as solitary males having ephemeral contact with groups of unrelated females (state 6). This may reflect that such a state is evolutionarily unstable, with males who established longer-term associations with groups of females able to secure more matings.

Foley and Lee (1989) applied this framework to study paths through "socioecological space" needed to arrive at humans. This required examining the clade to which hominins belong, specifically African apes, where state 2 is the gorilla and state 3 is the chimpanzee. With shifts to more open savanna environments, the authors proposed that early hominins probably lived in larger groups, which would most efficiently build on male kin alliances. Thus, they would have been likely to fill social state 4, with stable associations among males and females, females grouping with non-kin, and males living in long-term association with kin. This state could be reached with only one step from states 3, 8 and 12 (see figure 12.2). Thus, the model gives some insights to

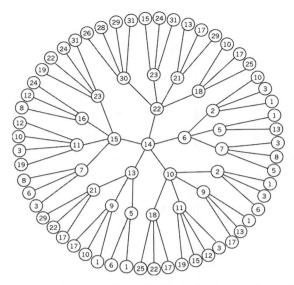

Figure 12.3. Parsimonious evolutionary pathways. Parsimoniously adjacent pathways of the social state variables in figure 12.2 are mapped graphically. This graph shows state 14 at the center, but any of the states could be placed there. From any node, the distance to adjacent nodes involves only one step.

the probable paths of hominin social evolution based on comparative data in non-human primates and assuming parsimonious evolution.

Nunn and Van Schaik (2002) provided another approach to reconstructing ancestral states using trait covariation. We reasoned that some behavioral and ecological traits can be estimated directly from fossil remains; these "knowable" variables include aspects of diet, body mass, substrate use (arboreal versus terrestrial), and activity period (nocturnal versus diurnal). If these knowable character states covary predictably with other traits that leave a fainter mark in the fossil record, such as group size, it is possible to make inferences about these "unknowable" traits based on patterns of trait correlations among knowable and unknowable variables in living species. Moreover, if we have multiple knowable traits—a suite of traits discernible from the fossil record—this should help to narrow the estimates of the unknowable traits. For example, if we reconstructed a particular fossil taxon as nocturnal, small bodied, and having a diet of insects, we should be able to make better inferences about its home range size than if we only knew that the fossil taxon was small bodied. The soundness of any such conclusion depends on the strength of the association in extant species and our understanding of its causality; hence, this approach couples aspects of strategic modeling with a comparative perspective.

We used this comparative approach to investigate patterns of variation in primates. First, we examined patterns of variation among the character states in the variables that can be inferred directly from fossil material. We termed the set of character states different "syndromes," where syndrome refers to a characteristic pattern of behavioral variation (see also the ecological categories of Clutton-Brock and Harvey 1977). Because not all character state combinations are equally likely, one can use information on patterns of character states in extant taxa to check the "internal consistency" of a fossil taxon's reconstructed syndrome.

Second, we examined the association between the syndromes and traits that are less easily inferred in the fossil record—such as group size, home range size, and day range—using data on living species. We found that using syndromes (rather than a single trait) helped to narrow the ranges of unknowable variables. When putting confidence intervals on the estimates, however, we found that many of the estimates were still remarkably wide. We provided tables that make it possible to predict socioecological features of an extinct taxon when this taxon can be assigned to a particular syndrome, and we applied the method to several fossils as case studies (Nunn and Van Schaik 2002). This approach could be applied to study human evolution.

Another example of a comparative model was discussed in chapter 6, in the context of research aiming to reconstruct levels of intrasexual competition in hominins by comparing dimorphism in the fossil record to that found in extant primates (see figure 6.3; Plavcan and Van Schaik 1997a). In this case, different measures of dimorphism produced different estimates of male intrasexual competition.

As a final example, it is important to note comparative research by Robin Dunbar and colleagues in which they used variation in neocortex size, group size and other variables in living species to infer hominin behavior (Aiello and Dunbar 1993; Dunbar 2003). As described in Chapter 6, comparative research has demonstrated a positive correlation between group size and neocortex size after controlling for body mass (figures 6.1 and 6.4). Although a species is likely to exhibit variation in group size around the mean values used in comparative research, this pattern suggests that with neocortex data in hand, it is possible to make estimates of group size for a fossil taxon. Because neocortex size cannot be estimated directly from hominin fossil remains, a method was needed to convert estimates of cranial capacity (a volume) to neocortex ratio (i.e., the volume of the neocortex relative to the rest of the brain). Aiello and Dunbar (1993) achieved this by regressing brain volume on cranial capacity

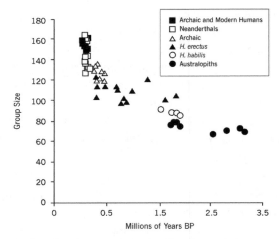

Figure 12.4. Group size throughout human evolution. Based on the associations among variables reflecting brain characteristics and group size, Aiello and Dunbar (1993) and Dunbar (2003) made predictions for group size (and other variables) based on hominin fossil cranial capacity. (Redrawn from Dunbar 2003.)

among living species to estimate brain volume for a given fossil and then using estimated brain volume for the fossil specimen to predict the neocortex ratio (based on the association between brain volume and neocortex ratio among extant species).

With this general comparative approach, Aiello and Dunbar (1993) and Dunbar (2003) examined variation in estimated group size over time (figure 12.4). Their models predicted that group sizes in early hominin evolution were relatively stable through time and similar to those observed in living apes. With the origin of the genus *Homo*, however, group size likely increased. Shortly after the evolution of *Homo erectus*, the authors predicted that group size rose at an exponential rate. From these and other analyses, inferences were drawn about the dates at which language arose (around five hundred thousand years ago), as well as transitions in cognitive capacity (e.g., levels of intentionality) and even the origins of musical ability. Clearly caution is required to avoid putting too much confidence in these estimates. For example, estimates of cranial capacity are imperfect in fossils, and the analysis builds from one association to another, which should result in wide confidence intervals on estimates of group size and other traits (e.g., Smith 1996). With this comparative approach, however, it becomes possible to make specific predictions for the origins of major cognitive transitions, which can then be tested using other types of fossil or archeological data.

Statistical Approaches for Probing Singularities

The previous section summarized the conceptual issues involved in reconstructing our recent evolutionary past. In what follows, I shift focus by considering four methodological approaches that can be used to make statistically rigorous inferences about evolutionary singularities.

Reconstructing ancestral states. Many of the questions described above involve reconstructing behavioral patterns based on a combination of fossil evidence and patterns of variation among living primates. For example, did males of the last common ancestor of African apes exhibit territorial behavior, and what was the probable group size of *Homo habilis*? Many similar questions also emerge concerning the origins of morphology associated with food processing (Organ et al. n.d.) or the origins of pair bonding in human evolution.

Most studies of human evolution have yet to use statistically based phylogenetic methods to make inferences about behavior in extinct lineages, yet the methods discussed in chapters 3 and 4 provide great opportunities for such research. Importantly, new phylogenetic methods can be used to place confidence intervals on reconstructions. Thus, it is now possible to reconstruct traits in the common ancestor of humans and *Pan* using these methods and to place *quantitative measures of confidence* on the reconstructions using maximum likelihood and Bayesian techniques. This is in striking contrast to the more conservative approach taken by Wrangham (1987) and Ghiglieri (1987), who used only direct descendants of the common ancestor to reconstruct behaviors, and limited their reconstructions to only those traits that were found in all the descendants. Such methods could also be used to rigorously investigate ancestral states in newly discovered fossil taxa, such as *Ardipithecus ramidus* (Lovejoy 2009; White et al. 2009b; Whiten et al. 2010).

In addition to reconstructions of a single trait using only phylogeny, it is now possible to use both phylogenetic information and the association among traits in living species to make more informed inferences about extinct organisms. In chapter 9, for example, I reviewed recent work by Organ et al. (2007), in which the authors investigated genome size in dinosaurs based on the association between cell size and genome size among living species. With this association and data on the sizes of bone cells (i.e., osteocytes) in living and fossil organisms as a predictor of genome size, Organ et al. (2007) were able to make phylogenetically and statistically informed inferences about genome size in extinct species (see AnthroTree 12.1).

Two important aspects of this approach are relevant to studying hominin evolution. First, the user ends up with a distribution of estimates rather than a single-point estimate. Thus, the Bayesian approach provides a way to place confidence intervals on the reconstructions. Second, the method combines the advantages of phylogenetic referential models with comparative models; it makes use of both the phylogenetic position of the fossil of interest and the correlation among traits (see also Garland and Ives 2000). With a hominin phylogeny, this method could be used to make inferences about hominin behavior. To this end, it would be useful to have a Bayesian inference of hominin phylogeny from which a set of trees could be used to control for phylogenetic uncertainty (Organ et al. n.d.).

The model of evolutionary change in hominin lineages. Another important set of questions involves the tempo and mode of evolution. For example, does the temporal pattern of trait variation in hominins indicate stabilizing selection? Do the traits show phylogenetic signal? It is widely believed that lineages of hominins radiated into niches that are very different from those found on the lineage leading to modern humans, as indicated by the extreme chewing morphology found in robust australopithecines. Does trait variation actually support an adaptive radiation model for these and other hominin lineages? Or, are there periods of rapid evolutionary change associated with speciation events, and stasis at other times?

New phylogenetic methods can be used to investigate trait evolution in relation to different evolutionary models, including Brownian motion, an Ornstein-Uhlenbeck (OU) model for stabilizing selection, speciational change, and the ecological niche-filling model, in which trait evolution is more rapid early in the radiation of a group of organisms and slows as niche space becomes more filled (see chapter 5). In one study, Hansen (1997) fitted an OU model to dental evolution in fossil horses. He was able to estimate optimal tooth design for grazers and browsers, with the model accounting for 72 percent of the variation in tooth morphology (hypsodonty). Such an approach could be used to model selective optima in robust versus non-robust australopithecines, or in *Homo* versus *Australopithecus* more generally. Speciational and punctuational models of evolution also can be studied, for example, by fitting the parameter κ (see chapter 5, Pagel 1999a).

The general point is that reconstructing hominin evolution is not only about probing a single key node in the evolutionary past, such as the common ancestor of *Pan* and *Homo*, the first fully bipedal hominin, or neocortex size in the first members of the genus *Homo*. Rather, we should also be interested in investigating the evolutionary process itself. In addition to general evolutionary models for describing evolutionary change, it is also important to investigate trends towards increases or decreases in key characters along the human evolutionary lineage, a point made by Tooby and DeVore (1987) and illustrated by Dunbar's (2003) analysis of group size in relation to cranial capacity (see figure 12.4). In addition, with information on evolutionary trends one can use the trend itself when reconstructing ancestral states, thus allowing the trait values to differ from what is found in a Brownian motion model without evolutionary trends (Pagel 1999a).

In this regard, new methods provide a way to detect directional trends in evolution (Pagel 1997, 1999a). As described in chapter 4, a directional model can be assessed statistically by examining whether longer distances along the root to tips paths of a tree are associated with larger trait values (this is the expectation for a positive directional trend; the opposite is expected for directional trends towards smaller values; see figure 4.7 and AnthroTree 4.7). The method requires that the path lengths from root to tips vary across the tree, as would be the case for a dated phylogeny that includes fossils. This model has been used to study brain evolution in hominins (Pagel 2002). As expected, the directional model obtained a significantly higher likelihood than a random model, indicating that brain size underwent directional evolution in hominins (see figure 4.8). Although this pattern is well known from the fossil record, it highlights that the method has the statistical power to detect known trends, giving us greater confidence when it is used to study other traits for which the fossil record is less complete.

These examples arise directly from within comparative biology, in which interspecific variation is mapped onto phylogeny. Other comparative approaches come from the related (but rarely overlapping) area of quantitative genetics. In one study, for example, Ackermann and Cheverud (2004b) used methods from quantitative evolutionary theory to investigate whether morphological variation in hominin cranial remains can be attributed to selection or genetic drift. They tested their hypothesis in Pliocene and Pleistocene hominins by comparing patterns of within-species variability in living apes to variability found across hominin samples; proportionality among these levels of variation would be consistent with drift. They found that drift can explain variation among samples of *Homo* fossils, while australopithecine variation was less consistent with an explanation based on genetic drift.

Examining evolutionary change along a single branch of a tree. If we are studying evolutionary singularities, we are by definition faced with analyzing change on a single branch of a phylogeny. A variety of approaches are available to compare trait evolution on one branch to other branches (Garland et al. 1993). Such tests can be used to discern if the trait change is somehow exceptional, compared to evolutionary changes along other branches on the tree. These tests also can be used to identify other traits that show large amounts of change (and thus might be related to the trait of interest). However, it is important to focus on traits for which an association is expected a priori, rather than undertaking a "fishing expedition" that examines as many traits as possible.

A method developed by Mark McPeek (1995) provides one way to investigate change along a single branch in a statistically rigorous way. His method is based on independent contrasts, and he provides a nice overview of the method and how it might be used to study change on single branches (although it is worth noting that the method also can be applied to more than one branch). The central proposition of the method is that a constant-rate model of character evolution by Brownian motion might be incorrect; larger amounts of change might occur, for example, when a lineage of organisms enters a new habitat or when a key innovation arises along a lineage. In such cases, we expect to see "large and directed" character change on one or more branches of the tree, and of course we would like to identify such cases.

McPeek proposed a way to investigate whether exceptional change occurred along a phylogenetic branch of interest. The procedure starts by "pruning" the single branch from the tree and then calculating standardized contrasts on the tree using standard implementations of independent contrasts. The resulting contrasts are saved. Next, the user "grafts" the pruned branch back onto the tree, estimates ancestral values for the nodes of the pruned branches, and then calculates a contrast along the pruned branch itself. This single contrast is calculated so that it covers just the one branch (rather than two branches in the standard implementation of independent contrasts). Like the other contrasts, however, it is standardized (see chapter 7). It then becomes possible to examine whether the change in the single branch is drawn from a different distribution than the change (contrasts) in other portions of the tree using, for example, a *t* test. The method can be applied to any single branch

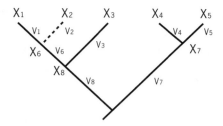

Figure 12.5. Testing for exceptionally large amounts of change along a single branch. McPeek (1995) developed an approach based on independent contrasts (see Chapter 7) to investigate whether a trait has undergone an exceptional amount of evolution along one or more branches of a phylogenetic tree. In this case, we are interested in the change from X_6 to X_2, where X_6 is reconstructed. (Redrawn from McPeek 1995. See also O'Meara et al. 2006; Thomas et al. 2006; and Revell 2008.)

except for those leading to the root, and for multiple branches in most (but not all) cases (see McPeek 1995 for more details).

McPeek provides a hypothetical example involving a single branch leading from a node to a tip (figure 12.5), which is thus similar to the situation we face in examining evolutionary change along the branch leading from the last common ancestor of *Homo-Pan* to modern humans. Notice in this case that the contrasts calculated across the whole tree are X_1-X_3, X_4-X_5, and X_7-X_8, while the putatively "exceptional" contrast runs from X_2-X_6, where X_6 is a reconstructed node. Thus, in calculating the standardized contrast, we must take into account that we are using an estimated value for X_6 rather than real data (see McPeek 1995). This is accomplished by adding an extra "burst" of evolution along the branch, as described in chapter 7 when standardizing independent contrasts.

To relate this method to human evolution, I applied McPeek's (1995) method to study the intermembral index (IMI), which measures the ratio of the forelimbs to hindlimbs and covaries strongly with locomotor behavior (Napier and Walker 1967; Napier 1970; Martin 1990). As shown in chapter 4, the reconstruction of the IMI in apes suggested that change in the human lineage is accelerated (in a negative direction) relative to the other apes (see figure 4.1), as expected given the large change in morphology associated with bipedal locomotion. Importantly, this was only a suggestion rather than a formal test, as it was based on the fact that the human value fell outside the 95% confidence interval on the root estimate. Using McPeek's (1995) method, I found that change along the branch leading to modern humans is indeed significantly larger than the other contrasts ($t_{10} = 5.01$; $p = .0005$; see Anthro-Tree 12.2).

More recently, Revell (2008) conducted numerical simulations to compare McPeek's method to two recent (and very similar) maximum likelihood approaches for detecting change in evolutionary rates on portions of a phylogeny (O'Meara et al. 2006; Thomas et al. 2006). Revell found that the methods produce type I error rates in line with expectations, but the maximum likelihood methods have higher statistical power. In addition, maximum likelihood provides an estimate of the heterogeneity in evolutionary rates. Because we are often dealing with small sample sizes in human evolution, it is thus preferable to use the maximum likelihood approach, which was described in chapter 5 (see AnthroTree 5.8). Revell (2008) provides a useful overview of all the methods and a worked example of the maximum likelihood method. Similarly, Garland et al. (1993) give further perspective on detecting exceptional amounts of evolutionary change, specifically in the context of phylogenetic ANCOVA (see chapter 9).

Quantifying transition rates. One of the major weaknesses of studying evolutionary singularities stems from the fact that most comparative methods make use of convergence in testing adaptive hypotheses. For example, it is generally through repeated evolutionary increases in group size and corresponding increases in neocortex size that independent contrasts approaches can assess the links between group size and brain evolution (e.g., figure 6.4). Thus, a method that makes use of evolutionary transition rates—rather than multiple instances of change—would offer some hope to assess correlated change in the context of evolutionary singularities. We should keep in mind, however, that such methods are more convincing when multiple origins are used to estimate the necessary evolutionary parameters, even if these multiple origins are not directly assessed and taken into account in the statistical tests that use the rates.

Pagel's (1994a) method for studying correlated evolution in discrete traits was described in chapter 7. The method works by quantifying transition rates under two models of evolution: one in which two traits evolve independently of one another (i.e., an independent model) and another in which the traits exhibit correlated evolution (a dependent model), such that the value of one trait determines the probability that the other trait changes (and vice versa). The likelihood of these models can be compared statistically using an LRT to assess whether the correlated model is supported over the uncorrelated model (see AnthroTree 3.5 and 4.7).

In describing the method, Pagel (1994a) used an example with only a single evolutionary origin, specifically involving the evolution of exaggerated sexual swellings in apes and whether this trait is associated with females mating with multiple males (Clutton-Brock and Harvey 1976). In the apes, exaggerated swellings are found only in *Pan troglodytes* and *Pan paniscus*, which may reflect a single origin of these traits (see Nunn 1999a). The correlated model produced a higher likelihood (log likelihood = −8.43) than the uncorrelated model (−11.91) in Pagel's example. However, this result was not statistically significant relative to a simulated null distribution ($p > .12$).

Pagel's method has been applied to study variation cross-culturally (Holden and Mace 2003; Fortunato and Mace 2009) and across primates (Nunn 2000), and it offers a promising possibility for studying evolutionary singularities. However, more work is needed to assess the sensitivity of the method to the sample of species in the test, especially when focusing on single evolutionary origins (e.g., Lindenfors 2006) and to more formally investigate the statistical power of the method in relation to the number of evolutionary origins.

Summary and Synthesis

Because most comparative methods use convergence to test adaptive hypotheses in a statistically rigorous way, single evolutionary events—whether found in only a single species or a monophyletic group of species—pose serious challenges for the comparative method. Human evolution is characterized by a suite of unique characters relative to other primates, including bipedal locomotion, complex language, and cumulative cultural evolution. We would obviously like to study these traits in broad comparative perspective, but how can this be achieved without falling into the trap of adaptive storytelling? In other words, how can we investigate evolutionary singularities in a statistically rigorous way?

This chapter focused on statistically based phylogenetic methods that can be used to investigate single evolutionary events, drawing on methods that were presented in more detail in previous chapters. These include methods to place confidence intervals on reconstructed ancestral states, to investigate evolutionary models, and to test whether evolutionary change along a branch is greater than expected (as based on evolutionary changes in other branches of the tree). Obviously, we can generate a more comprehensive understanding of trait evolution if we also have genetic data to detect selection, information on past selective regimes, and details on the performance of different traits in different environments (e.g., Bock 1977; Leroi et al. 1994; Doughty 1996).

The comparative method forms a central pillar in linking multiple lines of evidence to probe the adaptive basis of a trait, yet caution is needed when only single evolutionary origins are available for study.

A critical issue for the future of modeling hominin evolution involves inferring the phylogeny of fossil hominins. Ideally, this would be achieved using Bayesian methods to control for uncertainty in the evolutionary relationships so that subsequent comparative tests are not conditioned on a single tree (Pagel and Lutzoni 2002). Research is underway in this regard (Organ et al. n.d.). We also need to decide on the sample of primates to use in the reconstruction. Most work has focused on simply reconstructing the ancestor of the great apes using intuition or parsimony; many researchers have thus limited the scope of sampling to the great apes. The newer methods discussed in this book, such as maximum likelihood estimates of transition rates for discrete traits, estimate rates of evolution across the tree. By estimating these rates across primate phylogeny rather than only among great apes, we should end up with better inferences of ancestral states (provided of course that the rates are constant across the tree). In particular, the confidence intervals should narrow. Thus, the methods described here provide a strong case for primatological research at a broad scale, including monkeys and possibly even tarsiers and strepsirrhines, rather than focusing solely on great apes (see also Martin 2002).

13 Developing a Comparative Database and Targeting Future Data Collection

A comparative analysis is only as good as the data that go into it. Unlike experimental approaches that generate new data, comparative approaches typically amalgamate existing data, often from diverse sources. Moreover, the data that are compiled are often reused in later comparative studies, making it essential that later users can track down the original sources. It is therefore important to consider how one should build a comparative database and how best to share it with others.

With the advent of so many new comparative methods in the past twenty years or so and greater numbers of comparative databases available (Anthro-Tree 13.1), I have the sense that many researchers fail to appreciate the importance of data building for a comparative analysis; some even seem to view it more as a "chore" than part of the "real" research. Yet building a database also involves interesting technical aspects, and getting to know the data based on reading the original sources is essential for understanding both expected and unexpected patterns in the overall dataset. Indeed, the database is often the most time consuming step in comparative research on new questions. Moreover, once a database is built, it can be used to identify sampling gaps and to target particular species for future data collection. Data compilation is therefore so essential to comparative research that it justifies having its own chapter.

When building a comparative database, many issues need to be considered, including the following. How will you record the data in a database so that you or other researchers can easily validate the data, for example by locating the original reference or resource? Can data from a field or laboratory study be coded in a way that facilitates the extraction of intraspecific variation? How will you share the data with others after publication of the results, and how much of the data should be shared? Which species will be included in the database? Is it possible to build the database so that it can flexibly grow as new data and research questions emerge?

This chapter provides a brief overview of databases in the context of comparative research (see also Jones and Teeling 2009). A major point is that as the comparative database grows in size and complexity, a *relational database* structure provides a more efficient and flexible way to organize the data, compared to "flat file" representation, such as spreadsheets. A relational database is usually composed of two or more tables in which particular fields in one table are linked to particular fields in other tables. A relational database provides an effective way to maintain data quality, transparency, and completeness; to store the data efficiently; and to enable the database to grow as the research project expands beyond its original goals. In addition, many relational database programs provide an easy way to share the database with collaborators over the Internet. Many familiar software programs also use relational databases, such as those that run geographic information systems (GIS) or statistics packages. Thus, a small investment in learning even basic database concepts can offer tremendous payoffs over the longer term.

The chapter also considers a related question: how can we combine existing comparative data and phylogenetic information to guide future data collection? In other words, which species are most valuable to study and add to comparative data compilations? I briefly review efforts to use phylogenies to identify the species that offer the most powerful tests of adaptive hypotheses, including a recent approach known as "phylogenetic targeting" (Arnold 2008; Arnold and Nunn 2010). Phylogenetic targeting is important for deciding which species to include in a comparative database, for example when data compilation from the literature requires significant amounts of research effort. More importantly, this approach can be used to identify species to study in the field, laboratory or museum, that is, when the researcher needs to build a comparative database through original data collection on a range of different species.

Using Relational Databases in Comparative Research

Until you have firsthand experience with building and using a database, the term "relational database" is abstract and perhaps even intimidating. In this context, it is surprising how easy relational databases can be to design and use. A brief introduction to some of the major *relational database management systems* (RDBMSs) is provided in AnthroTree 13.2, which focuses on freely available systems that run on multiple platforms and allow for secure sharing and compilation of data by collaborators over the Internet.

At the outset, I would like to emphasize two points. First, this chapter is only meant as an entry portal for learning about relational databases, in essence a "teaspoon of sugar" to help ease the reader into sometimes arcane terminology involving foreign keys, queries and graphical user interfaces (more on these later). Second, it is important to think critically about the database structure that is right for the task at hand. Indeed, it is essential at the earliest stages of data collection to make a *data model* or *schema* that describes the variables, their units, and the relationships among variables in different tables. Knowing some of the general characteristics and advantages of relational databases will help users decide when to shift from spreadsheets, which are typical of so many comparative databases, to a relational database structure.

As already noted, a relational database usually involves two or more tables that contain multiple fields (technically, however, a single table can be relational when fields are linked within that table, or when the underlying management system itself is relational). Typically we think of the tables as having rows and columns. The different columns represent different fields, while the rows represent different species, studies or cultural groups. Figure 13.1 shows a simple example involving diet in several species of lemurs, where the units in rows are different findings for particular species. The "data" table at the top provides information on the species studied, its group size, body mass, and the population of animals that was sampled ("location"). The "data" table also gives a citation to the reference that provided the information (this table is for illustrative purposes and obviously is far from complete). With data on the species and location (in the "locality" table), data related to intraspecific variation can be readily assembled (e.g., variation among locations for the same species).

Tables can be linked together. Thus, the "citation" field of the "data" table is joined to the "citation" field of the "references" table at the bottom left of the figure, which provides more details on the reference. Notice that because one reference provided data on three species, fewer records are needed in the "references" table, which makes for more efficient data entry and storage (i.e., compared to a "flat file" that would require entry of the full reference on multiple rows). This example database has a third table that provides information on the sampling location, which again has fewer records than the "data" table because there are fewer distinct elements in this field in the "data" table.

All three tables each contain one *primary key*, and two of the tables have *foreign keys* ("references" and "locality"). The primary key provides a unique identifier for each record in a table. Typically, the primary key is a single

Table: *Data*

ID	Citation	Species	Group Size	Body Mass	Location
1	Tan 1999	*Hapalemur simus*	7	2450	Ranomafana
2	Tan 1999	*Hapalemur aureus*	3.5	1548	Ranomafana
3	Tan 1999	*Hapalemur griseus*	5.5	935	Ranomafana
4	Mutschler et al. 2000	*Hapalemur griseus alaotrensis*	3	1239	Lac Alaotra

Join   Join

Table: *Locality*

foreign key

ID	Location	Lat.	Long.
1	Ranomafana	-21.26	47.33
2	Lac Alaotra	-17.5	48.5

Table: *References*

foreign key

ID	Citation	Reference
1	Tan 1999	Tan, C.L. 1999. Group composition, home range size, and diet of three symatric baboo lemur species (genus *Hapalemur*) in Ranomafana National Park, Madagascar. *International Journal of Primatology* 20:547-566.
2	Mutschler et al. 2000	Mutschler, T., Nievergelt, C.M. & Feistner, A. T. C. 2000. Social organization of the Alaotran gentle lemur (*Hapalemur griseus alatrensis*). *American Journal of Primatology* 50:9-24.

Figure 13.1. Example of a relational database. This simplified example shows data for group size and body mass for three species of bamboo lemurs (genus *Hapalemur*). Three tables are shown, one containing the data ("Data"), another containing the detailed reference information ("References"), and another giving details on geographic location of the sample ("Locality"). Notice that one of the references (Tan 1999) provides data for three different species of lemurs, and tables can have primary keys (ID) and foreign keys (e.g., "Citation" for the "References" table). The foreign key limits entry into the "Data" table to only those references found in the "References" table. In an actual database, further tables might provide details on taxonomy (see figure 13.2).

field (column), but it can comprise multiple fields if this is needed to ensure uniqueness. The foreign key is a referential constraint between two tables; it identifies a column (or a set of columns) in one table that refers to a column (or set of columns) in another table. A foreign key is important, because it ensures that values of data entered in one table are also found in another linked table. A foreign key "constraint" provides the scaffolding to use all of the data that are entered, that is, to link among the different tables and ensure that all data are consistent and linked correctly among tables (*relational integrity*). In figure 13.1, for example, all "citation" fields of the "data" table are required to be present in the "citation" field of the "references" table. To ensure relational integrity, the user cannot delete one of the citations from the "references" table if that citation was also found in the "data" table, which guarantees that we will not lose details on the reference at some later time. Importantly, however, this depends on specific settings for a join between fields, and a variety of flexible options are possible for ensuring relational integrity.

Fields in the tables also can have other kinds of constraints, such as predefined sets of words to choose from for entering data (i.e., *enumerated text*). Constraints also can limit the types of data that can be entered in a field, for example, to require that data are integers rather than "floating point" (decimal) values or to limit the number of characters in a text field. These limits can help to minimize the overall size of the database in terms of memory allocation, thus saving hard drive space. The constraints also prevent users from inadvertently entering an incorrect data type, which helps to ensure the overall quality and usability of the database. From the perspective of comparative data, constraints keep the database consistent and complete in all the key fields—a goal that often fails when users attempt to manually link up information across flat files, such as Excel spreadsheets.

Data is extracted from a database using *queries*. A key advantage of most RDBMSs is that they make use of structured query language (SQL, pronounced "S-Q-L" or "sequel"), and its variants (see AnthroTree 13.3; Date and Darwen 1996). SQL is a programming language that was designed to insert and extract data from relational databases. SQL can also be used to manage databases, for example to add new users, change the access that different users have to the database, update fields, or add new tables or constraints among tables. Basic SQL can be learned intuitively by using a variety of graphical user interfaces (GUIs, pronounced "gooey") that run queries on databases by generating SQL scripts, and by perusing a number of websites and other resources that provide introductions to SQL (see AnthroTree 13.3). The GUIs typically have graphical drag-and-drop functions, but often they also generate actual SQL commands that the user can view and edit. While GUIs are often sufficient for most basic data management and extraction purposes, greater familiarity with SQL code itself has tremendous benefits for dealing with larger databases, to run update queries, and for more complex data extraction needs.

What about entering data into a relational database? This critical task may seem challenging and abstract, especially in the context of foreign key and other constraints that are implemented to maintain data integrity. Again, however, there are many options for entering data in a user-friendly and efficient way subject to data constraints, including various GUIs. Given that most databases can be shared easily over the web, another option is to develop a password-protected, web-based application so that users can insert, edit, or obtain data on a dedicated web page from anywhere in the world. Although many options are available, a scripting language known as PHP is particularly helpful in this regard. It provides a means to produce dynamic web pages and interfaces with MySQL (typically pronounced "my-S-Q-L") and other RDBMSs very effectively. In one widely used application, for example, Wikipedia runs the software MediaWiki, which is written in PHP and uses MySQL to store data. Relational databases are everywhere on the web!

Building a fully referenced database. One of the most important characteristics of a comparative database is that it should provide a reference to the data source for every data point. We might want to track down, for example, why *Cebus capucinus* was coded as a frugivore in a database on primate behavior, or, for a database on cultural variability, why the Macedonians are listed as having marriage practices involving dowry. We might also want to know whether data on lemur body masses were from captive animals, or from a population of free-ranging animals that are not provisioned with additional food. Were the data obtained from a primary research publication, a review paper, or a personal communication from an expert on that species or human population? All of these questions can be dealt with by having specific fields in the database for references, information on the reference, or factors relevant to using the data.

At least three groups of people might be interested in validating a particular data point: the original researcher and his or her collaborators as they use the data in analyses; a future user of the database, including someone who might want to expand the database with new data; or a reader of the published research who is simply curious about why a particular taxon or group was coded as it was. If each datum has a citation to a complete reference, it becomes possible to locate the original source for research purposes and to easily print out full reference information for each species (e.g., for publications and associated online material).

Often it is relatively straightforward to maintain a linkage between the original references and data, but a few rules of thumb are worth noting. First, it is often useful to enter the full reference information in a separate table in the database that is linked via a foreign key (e.g., as shown in figure 13.1). With such a database design, it is easy to record the reference when a study provides data on multiple species—that is, multiple "rows" in the database—because only a unique citation, and not the full reference information, is needed. Second, it is important to have one or more "notes" fields in the database to explain how and why a particular value was assigned to a species, society, population, or any other field. When possible, however, it is best to use enumerated text to describe specific aspects of the study that are of interest, such as whether the animals were captive or not, and reserve the actual notes field for general notes. I also find it useful to print the paper and write directly on it, with notes indicating why a particular value was used (this also can be done by adding "comments" to the electronic version of the paper).

Incorporating taxonomic ambiguity and mismatches. One of the biggest challenges in comparative biology is to match up the taxonomy used in the phylogeny to various taxonomies that different researchers use when studying species (Isaac et al. 2004; Jones and Teeling 2009). In my career as a comparative biologist studying primates, I have spent literally hundreds of hours poring over taxonomic synonyms to try to match up data from publications to the taxonomic structure used in the comparative analysis (which often follows the taxonomic structure of the primate phylogeny that I am using at the time). One new online resource—the website 10kTrees—provides a taxonomic translation tool to ease the transition among different taxonomic schemes (Arnold et al. 2010). Similar naming incongruities occur in the context of human cultural or morphological variation. More generally, "taxonomic ambiguity" can occur when different researchers organize their sampling points in different ways. In biological systems, the ambiguity often stems from taxonomic revisions and differences in deciding what constitutes separate species. In cross-cultural studies, ambiguity arises when data for different variables in a comparative study are collected at different organizational levels, for example, by language group for one variable (e.g., marriage system) and by province or nation for another variable (e.g., malaria prevalence).

Comparative biologists and anthropologists therefore need a way to deal with taxonomic ambiguity in a consistent and transparent way. In my research on primate parasites, we had to deal with taxonomic ambiguity for two groups of organisms—the primate hosts and their parasites—and we developed a set of translation procedures that has proved useful for many other databases that I have developed since then. More specifically, we made use of what we called *translation tables*. A translation table is simply another table in the relational database that provides an explicit linkage between the name of a taxon given by the authors in the original source material (e.g., a journal article), and the "translation" of the taxon into the taxonomy that we are using in the comparative analysis. It works as follows: First, we enter the species name in the main data table exactly as the authors spelled it. If the authors used a scientific name, such as *Macaca fascicularis*, we enter that precisely as it was spelled (even if misspelled). If the authors used only the common name, such as "crab-eating macaque," we again enter it in the main data table exactly as the authors gave it in the original paper. Second, we create a translation table that converts the authors' names for the different species into one or more taxonomic schemes (e.g., Corbet and Hill 1991; Groves 2001b). The original data are then joined to the translation table via the reported species name as a foreign key, making it easy to translate from the original papers to any taxonomy that is desired (figure 13.2).

This procedure is easy to implement and has several advantages over the more typical approach of entering the "corrected" name directly into the database (i.e., entering what you think the authors should have called the species). Importantly, it provides greater transparency between the published literature and where we place the species on the phylogeny used in the analysis; this transparency is achieved by simply providing the translation table over the web or as supplementary material in a publication. In addition, it provides a way to easily update the taxonomy should new taxonomic schemes become available; one simply edits the translation table to reflect how the original name should translate to the new taxonomy. A translation table also provides a way to quickly deal with the same taxonomic question if it comes up again later in data collection, rather than having to remember or reconstruct what you did earlier. Last, this approach produces a core database that matches exactly what is in the published sources used to construct the database (i.e., the "reporteddata" table figure 13.2), with all interpretations by the comparative biologist occurring in other tables ("translation" table in figure 13.2). This makes it easier to double-check data and track down any errors to either a data-entry issue, or to interpretations of the data by the comparative research team.

As a final point, it is worth mentioning an underappreciated issue involved with using translations to the most recent taxonomies. The general trend in biological classification is toward the splitting of species into new species, and this is certainly true for primates (Groves 2001b, 2004; Isaac et al. 2004). This

Table: *Reported Data*

PK	Species-Reported-Name	Male-Body-Mass	Reference
1	Macaca fascicularis	5.785	Smith 2008
2	crab-eating macacque	5.2	Jacobs 1942
3	Alouatta seniculus straminea	6.51	Robers 2007
4	anubis baboons	28.9	Martins 1976

Join

Table: *Translation*

foreign key

PK	Species-Reported-Name	Corrected-CorbetHill	Corrected-Groves
1	Macaca fascicularis	Macaca fascicularis	Macaca fascicularis
2	crab-eating macacque	Macaca fascicularis	Macaca fascicularis
3	Alouatta seniculus straminea	Alouatta seniculus	Alouatta seniculus
4	anubis baboons	Papio anubis	Papio hamadryas

Figure 13.2. A taxonomic translation table. By including a table that translates from "reported" names to "corrected" names, it becomes possible to easily update taxonomic information and to convert the raw data into multiple taxonomies. Here two "corrected" columns refer to two different taxonomic schemes (Corbet and Hill 1991; Groves 2001b).

means that an article published in, say, 1970 could represent any of several new species recognized in the present. An older taxonomy will tend to have fewer species and thus fewer data points. Hence, we might wish to use a more recent taxonomy to increase the sample size (and thus statistical power) of a comparative test. On the other hand, we can increase the certainty with which we assign a value to a species if we use an older taxonomy; in such cases it is more straightforward to "bin" the species from the present highly split taxonomy into the older, more lumped taxonomy. If information is available on the location of sampling or if subspecific information is provided, it may be possible to identify which "newer" species the data corresponds to in a more up-to-date taxonomy. Hence, it is important to include the full subspecies name and geographic location when developing a comparative database, as this will allow for greater flexibility when the taxonomy is revised in the future.

Database errors and double-checking. Many of the most exciting advances in phylogenetic comparative methods described in earlier chapters involve controlling for phylogenetic uncertainty (Huelsenbeck et al. 2000; Lutzoni et al.

2001; Pagel and Meade 2006a). This is obviously an important issue, but why have comparative biologists not given similar attention to dealing with uncertainty in the data used in comparative analyses? The values for ecological, behavioral, and morphological data are only estimates of true underlying values (Purvis and Webster 1999), and researchers often spend too little time ensuring that their data are of high quality (Freckleton 2009). In addition, errors can be introduced inadvertently during the collation of data from the literature by the comparative biologist, and we should keep in mind that the researchers who studied a given species may have introduced errors, either through their research procedures or through typographical errors during the publication process.

Comparative biologists usually assume that the data in the primary literature are correct, unless of course the values are clearly in error (e.g., mass labeled as kilograms that are actually in grams, or values that are inexplicably very far outside the range of other estimates for that taxon). A "notes" field can provide a way for the person entering data to explain adjustments to a published value, or to justify not including the data (although it is again worth remembering that text fields such as this should be used sparingly, as the information they contain usually cannot be easily analyzed). More generally, comparative biologists must often take the data as given and can only be certain that the data were entered correctly in their database; thus "double checking" of data is an important step in developing a comparative database. In this regard, it is important to include a field in the database that indicates whether the data have been double-checked, and then to only use data that fulfill this requirement. Even with a field that confirms the status of data, occasional spot-checks are important, especially when the database is available to multiple users who might inadvertently change a record. Whenever possible, someone other than the person who entered the data should do the double-checking.

It seems likely that enterprising scientists will develop a suite of more principled approaches to dealing with data coding errors in future work, using, for example, a resampling procedure to deal with data that are misclassified (i.e., assigned to the wrong species) or erroneously entered in the database. Even if such methods are developed, however, it will always be better to take the time to ensure that the comparative data are as close to the original records as possible. For this, only one solution exists, namely, to double-check the data, provide notes in the database on why species were assigned particular values, and include a field indicating that the data have been double checked relative to the original source.

Sharing data. Data can be shared in a variety of formats. If the databases are not too large, the author can publish them as part of an article that describes the results of tests that use the data. This can be achieved with a table in the main text of a scientific paper. For larger databases, the values can be placed in an appendix or as online supplementary files, or even in some journals specifically focused on publishing data (e.g., *Ecological Archives*).

How much of the data should be provided? It is important to include the data that are essential for replicating the study, or at least references to databases that are published (or available in some other way) so that the results of the study can be replicated. If space permits in the publication, authors should include references that were used to construct the database; if space is not available, then it might be worth providing the entire database over the Internet. In my research on primate disease ecology, for example, my colleagues and I put a database of primate parasites online (http://www.mammalpara sites.org; Nunn and Altizer 2005). We placed the key fields of the relational database on a separate server and built a "front end" so that users could flexibly search the database according to parasite taxonomy, host taxonomy, or geographic location. Thus, rather than having to design their own queries, users interact with a familiar web-based interface to extract the data in which they are interested.

A related issue involves sharing data among collaborators during the research itself, especially in the process of building and refining the comparative database. Collaborators often shuttle a file around by email and work on it offline. The obvious drawback of this approach is that two people might alter the files at the same time, making it difficult to keep track of the most up-to-date version of the database. It would help to share a database online, and there are several mechanisms for achieving this, including online collaboration tools such as Wikis or GoogleDocs. In my view, however, the most efficient way to build a database collaboratively is to create a relational database for a project that can be put on a server and securely accessed by other members of the research team. With such a set-up, it becomes possible to devise web-based tools for entering data with, for example, PHP, and to give limited access rights to some users, such as students who are entering data, so that data are not deleted or altered accidentally. In this way, all members of the research team can feel that they are playing a direct role in constructing the database without the risk of overwriting or losing data, and without waiting for another person to finish their tasks, as can commonly happen when researchers attempt to share a single file via email.

Summary. This introduction to relational databases is far too brief, but I hope it will ease readers into pursuing further reading and experimentation with RDBMSs (see also Jones and Teeling 2009). It is also important to keep in mind a point made earlier, namely, that not all databases need to start in relational format; often a "flat file," such as a spreadsheet, is sufficient for the initial stages of data entry. Staff can then be hired or experts consulted to build a relational database as the flat-file grows in size and complexity, and the guidelines given above can help to shepherd hired staff towards building a database that is most suited for the project at hand. If you find that you are typing in the same text multiple times or having difficulty keeping track of which version of a file has the latest data, it is probably worthwhile to develop a relational database that can be shared with your collaborators via the Internet.

Phylogenetic Targeting

The examples in this book follow the typical approach in comparative biology and cross-cultural research: they make use of data that are already available. My own comparative research is no exception. In our study of primate ranging patterns, Robert Barton and I did not go to the field to obtain new data on species that had not yet been studied (Nunn and Barton 2000); instead, we used what was available in the literature. Similarly, in my studies of primate disease ecology (Nunn et al. 2003, 2004, 2005) and the resulting databases (Nunn and Altizer 2005), my colleagues and I used the data that were available in the literature or in existing databases (e.g., the Natural History Museum's Host-Parasite Database, see Vitone et al. 2004).

Wouldn't it be nice to use the comparative approach to also guide future data collection? In the case of home range size, Robert Barton and I could have proposed, for example, ten species most in need of studying to further test hypotheses regarding the scaling of home range size in relation to ecological factors, while my other colleagues and I could have proposed the species or areas most in need of parasite sampling (which we did in later papers, Hopkins and Nunn 2007, 2010). More specifically, with a phylogeny and data for a broad range of organisms, one could in principle identify those species that offer the strongest test of a particular set of hypotheses. In the case of continuous traits, for example, one could select species that differ by some threshold in the predictor variable, the assumption being that a larger difference in a predictor variable should offer the best chances to detect a difference in the dependent variable (see Ackerly 2000).

Phylogeny can shed light on two issues that are important when selecting species for inclusion in a comparative study. First, with regard to the taxonomic scope of the study, it is important to include multiple evolutionary transitions in the traits of interest (Pagel 1994b). In the case of sexual swellings in primates, for example, it would be better to study Old World monkeys and apes, rather than studying only apes, because multiple origins of this trait probably have occurred in the broader set of taxa, while only one transition probably occurred among the apes (Nunn 1999a). Second, it is important to choose taxa that are not too different in other traits that might confound the analysis, whether these potential confounds are quantitative traits, such as body mass or life history characteristics, or more fundamental aspects of organismal biology, such as activity period or visual acuity. Because more closely related species will tend to share many other characters in common, phylogeny can serve as a useful surrogate for similarity, with comparisons among close relatives thus tending to control for potentially confounding factors (when possible, of course, it may be preferable to control for confounding variables statistically).

These rules of thumb are useful, but it would be desirable to have a more quantitative approach for deciding on the species to study. With this goal in mind, Christian Arnold and I developed a systematic approach to using phylogenetic and comparative data to guide the collection of new data. We call this method *phylogenetic targeting* (Arnold 2008; Arnold and Nunn 2010). Imagine that you have a hypothesis, a phylogeny for a group of species, and a dataset for those species that can serve as the independent variable when testing the hypothesis. Further, imagine that you lack data on the dependent variable that would be necessary to conduct the comparative test; in other words, you have data for a predictor variable, but not for the dependent variable, in a regression model. Last, assume that a greater amount of change in the predictor variable should offer greater power to detect any changes in the dependent variable.

With these data and assumptions, it is possible to identify the species that should be studied with respect to the dependent variable (Arnold 2008; Arnold and Nunn 2010). Essentially, given that it is possible to collect data for n species, which n species with data on the predictor variable(s) would offer the greatest power to test the hypothesis? The idea is to generate a series of paired comparisons in which closely related species are compared with respect to two or more variables and phylogenetic relatedness (Moller and Birkhead 1992; Mitani et al. 1996b). To maintain independence among the mul-

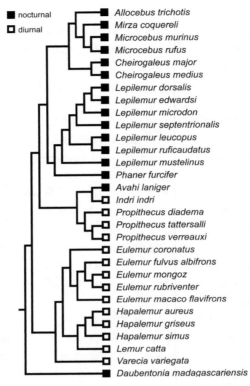

Figure 13.3. Paired comparisons. Once a branch is used in a comparison, it cannot be reused for any further comparisons. The tree comes from an early version of the 10kTrees website (Arnold et al. 2010).

tiple comparisons, each branch on the tree can be used only one time (Burt 1989; Moller and Birkhead 1992; Maddison 2000). Thus, if we are making comparisons among nocturnal and diurnal lemurs, we might compare *Avahi* with *Indri*, but subsequently we could not compare *Avahi* with *Propithecus* or *Indri* with *Phaner* (figure 13.3). Our approach to phylogenetic targeting can, in principle, be extended to calculate contrasts at deeper nodes in the tree (e.g., using independent contrasts; see chapter 7) or with GLS approaches (see Arnold and Nunn 2010).

An example involving animal cognition helps to make the idea of phylogenetic targeting more concrete (E. MacLean, B. Hare, L. Matthews, C. Nunn and others, in prep.). Animal cognition has been an area of intense research interest over the past decade, with a variety of hypotheses proposed for the factors that influence social and other forms of cognition (Byrne and Whiten 1989; Shettleworth 1998). Measures of cognitive performance are difficult and

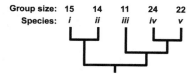

Figure 13.4. **Phylogenetic targeting.** Hypothetical data on group size are shown for five species. Based on the reasoning that a larger amount of evolutionary change in one trait should select for greater change in another, as yet unmeasured trait, *species iii* and *iv* should be studied (i.e., the contrast is largest in absolute magnitude). To maintain the independence of the paired comparisons, once branches are used they cannot be re-used; thus, species *v* would no longer be available because there is no way to connect that species to other species in the tree without crossing a used branch.

expensive to obtain, however, in large part because cognitive studies rely on having access to captive animals, and rarely are the same sets of cognitive tasks given to a wide range of species under comparable experimental conditions.

In the context of phylogenetic targeting, which species' cognitive performance should we study? Given the high costs of housing animals and conducting the experiments, we might have funds available for only a small number of comparisons, and we want to choose species that offer the strongest tests of the hypotheses of interest. With a set of hypotheses about cognitive performance, a phylogeny, and data relevant to the hypotheses, it is possible to pinpoint which species would best test a hypothesis. By "best," I am referring to having the highest power to detect an effect (i.e., the biggest evolutionary change in one of the traits of interest), and also able to control for alternative explanations (i.e., the smallest possible change in other traits that represent alternative hypotheses).

In the case of cognition, let us start with a single hypothesis, namely, that living in a larger group favors greater cognitive ability, for example in the context of social or Machiavellian intelligence (Byrne and Whiten 1989, 1998). One widely used predictor variable for this hypothesis is group size (Dunbar 1995, 1998; Deaner et al. 2000).

What about the dependent variable, which involves collecting new data on cognitive performance? For this, we need to decide on the domain of cognition to study, with possibilities that include social or spatial cognition (Shettleworth 1998); for the Machiavellian hypothesis, we would probably opt for tests of social cognition. Imagine data on group size for a hypothetical set of species in figure 13.4. If we were to compare only two species, the strongest test would involve comparison of *species iii* and *iv*; this pair differs to the greatest extent in group size and, thus, under the social intelligence hypothe-

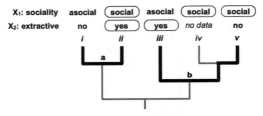

Figure 13.5. Selecting pairs of species to control for alternative hypotheses. Hypothetical data are shown for sociality and extractive foraging for five species, with *species iv* missing data on extractive foraging (and thus excluded from further consideration). For simplicity, data are scored as binary traits indicating social versus asocial categories, and extractive versus nonextractive foraging, and two possible pairwise comparisons are indicated. Circled character states indicate the species that are expected to have greater cognitive performance under the sociality and extractive foraging hypotheses, respectively. Thus, only contrast b provides a means to distinguish among the hypotheses: greater cognitive performance by *species v* would support the sociality hypothesis, while greater cognitive performance by *species iii* would support the extractive foraging hypothesis.

sis should also differ to the greatest extent in cognitive performance in the social domain (note, however, that we may also want to control for branch lengths, as was done when calculating independent contrasts in chapter 7; see Arnold and Nunn 2010).

To make a statistically valid comparative test of the hypothesis, we would need at least 6 such comparisons across a larger phylogeny, requiring the collection of data on 12 or more species. In fact, considering a larger phylogeny emphasizes the importance of having a systematic way of targeting species for comparison: in the case of primate phylogeny, with 233 species (Bininda-Emonds et al. 2007), there are 27,028 possible pair-wise comparisons to consider! Phylogenetic targeting provides a means to systematically identify a set of independent comparisons from among this full set of possibilities (Anthro-Tree 13.4).

Importantly, phylogenetic targeting also can be implemented to control for alternative hypotheses for variation in the dependent variable or even to test mutually exclusive hypotheses. Thus, returning to the example of cognitive evolution, many factors other than sociality could influence cognitive ability, including home range size, life history traits, metabolic rate, and diet (e.g., Deaner et al. 2000, 2003; Dunbar and Shultz 2007a). For example, imagine a second hypothesis involving extractive foraging, with the prediction that extractive foragers require greater cognitive ability to facilitate complex processing skills, including the use of tools (Gibson 1986). With information on extractive foraging, we often can devise comparisons that control for this alternative explanation, as shown in figure 13.5. In this case, I coded species as social versus nonsocial and extractive versus nonextractive foragers (indicated by a "yes" or "no," respectively). The values for sociality are not meant to match those for group size in figure 13.5, and I assumed that data on extractive foraging are available for only four of the five species (hence *species iv* would be excluded from the analysis).

Two paired contrasts are indicated "a" and "b," and circled values indicate where greater cognitive performance would support different hypotheses. For example, if we compared *species iii* and *v* and found that *species v* performed better on the cognitive task, this would indicate support for the sociality hypothesis, but if *species iii* performed better, it would support the extractive foraging hypothesis. The point is that the hypotheses can be distinguished with contrast "b," but they cannot be distinguished with contrast "a"; contrast "b," therefore, is given a higher score by the algorithm used in phylogenetic targeting (see Mitani et al. 1996b for a similar approach to dealing with competing hypotheses in a pair-wise comparison framework).

These figures are meant to illustrate a more general process of scoring comparisons according to various criteria involving competing hypotheses, confounding variables, and availability of data already collected for the dependent variable. Implementation of the method typically involves much larger numbers of species than in figures 13.4 and 13.5, and for these more complicated scenarios with many species, a "maximal pairing" algorithm can be used to select a particular number of comparisons that maximize the power to test a particular hypothesis (Arnold 2008; Arnold and Stadler 2010).

As noted above, these general principles also are applicable to methods other than pair-wise comparisons, such as phylogenetically independent contrasts or GLS. Nonetheless, in the context of difficult to measure dependent variables, such as cognitive performance on a particular task, the paired comparisons approach has many advantages. For example, the method of paired comparisons does not require reconstruction of ancestral states, it is less vulnerable to phylogenetic errors, and it works equally well with discrete or continuously varying traits (e.g., Read and Nee 1995; Maddison 2000). By comparing close relatives, it becomes possible to tailor the measurement variable to the species of interest. Returning to the example of cognitive tasks, for example, we might need to use a different cognitive experiment for lemurs than for monkeys; the lemur task might be based more on olfactory cues or easily manipulated objects, while the monkey task might make use of visual cues or behaviors that require greater manual dexterity. Thus, it becomes possible to assess the significance of a general pattern by comparing lemurs with lemurs and monkeys with monkeys using the same tasks within these groups. Last, paired comparison does not rely on explicit evolutionary models.

Simulations revealed that phylogenetic targeting offers significantly higher power to detect correlated evolution than simply plucking data at random from the species in the clade of interest (Arnold and Nunn 2010). Importantly, the simulations revealed that even when the number of comparisons is held constant, the power of tests increases by using phylogenetic targeting across a larger clade of organisms. Thus, it is best to apply the method across as many species as possible.

Summary and Synthesis

The informatics revolution has provided a number of tools that make comparative research easier to conduct. Importantly, these methods not only save time but also increase the quality of comparative research by making it easier to share larger amounts of data, retaining explicit links to citations for each datum in the database, and providing opportunities to flexibly expand databases as new questions arise. In addition, many of the RDBMSs work flexibly with other programs, including programs for GIS and statistical packages.

This chapter also covered a new set of methods that can be used to expand databases by systematically identifying species to study. One of these methods, phylogenetic targeting, provides a flexible and systematic approach to target future data collection, and is already being used to study primate cognition. Importantly, this method and others like it can be applied in the funding process and research planning stage to identify the species most valuable for future study, and it has wide applicability in the context of collecting data on wild species before they go extinct.

14 Conclusions and Future Directions

Over fifteen years ago, Paul Harvey and Mark Pagel published their influential *The Comparative Method in Evolutionary Biology*. Their book was foundational in many ways, and it is largely responsible for converting many people—including me—to using phylogeny-based comparative methods. In their concluding chapter, the authors noted presciently that, "New models, based on undoubtedly wicked mathematics, will gradually emerge" (1991, 203). Even a cursory glance at the comparative methods literature reveals that their prediction has become reality; the methodology is challenging to understand and often difficult to implement.

In this book, I focused on providing a basic introduction to several key sets of comparative methods from the standpoint of a user of the methods, rather than giving a comprehensive review of all the methods and their mathematical underpinnings. I had to make many difficult decisions when considering the methods and examples to include in this book and in deciding how much detail to provide. It is important to emphasize that I focused only on a subset of methods and a handful of examples, many of which were appealing to me because they fell close to my research interests. In addition to helping others understand the methods and apply them, I hope this book provides an entry point for those interested in learning more about the statistical nuts and bolts of phylogeny-based comparative methods. The references contained herein and on AnthroTree provide a road map to learn more about the methods and their application.

Within evolutionary anthropology, we have reached a point where we can synthesize the vast amounts of data that have accumulated into a more coherent picture of human diversity and primate evolution. Detailed data on the behavior, morphology, and ecology of different primate species have filled our libraries, electronic databases, and file drawers; similar advances are occurring (or have occurred) for linguistic and cultural data. Genetic data have accumulated to the point where it is possible to compile a dataset of aligned sequences for over half of the extant primates species, and from that dataset it

317

is possible to infer phylogeny for the majority of primate species (see Anthro-Tree 2.7; Arnold et al. 2010). Although broad-scale patterns have long been of interest to linguists, cross-cultural anthropologists, and biological anthropologists (e.g., Swadesh 1952; Crook and Gartlan 1966; Murdock and White 1969; Clutton-Brock and Harvey 1977; Cavalli-Sforza et al. 1994), never before have we had so much data at our fingertips or access to methods that can integrate these data into broad-scale quantitative tests of particular hypotheses.

The methods reviewed in this book are likely to play a pivotal role in synthesizing and understanding the growing tower of comparative data. Many of these methods can achieve important goals that Harvey and Pagel and other early practitioners could only outline as interesting for the future, such as systematically incorporating phylogenetic uncertainty into comparative studies and addressing violations of standard evolutionary assumptions (e.g., Brownian motion). Importantly, we should use methods and data that allow us to place measures of confidence on our estimates, for example when reconstructing ancestral states. Nowhere is this more essential than in studies of evolutionary singularities, where quantitative, statistically rigorous tests are sorely needed, as is caution against over-interpreting the output from such tests (see chapter 12). These methods are also already playing a role outside evolutionary anthropology and biology, for example, in studying historical manuscripts (Platnick and Cameron 1977; Barbrook et al. 1998; Howe et al. 2001). Phylogenetic comparative methods could be focused on other subjects, such as providing a more rigorous framework to study historical, economic, and environmental questions in our more recent past (Diamond and Robinson 2010).

In addition to this plea for greater synthesis of existing data, however, we must continue to collect new cultural, biological, and linguistic data. All of the comparative approaches described in this book require original data collected by field biologists, anthropologists, linguists, paleontologists, and laboratory researchers. All comparative research stands on the shoulders of many giants, and the way forward will involve a combination of synthetic comparison and original data collection—ideally conducted through collaborations between these two types of researchers using, for example, approaches based on phylogenetic targeting (chapter 13). Indeed, original data collection can be a great gig for researchers interested in comparative studies, because it helps to instill an appreciation of the strengths and weaknesses of existing data, and hands-on experience with the data can guide the development of new methods for collecting and analyzing data. Data collection also can inspire new ideas for comparative research, especially when acquiring data takes the researcher into the field to see the cultural traditions, environment, morphology, or behavior of the subjects of interest, whether the subjects are people, fossils, or nonhuman primates.

From a practical perspective, we are losing biological and cultural diversity at a faster rate than ever before. To maximize the future potential of comparative research, we should document as much biological and cultural diversity as possible before it disappears; indeed, research activity aimed at underrepresented cultures, languages, or organisms can even reduce the loss of diversity, for example, when a long-term research presence reduces the rate of hunting in biological systems (Koendgen et al. 2008). This recommendation does not rule out conducting broad-scale analyses with the data we have on hand. As described in previous chapters, comparative approaches can play a role in understanding extinction processes (chapter 11) or in deciding which species to study (chapter 13), again suggesting that we will benefit from integrating comparative and field research to a greater extent.

Where is future effort needed in terms of methodology, data collection, and putting the methods into practice? In linguistics research, the goals of the lexicostatisticians (Swadesh 1952; Kruskal et al. 1971) are being realized with the help of quantitative methods from evolutionary biology that allow for different rates of evolution across the lexicon (Pagel 2000b; Gray and Atkinson 2003; Pagel et al. 2007). Similarly, new methods are providing a means to assess evolutionary history in the context of linguistic contact, such as borrowing (Bryant et al. 2005; McMahon and McMahon 2005; Nakhleh et al. 2005). We need cognate classifications for more languages, including for more American Indian and Asian populations, so that these methods can be used to uncover human population origins and movements at a global scale and to assess whether patterns of change found in well-studied linguistic groups, such as Indo-European, represent general patterns.

In terms of cross-cultural research, recent computer simulation studies have begun to shed light on the impact that horizontal transmission has on tests of correlated evolution, measures of phylogenetic signal, and reconstruction of ancestral states (Nunn et al. 2006, 2010; Currie et al. 2010). Although results from these analyses are not always perfectly congruent in their recommendations, they help to triangulate on the factors that can lead cross-cultural research astray, and they may be used to identify new methods to better detect and accommodate horizontal transmission in comparative tests. We should apply simulation approaches to other methods that are starting to be used in cross-cultural research, including methods based on host-parasite congruence (Riede 2009; Tehrani et al. 2010), split decomposition (Bryant et al. 2005), and Bayesian phylogenetic methods to detect incongruence in cultural trait histories (Matthews et al. forthcoming). While simulations are sure to play a major role in cross-cultural research (Nunn et al. 2010), just as they did in biological research (Martins and Garland 1991; Diaz-Uriarte and Garland 1996; Harvey and Rambaut 2000), we clearly also need better data on cross-cultural variation, and we need an easy way to link these data to geographic, linguistic, and genetic data to control for spatiotemporal non-independence of cultural data. Basic questions about the transmission process at the group level—such as the degree of vertical versus horizontal transmission—still have not received adequate answers, in large part due to the challenges of disentangling these processes from the comparative patterns (Borgerhoff Mulder et al. 2006). An important direction for the future is to investigate whether and why suites of traits have similar transmission histories (e.g., Boyd et al. 1997; Jordan and Shennan 2009; Matthews et al. forthcoming).

My own research has focused mainly on comparative studies of nonhuman primates, and I see a number of methodological directions to expand primate comparative studies in the future. First, as described in previous chapters, we need to apply methods that take into account uncertainty in both the evolutionary relationships among the taxa and the branch lengths that connect the nodes on the tree. One such phylogeny has recently been provided (see AnthroTree 2.7; Arnold et al. 2010) and applied to study primate diversification (Matthews et al. 2010). Second, methods are available for incorporating intraspecific variation (Ives et al. 2007; Felsenstein 2008), but they have yet to be applied in many comparative studies of primates (e.g., Lindenfors et al. 2010). User-friendly programs are needed to implement these methods, for example, in the context of estimating allometric slopes, and comparative databases are needed to provide common metrics of intraspecific variation. Last, it is important to revisit many classic comparative studies of primates that have yet to be investigated using recent advances in phylogenetic tools, including studies of dentition in relation to diet (Kay and Simons 1980; Kay 1984), primate range defense (Mitani and Rodman 1979), and the intermembral index (Napier and Walker 1967; Napier 1970; Martin 1990).

Many (if not all) of the methods discussed here also apply to biological research outside of human populations and primate species. One of my goals in writing this book was to make these methods more accessible to biologists by focusing on a system and questions that should be familiar to most readers, since many biologists share a basic interest and familiarity with our evolutionary origins, human cultural variation, and the biology of our closest evolutionary relatives, the nonhuman primates. I also wish to share with biologists many of the exciting results that are emerging from the application of phylogenetic methods to anthropological and linguistic data. A valuable outcome from this book would be greater collaboration between evolutionary biologists on the one side and linguists, anthropologists and primatologists on the other. Moreover, as new methods develop within anthropology to study horizontal transmission (Dow 2007), these methods may prove useful to biologists in the context of horizontal genetic transmission or ecological underpinnings of biological variation (e.g., Freckleton and Jetz 2008).

Last, I am very excited by the prospect of using comparative data and phylogeny to guide future data collection, both in nonhuman primates and in human cultural assemblages (including languages). Which species, cultural entities, and languages are missing from our comparative databases, and based on their known characteristics, where should we place funding priorities for future research? The phylogenetic targeting approach discussed in chapter 13 provides a new way to make use of comparative data and phylogeny (Arnold 2008; Arnold and Nunn 2010). Importantly, phylogenetic targeting can be used to systematically prioritize those species that should be studied in the future, and thus enables comparative databases to grow in the most productive directions before cultural and biological diversity is lost.

Writing this book has had a major impact on my comparative research. It led me to sharpen my phylogenetic toolkit in ways that I never expected, such as controlling for phylogenetic uncertainty and making greater use of flexible GLS and evolutionary modeling approaches. I also became even more excited about the future of comparative perspectives across evolutionary anthropology. While the "wicked mathematics" of comparative biology is likely to continue apace (Harvey and Pagel 1991), we should not shy away from new and improved methods that are undoubtedly on the horizon. Similarly, newer, open-source software, such as R (R Development Core Team, 2009), has a steep learning curve, but it offers incredible opportunity for the growth and application of phylogeny-based comparative methods (Freckleton 2009). I look forward to seeing the future shape of phylogenetic comparative methods and their impact on our understanding of what makes us human.

References

Aberle, D. 1961. Matrilineal descent in cross-cultural perspective. In *Matrilineal kinship*, edited by D. Schneider and K. Gough, 655–730. Berkeley: University of California Press.

Abouheif, E. 1998. Random trees and the comparative method: A cautionary tale. *Evolution* 52:1197–204.

———. 1999. A method for testing the assumption of phylogenetic independence in comparative data. *Evolutionary Ecology Research* 1:895–909.

Abouheif, E., and D. J. Fairbairn. 1997. A comparative analysis of allometry for sexual size dimorphism: Assessing rensch's rule. *American Naturalist* 149:540.

Acerbi, A., P. M. McNamara, and C. L. Nunn. 2008. To sleep or not to sleep: The ecology of sleep in artificial organisms. *BMC Ecology* 8:10.

Ackerly, D. D. 2000. Taxon sampling, correlated evolution, and independent contrasts. *Evolution* 54:1480–92.

Ackerly, D. D., and M. J. Donoghue. 1998. Leaf size, sapling allometry, and Corner's rules: Phylogeny and correlated evolution in maples (*Acer*). *American Naturalist* 152:767–98.

Ackermann, R. R., and J. M. Cheverud. 2004a. Morphological integration in primate evolution. In *Phenotypic integration: Studying the ecology and evolution of complex phenotypes*, edited by M. Pigliucci and K. Preston, 302–19. Oxford: Oxford University Press.

———. 2004b. Detecting genetic drift versus selection in human evolution. *Proceedings of the National Academy of Sciences of the United States of America* 101:17946–51.

Ackermann, R. R., and R. J. Smith. 2007. The macroevolution of our ancient lineage: What we know (or think we know) about early hominin diversity. *Evolutionary Biology* 34:72–85.

Adams, D. 2008. Phylogenetic meta-analysis. *Evolution* 62:567–72.

Agapow, P. M., O. R. P. Bininda-Emonds, K. A. Crandall, et al. 2004. The impact of species concept on biodiversity studies. *Quarterly Review of Biology* 79:161–79.

Agapow, P. M., and N. J. B. Isaac. 2002. MacroCAIC: Revealing correlates of species richness by comparative analysis. *Diversity and Distributions* 8:41–43.

Agapow, P. M., and A. Purvis. 2002. Power of eight tree shape statistics to detect nonrandom diversification: A comparison by simulation of two models of cladogenesis. *Systematic Biology* 51:866–72.

Aiello, L. C., and R. I. M. Dunbar. 1993. Neocortex size, group size, and the evolution of language. *Current Anthropology* 34:184–93.

Alberch, P., S. J. Gould, G. F. Oster, and D. B. Wake. 1979. Size and shape in ontogeny and phylogeny. *Paleobiology* 5:296–317.

Alcock, J. 1998. *Animal behavior.* Sunderland, MA: Sinauer Associates.

Ali, F., and R. Meier. 2008. Positive selection in ASPM is correlated with cerebral cortex evolution across primates but not with whole-brain size. *Molecular Biology and Evolution* 25:2247.

———. 2009. Primate home range and GRIN2A, a receptor gene involved in neuronal plasticity: Implications for the evolution of spatial memory. *Genes, Brain and Behavior* 8:435–41.

Alroy, J. 1998. Cope's rule and the dynamics of body mass evolution in North American fossil mammals. *Science* 280:731–34.

Altmann, J. 1980. *Baboon mothers and infants.* Cambridge, MA: Harvard University Press.

———. 1990. Primate males go where the females are. *Animal Behaviour* 39:193–95.

———. 2000. Models of outcome and process: predicting the number of males in primate groups. In Kappeler 2000, 236–47.

Ammerman, A. J., and L. L. Cavalli-Sforza. 1984. *The neolithic transition and the genetics of populations in Europe.* Princeton, NJ: Princeton University Press.

Andelman, S. J. 1986. Ecological and social determinants of cercopithecine mating patterns. In *Ecological aspects of social evolution: Birds and mammals,* edited by D. I. Rubenstein and R. W. Wrangham, 201–16. Princeton, NJ: Princeton University Press.

Anthony, D. 2008. *The horse, the wheel, and language: How bronze-age riders from the Eurasian steppes shaped the modern world.* Princeton, NJ: Princeton University Press.

Antonovics, J., and P. H. Van Tienderen. 1991. Ontoecogenophyloconstraints? The chaos of constraint terminology. *TREE* 6:166–67.

Arnold, C. 2008. Phylogenetic targeting: A systematic approach and computer program for targeting research effort in comparative evolutionary biology. Master's thesis, University of Leipzig.

Arnold, C., L. J. Matthews, and C. L. Nunn. 2010. The *10kTrees website*: A new online resource for primate phylogeny. *Evolutionary Anthropology* 19:114–18.

Arnold, C., and C. L. Nunn. 2010. Phylogenetic targeting of research effort in evolutionary biology. *American Naturalist* 176:601–12.

Arnold, C., and P.F. Stadler. 2010. Polynomial algorithms for the Maximal Pairing Problem: Efficient phylogenetic targeting on arbitrary trees. *Algorithms for Molecular Biology* 5:25.

Ashton, K. G. 2004. Comparing phylogenetic signal in intraspecific and interspecific body size datasets. *Journal of Evolutionary Biology* 17:1157–61.

Atkinson, Q. D., and R. D. Gray. 2005. Curious parallels and curious connections: Phylogenetic thinking in biology and historical linguistics. *Systematic Biology* 54:513–26.

Atkinson, Q. D., A. Meade, C. Venditti, S. J. Greenhill, and M. Pagel. 2008. Languages evolve in punctuational bursts. *Science* 319:588.

Atkinson, Q., G. Nicholls, D. Welch, and R. Gray. 2005. From words to dates: Water into wine, mathemagic or phylogenetic inference? *Transactions of the Philological Society* 103:193–219.

Baerends, G. 1958. Comparative methods and the concept of homology in the study of behaviour. *Archives Neerlandaises de Zoologie* 13:401–17.

Bandelt, H.-J., and A. W. M. Dress. 1992. Split decomposition: A new and useful approach to phylogenetic analysis of distance data. *Molecular Phylogenetics and Evolution* 1:242–52.

Barbrook, A. C., C. J. Howe, N. Blake, and P. Robinson. 1998. The phylogeny of *The Canterbury Tales. Nature* 394:839.

Barbujani, G., and R. R. Sokal. 1990. Zones of sharp genetic change in Europe are also linguistic boundaries. *Proceedings of the National Academy of Sciences of the United States of America* 87:1816–19.

Barelli, C., M. Heistermann, C. Boesch, and U. H. Reichard. 2008. Mating patterns and sexual swellings in pair-living and multimale groups of wild white-handed gibbons, *Hylobates lar. Animal Behaviour* 75:991–1001.

Barraclough, T. G., P. H. Harvey, and S. N. Nee. 1995. Sexual selection and taxonomic diversity in passerine birds. *Proceedings of the Royal Society B: Biological Sciences* 259:211–15.

Barton, R. A. 1996. Neocortex size and behavioural ecology in primates. *Proceedings of the Royal Society B: Biological Science* 263:173–77.

———. 1998. Visual specialization and brain evolution in primates. *Proceedings of the Royal Society B: Biological Sciences* 265:1933–37.

Barton, R. A., and R. I. M. Dunbar. 1997. Evolution of the social brain. In *Machiavellian Intelligence II: Extensions and Evalutations*, edited by A. Whiten and R. W. Byrne, 240–63. Cambridge: Cambridge University Press.

Barton, R. A., and P. H. Harvey. 2000. Mosaic evolution of brain structure in mammals. *Nature* 405:1055–58.

Basolo, A. L. 1990. Female preference predates the evolution of the sword in swordtail fish. *Science* 250:808–10.

Bateman, R., I. Goddard, R. O'Grady, et al. 1990. Speaking of forked tongues: The feasibility of reconciling human phylogeny and the history of language. *Current Anthropology* 31:1–24.

Baum, D. A., and A. Larson. 1991. Adaptation reviewed: A phylogenetic methodology for studying character macroevolution. *Systematic Zoology* 40:1–18.

Baum, D. A., S. D. Smith, and S. S. S. Donovan. 2005. The tree-thinking challenge. *Science* 310:979–80.

Beals, K. L., L. S. Courtland, and S. M. Dodd. 1984. Brain size, cranial morphology, climate, and time machines. *Current Anthropology* 25:301–18.

Bellwood, P. 2001. Early agriculturalist population diasporas? Farming, languages, and genes. *Annual Review of Anthropology* 30:181–207.

Belshaw, R., and D. L. J. Quicke. 2002. Robustness of ancestral state estimates: Evolution of life history strategy in ichneumonoid parasitoids. *Systematic Biology* 51:450–77.

Bentley, G. R., T. Goldberg, and G. Jasienska. 1993. The fertility of agricultural and non-agricultural traditional societies. *Population Studies* 47:269–81.

Bergsland, K., and H. Vogt. 1962. On the validity of glottochronology. *Current Anthropology* 3:115–53.

Bergsten, J. 2005. A review of long-branch attraction. *Cladistics* 21:163–93.

Berrigan, D., E. L. Charnov, A. Purvis, and P. H. Harvey. 1993. Phylogenetic contrasts and the evolution of mammalian life histories. *Evolutionary Ecology* 7:270–78.

Billing, J., and P. W. Sherman. 1998. Antimicrobial functions of spices: Why some like it hot. *Quarterly Review of Biology* 73:3–49.

Bininda-Emonds, O. R. P. 2004a. *Phylogenetic supertrees: Combining information to reveal the tree of life*. Dordrecht: Kluwer Academic Publishers.

———. 2004b. The evolution of supertrees. *Trends in Ecology & Evolution* 19:315–22.

Bininda-Emonds, O. R. P., M. Cardillo, K. E. Jones, et al. 2007. The delayed rise of present-day mammals. *Nature* 446:507–12.

Bininda-Emonds, O. R. P., J. L. Gittleman, and M. A. Steel. 2002. The (super)tree of life: Procedures, problems, and prospects. *Annual Review of Ecology and Systematics* 33:265–89.

Birdsell, J. B. 1953. Some environmental and cultural factors influencing the structuring of Australian aboriginal populations. *American Naturalist* 87:171–207.

Birkhead, T. R., and A. P. Moller. 1992. *Sperm competition in birds*. London: Academic Press.

———. 1998. *Sperm competition and sexual selection*. San Diego, CA: Academic Press.

Blomberg, S. P., and T. Garland. 2002. Tempo and mode in evolution: Phylogenetic inertia, adaptation and comparative methods. *Journal of Evolutionary Biology* 15:899–910.

Blomberg, S. P., T. Garland, and A. R. Ives. 2003. Testing for phylogenetic signal in comparative data: Behavioral traits are more labile. *Evolution* 57:717–45.

Bloom, G., and P. W. Sherman. 2005. Dairying barriers affect the distribution of lactose malabsorption. *Evolution and Human Behavior* 26:301–12.

Bloomquist, E., and M. Suchard. 2010. Unifying vertical and nonvertical evolution: A stochastic ARG-based framework. *Systematic Biology* 59:27–41.

Blust, R. 2000. Why lexicostatistics doesn't work: the "universal constant" hypothesis and the Austronesian languages. In Renfrew, McMahon, and Trask 2000, 311–32. Cambridge: The McDonald Institute for Archaeological Research.

Bock, W. T. 1977. Adaption and the comparative method. *Major patterns in vertebrate evolution*, edited by M. Hecht, P. C. Goody, and B. M. Hecht, 57–82. New York: Plenum.

Bollback, J. P. 2006. SIMMAP: Stochastic character mapping of discrete traits on phylogenies. *BMC Bioinformatics* 7:88.

Borgerhoff Mulder, M. 1988. Kipsigis bridewealth payments. In *Human reproductive behaviour: A Darwinian perspective*, edited by L. Betzig, M. Mulder Borgerhoff, and P. Turke, 65–82. Cambridge, Cambridge University Press.

———. 2001. Using phylogenetically based comparative methods in anthropology: more questions than answers. *Evolutionary Anthroplogy* 10:99–111.

Borgerhoff Mulder, M., M. George-Cramer, J. Eshleman, and A. Ortolani. 2001. A study of East African kinship and marriage using phylogenetically controlled comparison. *American Anthropologist* 103:1059–82.

Borgerhoff Mulder, M., C. L. Nunn, and M. C. Towner. 2006. Macroevolutionary studies of cultural trait transmission. *Evolutionary Anthroplogy* 15:52–64.

Boyd, R., M. Borgerhoff Mulder, W. H. Durham, and P. J. Richerson. 1997. Are cultural phylogenies possible? In *Human by nature: Between biology and the social sciences*, edited by P. Weingart, S. D. Mitchell, P. J. Richerson, and S. Maasen, 355–86. Mahwah, NJ: Erlbaum.

Boyd, R., and P. J. Richerson. 1985. *Culture and the evolutionary process*. Chicago: University of Chicago Press.

Brashares, J. S. 2003. Ecological, behavioral, and life-history correlates of mammal extinctions in West Africa. *Conservation Biology* 17:733–43.

Bromham, L., M. Woolfit, M. S. Y. Lee, and A. Rambaut. 2002. Testing the relationship between morphological and molecular rates of change along phylogenies. *Evolution* 56:1921–30.

Brooks, D. R., and D. R. Glen. 1982. Pinworms and primates: a case-study in coevolution. *Proceedings of the Helminthological Society of Washington* 49:76–85.

Brooks, D. R., and D. A. McLennan. 1991. *Phylogeny, ecology, and behavior*. Chicago, Chicago University Press.

Brown, C. H. 2006. Prehistoric chronology of the common bean in the New World: The linguistic evidence. *American Anthropologist* 108:507–16.

Brown, C. H., E. W. Holman, S. Wichmann, and V. Velupillai. 2008. Automated classification of the world's languages: A description of the method and preliminary results. *Language Typology and Universals* 61:285–308.

Brown, J. H. 1995. *Macroecology*. Chicago: University of Chicago Press.

Brown, P., T. Sutikna, M. J. Morwood, et al. 2004. A new small-bodied hominin from the Late Pleistocene of Flores, Indonesia. *Nature* 431:1055–61.

Bryant, D., F. Filimon, and R. D. Gray. 2005. Untangling our past: Languages, trees, splits and networks. In Mace, Holden, and Shennan 2005, 69–85.

Bryant, D., and V. Moulton. 2004. Neighbor-net: An agglomerative method for the construction of phylogenetic networks. *Molecular Biology and Evolution* 21:255–65.

Buchanan, B., and M. Collard. 2007. Investigating the peopling of North America through cladistic analyses of early paleoindian projectile points. *Journal of Anthropological Archaeology* 26:366–93.

Burnham, K. P., and D. R. Anderson. 1998. *Model selection and inference: a practical information-theoretic approach*. New York: Springer-Verlag.

Burrell, A., C. Jolly, A. Tosi, and T. Disotell. 2009. Mitochondrial evidence for the hybrid origin of the kipunji, *Rungwecebus kipunji* (Primates: Papionini). *Molecular Phylogenetics and Evolution* 51:340–48.

Burt, A. 1989. Comparative methods using phylogenetically independent contrasts. In *Oxford surveys in evolutionary biology*, edited by P. H. Harvey, and L. Partridge, 33–53. Oxford: Oxford University Press.

Burt, D. B. 2001. Evolutionary stasis, constraint and other terminology describing evolutionary patterns. *Biological Journal of the Linnean Society* 72:509–17.

Burton, M. L., and D. R. White. 1984. Sexual division of labor in agriculture. *American Anthropologist* 86:568–83.

Butler, M. A., and A. A. King. 2004. Phylogenetic comparative analysis: A modeling approach for adaptive evolution. *American Naturalist* 164:683–95.

Butler, M. A., and J. B. Losos. 1997. Testing for unequal amounts of evolution in a continuous character on different branches of a phylogenetic tree using linear and squared-change parsimony: An example using Lesser Antillean Anolis lizards. *Evolution* 51:1623–35.

Butler, M. A., T. W. Schoener, and J. B. Losos. 2000. The relationship between sexual size dimorphism and habitat use in Greater Antillean Anolis lizards. *Evolution* 54:259–72.

Byrne, R., and A. Whiten. 1989. *Machiavellian intelligence: Social expertise and the evolution of intellect in monkeys, apes, and humans*. Oxford: Oxford University Press.

———. 1998. Machiavellian intelligence. In *Machiavellian Intelligence II: Extensions and Evalutations*, edited by A. Whiten, and R. W. Byrne, 1–13. Cambridge: Cambridge University Press.

Calder, W. A. 1984. *Size, function, and life history*. Mineola, NY: Dover.

Calhim, S., and T. R. Birkhead. 2006. Testes size in birds: quality versus quantity—assumptions, errors, and estimates. *Behavioral Ecology* 18:271–75.

Call, J., M. Carpenter, and M. Tomasello. 2005. Copying results and copying actions in the process of social learning: chimpanzees (*Pan troglodytes*) and human children (*Homo sapiens*). *Animal Cognition* 8:151–63.

Campbell, L. 2004. *Historical linguistics: An introduction*. Cambridge, MA: MIT Press.

Campbell, L., and T. Kaufman. 1985. Mayan linguistics: Where are we now? *Annual Review of Anthropology* 14:187–98.

Cann, R. L., M. Stoneking, and A. C. Wilson. 1987. Mitochondrial-DNA and human-evolution. *Nature* 325:31–36.

Cannon, C., and P. Manos. 2001. Combining and comparing morphometric shape descriptors with a molecular phylogeny: the case of fruit type evolution in Bornean Lithocarpus (*Fagaceae*). *Systematic Biology* 50:860–80.

Capellini, I., R. A. Barton, P. McNamara, B. Preston, and C. L. Nunn. 2008. Ecology and Evolution of Mammalian Sleep. *Evolution* 62:1764–76.

Capellini, I., C. Venditti, and R. A. Barton. 2010. Phylogeny and metabolic scaling in mammals. *Ecology* 9: 2783–93.

Cardillo, M., J. Gittleman, and A. Purvis. 2008. Global patterns in the phylogenetic structure of island mammal assemblages. *Proceedings of the Royal Society B: Biological Sciences* 275:1549.

Cardillo, M., G. M. Mace, J. L. Gittleman, and A. Purvis. 2006. Latent extinction risk and the future battlegrounds of mammal conservation. *Proceedings of the National Academy of Sciences of the United States of America* 103:4157–61.

Carpenter, J. M. 1989. Testing scenarios: Wasp social evolution. *Cladistics* 5:131–44.

Cartmill, M. 1974. Pads and claws in arboreal locomotion. In *Primate locomotion*, edited by F. A. Jenkins, 45–83. New York: Academic Press.

Carvalho, P., J. A. Felizola Diniz-Filho, and L. M. Bini. 2006. Factors influencing changes in trait correlations across species after using phylogenetic independent contrasts. *Evolutionary Ecology* 20:591–602.

Cavalli-Sforza, L. L. 2000. *Genes, peoples, and languages.* New York: North Point Press.

Cavalli-Sforza, L. L., and M. W. Feldman. 1981. *Cultural transmission and evolution.* Palo Alto, CA: Stanford University Press.

Cavalli-Sforza, L. L., P. Menozzi, and A. Piazza. 1994. *The history and geography of human genes.* Princeton, NJ: Princeton University Press.

Cavalli-Sforza, L. L., A. Piazza, P. Menozzi, and J. Mountain. 1988. Reconstruction of human evolution: Bringing together genetic, archaeological and linguistic data. *Proceedings of the National Academy of Sciences of the United States of America* 85:6002–6.

Cavender-Bares, J., K. Kozak, P. Fine, and S. Kembel. 2009. The merging of community ecology and phylogenetic biology. *Ecology Letters* 12:693–715.

Chamberlain, A. T., and B. A. Wood. 1987. Early hominid phylogeny. *Journal of Human Evolution* 16:119–33.

Chan, K. M. A., and B. R. Moore. 2002. Whole-tree methods for detecting differential diversification rates. *Systematic Biology* 51:855–65.

———. 2005. SymmeTREE: Whole-tree analysis of differential diversification rates. *Bioinformatics* 21:1709–10.

Chapman, C., and J. Rothman. 2009. Within-species differences in primate social structure: Evolution of plasticity and phylogenetic constraints. *Primates* 50:12–22.

Charleston, M. A. 1998. Jungles: A new solution to the host/parasite phylogeny reconciliation problem. *Mathematical Biosciences* 149:191–223.

Cheverud, J. M. 1982. Phenotypic, genetic, and environmental morphological integration in the cranium. *Evolution* 36:499–516.

Cheverud, J.M. 1995. Morphological integration in the saddle-back tamarin (*Saguinus fuscicollis*) cranium. *American Naturalist* 145:63–89.

Cheverud, J. M., M. M. Dow, and W. Leutenegger. 1985. The quantitative assessment of phylogenetic constraints in comparative analyses: sexual dimorphism in body weight among primates. *Evolution* 39:1335–51.

Clubb, R., and G. Mason. 2003. Captivity effects on wide-ranging carnivores. *Nature* 425: 473–74.

Clutton-Brock, T. H., and P. H. Harvey. 1976. Evolutionary rules and primate societies. In *Growing points in ethology*, edited by P. P. G. Bateson and R. A. Hinde, 195–237. Cambridge: Cambridge University Press.

———. 1977. Primate ecology and social organization. *Journal of Zoology* 183:1–39.

———. 1984. Comparative approaches to investigating adaptation. In *Behavioural ecology*, edited by J. R. Krebs, and N. B. Davies, 7–29. Oxford: Blackwell.

Clutton-Brock, T. H., P. H. Harvey, and B. Rudder. 1977. Sexual dimorphism, socionomic sex ratio and body weight in primates. *Nature* 269:797–800.

Cochrane, E. E. 2004. Explaining cultural diversity in ancient Fiji: The transmission of ceramic variability. PhD diss., University of Hawai'i, Honolulu.

Coddington, J. A. 1988. Cladistic tests of adaptationist hypotheses. *Cladistics* 4:3–22.

———. 1994. The roles of homology and convergence in studies of adaptation. In *Phylogenetics and ecology*, edited by P. Eggleton and R. I. Vane-Wright, 53–78. London: Academic Press.

Coe, C. L. 1993. Psychosocial factors and immunity in nonhuman-primates—A review. *Psychosomatic Medicine* 55:298–308.

Collard, I. F., and R. A. Foley. 2002. Latitudinal patterns and environmental determinants of recent human cultural diversity: Do humans follow biogeographical rules? *Evolutionary Ecology Research* 4:371–83.

Collard, M., S. J. Shennan, and J. J. Tehrani. 2006a. Branching versus blending in macroscale cultural evolution: A comparative study. In Lipo et al. 2006b, 53–63.

———. 2006b. Branching, blending, and the evolution of cultural similarities and differences among human populations. *Evolution and Human Behavior* 27:169–84.

Collen, B., A. Purvis, and J. L. Gittleman. 2004. Biological correlates of description date in carnivores and primates. *Global Ecology and Biogeography* 13:459–67.

Combes, C. 2001. *Parasitism: The ecology and evolution of intimate interactions.* Chicago: University of Chicago Press.

Conroy, G. C. 2003. The inverse relationship between species diversity and body mass: Do primates play by the "rules"? *Journal of Human Evolution* 45:43–55.

Cooper, N., and A. Purvis. 2010. Body size evolution in mammals: Complexity in tempo and mode. *American Naturalist* 175:727–38.

Cooper, N., J. Rodriguez, and A. Purvis. 2008. A common tendency for phylogenetic overdispersion in mammalian assemblages. *Proceedings of the Royal Society B: Biological Sciences* 275:2031–37.

Corbet, G. B., and J. E. Hill. 1991. *A world list of mammalian species.* Oxford: Oxford University Press.

Coward, F., S. Shennan, S. Colledge, J. Conolly, and M. Collard. 2008. The spread of neolithic plant economies from the near east to northwest Europe: A phylogenetic analysis. *Journal of Archaeological Science* 35:42–56.

Cowlishaw, G., and R. I. M. Dunbar. 2000. *Primate conservation biology.* Chicago: University of Chicago Press.

Cowlishaw, G., and R. Mace. 1996. Cross-cultural patterns of marriage and inheritance: A phylogenetic approach. *Ethology and Sociobiology* 17:87–97.

Cracraft, J. 1992. Species concepts and speciation analysis. In *The units of evolution: Essays on the nature of species*, edited by M. Ereshefsky, 93–120. Cambridge, MA: MIT Press.

Croft, W. 2008. Evolutionary linguistics. *Annual Review of Anthropology* 37:219–34.

Crook, J. H. 1972. Sexual selection, dimorphism, and social organization in the primates. In *Sexual selection and the descent of man*, edited by B. G. Campbell, 231–81. Chicago: Aldine.

Crook, J. H., and J. C. Gartlan. 1966. Evolution of primate societies. *Nature* 210:1200–3.

Crozier, R. H. 1997. Preserving the information content of species: Genetic diversity, phylogeny, and conservation worth. *Annual Review of Ecology and Systematics* 28:243–68.

Cunningham, C. W. 1999. Some limitations of ancestral character-state reconstruction when testing evolutionary hypotheses. *Systematic Biology* 48:665–74.

Cunningham, C. W., K. E. Omland, and T. H. Oakley. 1998. Reconstructing ancestral character states: A critical reappraisal. *Trends in Ecology & Evolution* 13:361–66.

Currie, T., and R. Mace. 2009. Political complexity predicts the spread of ethnolinguistic groups. *Proceedings of the National Academy of Sciences of the United States of America* 106:7339.

Currie, T. E., S. J. Greenhill, and R. Mace. 2010. Is horizontal transmission really a problem for phylogenetic comparative methods? A simulation study using continuous cultural traits. *Philosophical Transactions of the Royal Society B: Biological Sciences* 365: 3903–12.

Darwin, C. 1859. *The origin of species*. London: John Murray.

———. 1871. *The descent of man and selection in relation to sex*. London: John Murray.

Date, C. J., and H. Darwen. 1996. *A guide to the SQL standard*. Reading, MA: Addison Wesley.

Davies, C. R., J. M. Ayres, C. Dye, and L. M. Deane. 1991. Malaria infection rate of Amazonian primates increases with body weight and group size. *Functional Ecology* 5:655–62.

Davies, T. J., and A. B. Pedersen. 2008. Phylogeny and geography predict pathogen community similarity in wild primates and humans. *Proceedings of the Royal Society B: Biological Sciences* 275:1695–701.

de Queiroz, A., and P. H. Wimberger. 1993. The usefulness of behavior for phylogeny estimation: Levels of homoplasy in behavioral and morphological characters. *Evolution* 47:46–60.

de Quiroz, K. 1996. Including the characters of interest during tree reconstruction and the problems of circularity and bias in studies of character evolution. *American Naturalist* 148:700–8.

Deaner, R. O., R. A. Barton, and C. P. van Schaik. 2003. Primate brains and life histories: Renewing the connection. In *Primate Life Histories and Socioecology*, edited by P. M. Kappeler and M. E. Pereira, 233–65. Chicago: University of Chicago Press.

Deaner, R. O., K. Isler, J. Burkart, and C. van Schaik. 2007. Overall brain size, and not encephalization quotient, best predicts cognitive ability across non-human primates. *Brain Behavior and Evolution* 70:115–24.

Deaner, R. O., and C. L. Nunn. 1999. How quickly do brains catch up with bodies? A comparative method for detecting evolutionary lag. *Proceedings of the Royal Society B: Biological Sciences* 266:687–94.

Deaner, R. O., C. L. Nunn, and C. P. van Schaik. 2000. Comparative tests of primate cognition: different scaling methods produce different results. *Brain, Behavior, and Evolution* 55:44–52.

Dediu, D., and D. R. Ladd. 2007. Linguistic tone is related to the population frequency of the adaptive haplogroups of two brain size genes, ASPM and microcephalin. *Proceedings of the National Academy of Sciences of the United States of America* 104:10944–49.

Delson, E., N. Eldredge, and I. Tattersall. 1977. Reconstruction of hominid phylogeny: A testable framework based on cladistic analysis. *Journal of Human Evolution* 6:263–78.

Demes, B., and W. L. Jungers. 1993. Long bone cross-sectional dimensions, locomotor adaptations and body size in prosimian primates. *Journal of Human Evolution* 25:57–74.

Desdevises, Y., P. Legendre, L. Azouzi, and S. Morand. 2003. Quantifying phylogenetically structured environmental variation. *Evolution* 57:2647–52.

DeVore, I. 1963. A comparison of the ecology and behavior of monkeys and apes.In *Classification and human evolution*, edited by S. Washburn, 301–19. Chicago: Aldine Publishing.

DeVore, I., and S. Washburn. 1963. Baboon ecology and human evolution.In *African ecology and human evolution*, edited by F. Howell and F. Bourliere, 335–67. New York: Aldine.

Dial, K. P., and J. M. Marzluff. 1989. Nonrandom diversification within taxonomic assemblages. *Systematic Zoology* 38:26–37.

Diamond, J. 1988. Express train to Polynesia. *Nature* 336:307–8.

———. 1997. *Guns, germs, and steel.* New York, W. W. Norton & Co.

Diamond, J., and P. Bellwood. 2003. Farmers and their languages: The first expansions. *Science* 300:597–603.

Diamond, J., and J. A. Robinson. 2010. *Natural experiments of history.* Cambridge, MA: Belknap.

Diaz-Uriarte, R., and T. Garland. 1996. Testing hypotheses of correlated evolution using phylogenetically independent contrasts: Sensitivity to deviations from Brownian motion. *Systematic Biology* 45:27–47.

———. 1998. Effects of branch length errors on the performance of phylogenetically independent contrasts. *Systematic Biology* 47:654–72.

Dickemann, M. 1982. Polygyny and Inheritance of Wealth [comments on Hartung's article in the same issue]. *Current Anthropology* 23:8–9.

Di Fiore, A., and D. Rendall. 1994. Evolution of social organization: A reappraisal for primates by using phylogenetic methods. *Proceedings of the National Academy of Sciences of the United States of America* 91:9941–45.

Diniz-Filho, J. A. F., C. E. R. De Sant'ana, and L. M. Bini. 1998. An eigenvector method for estimating phylogenetic inertia. *Evolution* 52:1247–62.

Disotell, T. R. 1996. The phylogeny of old world monkeys. *Evolutionary Anthropology* 5:18–24.

———. 2008. Primate phylogenetics. In *Encyclopedia of life sciences.* Chichester, UK: Wiley.

Dixon, R. M. W. 1997. *The rise and fall of languages.* Cambridge: Cambridge University Press.

Dixson, A. F., and M. J. Anderson. 2004. Effects of sexual selection upon sperm morphology and sexual skin morphology in primates. *International Journal of Primatology* 25:1159–71.

Doughty, P. 1996. Statistical analysis of natural experiments in evolutionary biology: Comments on recent criticisms of the use of comparative methods to study adaptation. *American Naturalist* 148:943–56.

Dow, M. M. 1984. A biparametric approach to network autocorrelation. *Sociological Methods and Research* 13:201–17.

———. 1993. Saving the theory: On chi-square tests with cross-cultural survey data. *Cross-Cultural Research* 27:247–76.

———. 2007. Galton's problem as multiple network autocorrelation effects—Cultural trait transmission and ecological constraint. *Cross-Cultural Research* 41:336–63.

Dow, M. M., M. L. Burton, D. R. White, and K. P. Reitz. 1984. Galton's problem as network autocorrelation. *American Ethnologist* 11:754–70.

Dow, M. M., and E. A. Eff. 2008. Global, regional, and local network autocorrelation in the standard cross-cultural sample. *Cross-Cultural Research* 42:148–71.

Dumont, E. 1997. Cranial shape in fruit, nectar, and exudate feeders: Implications for interpreting the fossil record. *American Journal of Physical Anthropology* 102:187–202.

Dunbar, R. I. M. 1988. *Primate social systems.* Ithaca, NY: Cornell University Press.

———. 1992. Neocortex size as a constraint on group size in primates. *Journal of Human Evolution* 20:469–93.

———. 1995. Neocortex size and group size in primates: A test of the hypothesis. *Journal of Human Evolution* 28:287–96.

———. 1998. The social brain hypothesis. *Evolutionary Anthropology* 6:178–90.

————. 2003. The social brain: Mind, language, and society in evolutionary perspective. *Annual Review of Anthropology* 32:163–81.

Dunbar, R. I. M., and G. Cowlishaw. 1992. Mating success in male primates: dominance rank, sperm competition and alternative strategies. *Animal Behaviour* 44:1171–73.

Dunbar, R. I. M., and P. Dunbar. 1988. Maternal time budgets of gelada baboons. *Animal Behaviour* 36:970–80.

Dunbar, R. I. M., and S. Shultz. 2007a. Understanding primate brain evolution. *Philosophical Transactions of the Royal Society B: Biological Sciences* 362:649.

————. 2007b. Understanding primate brain evolution. *Philosophical Transactions of the Royal Society B: Biological Sciences* 362:649–58.

Dunn, M., R. Foley, S. Levinson, G. Reesink, and A. Terrill. 2007. Statistical reasoning in the evaluation of typological diversity in Island Melanesia. *Oceanic Linguistics* 46: 388–403.

Dunn, M., A. Terrill, G. Reesink, R. A. Foley, and S. C. Levinson. 2005. Structural phylogenetics and the reconstruction of ancient language history. *Science* 309:2072–75.

Durham, W. H. 1991. *Coevolution: Genes, culture and human diversity*. Palo Alto, CA: Stanford University Press.

Dyen, I., J. B. Kruskal, and P. Black. 1992. An Indoeuropean classification: A lexicostatistical experiment. *Transactions of the American Philosophical Society* 82:iii–132.

Edwards, S. V., and M. Kot. 1995. Comparative methods at the species level: geographic variation in morphology and group size in grey-crowned babblers (*Pomatostomus temporalis*). *Evolution* 49:1134–46.

Eff, E. A. 2004. Does Mr. Galton still have a problem? Autocorrelation in the standard cross-cultural sample. *World Cultures* 15:153–70.

Eldredge, N., and J. Cracraft. 1980. *Phylogenetic patterns and the evolutionary process*. New York: Columbia University Press.

Elgar, M. A., M. D. Pagel, and P. H. Harvey. 1988. Sleep in Mammals. *Animal Behaviour* 36:1407–19.

Ember, C. R., and M. Ember. 1998. Cross-cultural research.In *Handbook of methods in cultural anthropology*, edited by H. R. Bernard, 647–87. Walnut Creek, CA: AltaMira Press.

Ember, M., and C. R. Ember. 1971. The conditions favoring matrilocal versus patrilocal residence. *American Anthropologist* 73:571–94.

Emlen, S. T., and L. W. Oring. 1977. Ecology, sexual selection, and the evolution of mating systems. *Science* 197:215–23.

Estes, S., and S. Arnold. 2007. Resolving the paradox of stasis: Models with stabilizing selection explain evolutionary divergence on all timescales. *American Naturalist* 169:227–44.

Etienne, R., and M. Apol. 2008. Estimating speciation and extinction rates from diversity data and the fossil record. *Evolution* 63:244–55.

Faith, D. P. 1992. Conservation evaluation and phylogenetic diversity. *Biological Conservation* 61:1–10.

————. 1994. Phylogenetic pattern and the quantification of organismal biodiversity. *Philosophical Transactions of the Royal Society B: Biological Sciences* 345:45–58.

————. 2002. Quantifying biodiversity: A phylogenetic perspective. *Conservation Biology* 16:248–52.

Falsetti, A. B., W. L. Jungers, and T. M. Cole. 1993. Morphometrics of the callitrichid forelimb: A case study in size and shape. *International Journal of Primatology* 14:551–72.

Feldman, M. W., and K. N. Laland. 1996. Gene-culture coevolutionary theory. *Trends in Ecology & Evolution* 11:453–57.

Felsenstein, J. 1979. Alternative methods of phylogenetic inference and their interrelationship. *Systematic Zoology* 28:49–62.

———. 1985a. Confidence limits on phylogenies: An approach using the bootstrap. *Evolution* 39:783–91.

———. 1985b. Phylogenies and the comparative method. *American Naturalist* 125:1–15.

———. 1988. Phylogenies and quantitative characters. *Annual Review of Ecology and Systematics* 19:445–71.

———. 2002. Quantitative characters, phylogenies, and morphometrics. In *Morphology, Shape and Phylogeny*, edited by N. MacLeod and P.L. Forey, 27–44. London: Taylor and Francis.

———. 2004. *Inferring phylogenies*. Sunderland, MA: Sinauer Associates.

———. 2008. Comparative methods with sampling error and within-species variation: Contrasts revisited and revised. *American Naturalist* 171:713–25.

Finarelli, J. A., and J. J. Flynn. 2006. Ancestral state reconstruction of body size in the Caniformia (Carnivora, Mammalia): The effects of incorporating data from the fossil record. *Systematic Biology* 55:301–13.

Fincher, C. L., and R. Thornhill. 2008. A parasite-driven wedge: Infectious diseases may explain language and other biodiversity. *OIKOS* 117:1289–97.

Fisher, D. O., and I. P. F. Owens. 2004. The comparative method in conservation biology. *Trends in Ecology & Evolution* 19:391–98.

FitzJohn, R. 2010. Quantitative traits and diversification. *Systematic Biology* 59:619–33.

Fleagle, J. G. 1988. *Primate adaptation and evolution*. New York: Academic Press.

Fleagle, J. G., and K. E. Reed. 1996. Comparing primate communities: A multivariate approach. *Journal of Human Evolution* 30:489–510.

———. 1999. Phylogenetic and temporal perspectives on primate ecology. In *Primate communities*, edited by J. G. Fleagle, C. Janson, and K. E. Reed, 92–115. Cambridge: Cambridge University Press.

Foley, R., and M. M. Lahr. 2003. On stony ground: Lithic technology, human evolution, and the emergence of culture. *Evolutionary Anthropology* 12:109–22.

Foley, R., and P. Lee. 1989. Finite social space, evolutionary pathways, and reconstructing hominid behavior. *Science* 243:901–6.

———. 1996. Comparative approaches to hominid socioecology. In *The archaeology of human ancestry*, edited by J. Steele and S. Shennan, 43–66. London: Routledge.

Forster, P., and A. Toth. 2003. Toward a phylogenetic chronology of ancient Gaulish, Celtic, and Indo-European. *Proceedings of the National Academy of Sciences of the United States of America* 100:9079–84.

Fortson, B. 2004. *Indo-European language and culture: An introduction*. Malden, MA: Blackwell.

Fortunato, L., C. Holden, and R. Mace. 2006. From bridewealth to dowry? A Bayesian estimation of ancestral states of marriage transfers in Indo-European groups. *Human Nature* 17:355–76.

Fortunato, L., and R. Mace. 2009. Testing functional hypotheses about cross-cultural variation: A maximum-likelihood comparative analysis of Indo-European marriage practices. In Shennan 2009, 235–50.

Freckleton, R. 2002. On the misuse of residuals in ecology: Regression of residuals vs. multiple regression. *Journal of Animal Ecology* 71:542–45.

———. 2009. The seven deadly sins of comparative analysis. *Journal of Evolutionary Biology*. 22:1367–75.

Freckleton, R. P., and P. H. Harvey. 2006. Detecting non-Brownian trait evolution in adaptive radiations. *PLoS Biology* 4:2104–11.

Freckleton, R. P., P. H. Harvey, and M. Pagel. 2002. Phylogenetic analysis and comparative data: A test and review of evidence. *American Naturalist* 160:712–26.

Freckleton, R. P., and W. Jetz. 2008. Space versus phylogeny: Disentangling phylogenetic and spatial signals in comparative data. *Proceedings of the Royal Society B: Biological Sciences* 276:21–30.

Freckleton, R. P., M. Pagel, and P. H. Harvey. 2003. Comparative methods for adaptive radiations. In *Macroecology: Concepts and consequences*, edited by T. M. Blackburn and K. J. Gaston, 391–407. Oxford: Blackwell.

Freckleton, R. P., A. B. Phillimore, and M. Pagel. 2008. Relating traits to diversification: a simple test. *American Naturalist* 172:102–15.

Friedman, E. M., and D. A. Lawrence. 2002. Environmental stress mediates changes in neuroimmunological interactions. *Toxicological Sciences* 67:4–10.

Fritz, J., J. Hummel, E. Kienzle, C. Arnold, C. Nunn, and M. Clauss. 2009. Comparative chewing efficiency in mammalian herbivores. *OIKOS* 118:1623–32.

Gangestad, S. W., and D. M. Buss. 1993. Pathogen prevalence and human mate preferences. *Ethology and Sociobiology* 14:89–96.

Garamszegi, L. Z., and C. L. Nunn. 2011. Parasite-mediated evolution of non-synonymous substitution rate at the functional part of the MHC in primates. *Journal of Evolutionary Biology* 24:184–95.

Garber, P. A. 1994. Phylogenetic approach to the study of tamarin and marmoset social systems. *American Journal of Primatology* 34:199–219.

Garland, T. 1992. Rate tests for phenotypic evolution using phylogenetically independent contrasts. *American Naturalist* 140:509–19.

Garland, T. and S. C. Adolph. 1994. Why not to do two-species comparative studies: Limitations on inferring adaptation. *Physiological Zoology* 67:797–828.

Garland, T., A. F. Bennett, and E. L. Rezende. 2005. Phylogenetic approaches in comparative physiology. *Journal of Experimental Biology* 208:3015–35.

Garland, T., A. W. Dickerman, C. M. Janis, and J. A. Jones. 1993. Phylogenetic analysis of covariance by computer simulation. *Systematic Biology* 42:265–92.

Garland, T., P. H. Harvey, and A. R. Ives. 1992. Procedures for the analysis of comparative data using phylogenetically independent contrasts. *Systematic Biology* 4:18–32.

Garland, T., and A. R. Ives. 2000. Using the past to predict the present: Confidence intervals for regression equations in phylogenetic comparative methods. *American Naturalist* 155:346–64.

Garland, T., K. L. M. Martin, and R. Diaz-Uriarte. 1997. Reconstructing ancestral trait values using squared-change parsimony: Plasma osmolarity at the origin of amniotes. In *Amniote origins*, edited by S. S. Sumida and K. L. M. Martin, 425–501. San Diego, CA: Academic Press.

Garland, T., P. E. Midford, and A. R. Ives. 1999. An introduction to phylogenetically based statistical methods, with a new method for confidence intervals on ancestral values. *American Zoologist* 39:374–88.

Garland, T., and M. R. Rose. 2009. *Experimental evolution*. Berkeley: University of California Press.

Gartner, G. E. A., J. W. Hicks, P. R. Manzani, et al. 2010. Phylogeny, ecology, and heart position in snakes. *Physiological and Biochemical Zoology* 83:43–54.

Gaulin, S. J. C., and J. S. Boster. 1990. Dowry as female competition. *American Anthropologist* 92:994–1005.

Gebo, D. L. 2004. A shrew-sized origin for primates. *Yearbook of Physical Anthropology* 47:40–62.

Genoud, M. 2002. Comparative studies of basal rate of metabolism in primates. *Evolutionary Anthropology* 11:108–11.

Ghiglieri, M. P. 1987. Sociobiology of the great apes and the hominid ancestor. *Journal of Human Evolution* 16:319–57.

Gibson, K. R. 1986. Cognition, brain size and the extraction of embedded food resources. In *Primate ontogeny, cognitive and social Behavior*, edited by G. Else and P. C. Lee, 93–105. Cambridge: Cambridge University Press.

Gilligan, I. 2008. Clothing and climate in aboriginal Australia. *Current Anthropology* 49:487–95.

Gilligan, I., and D. Bulbeck. 2007. Environment and morphology in Australian Aborigines: A re-analysis of the Birdsell database. *Yearbook of Physical Anthropology* 134:75–91.

Gimbutas, M. 1973. The beginning of the Bronze Age in Europe and the Indo-Europeans, 3500–2500 BC. *Journal of Indo-European Studies* 1:163–214.

Gingerich, P. D. 1977. Correlation of tooth size and body size in living hominoid primates, with a note on relative brain size in *Aegyptopithecus* and *Proconsul*. *American Journal of Physical Anthropology* 47:395–98.

———. 1983. Rates of evolution—Effects of time and temporal scaling. *Science* 222:159–61.

———. 1984. Primate evolution: Evidence from the fossil record, comparative morphology, and molecular biology. *American Journal of Physical Anthropology* 27:57–72.

Gittleman, J. L. 1989. The comparative approach in ethology: Aims and limitations. In *Perspectives in ethology*, edited by P. P. G. Bateson and P. H. Klopfer, 55–83. New York: Plenum.

———. 1994. Female brain size and parental care in carnivores. *Proceedings of the National Academy of Sciences of the United States of America* 91:5495–97.

Gittleman, J. L., C. G. Anderson, M. Kot, and H.-K. Luh. 1996. Phylogenetic lability and rates of evolution: a comparison of behavioral, morphological and life history traits. In *The comparative method in animal behavior*, edited by E. P. Martins, 166–205. Oxford: Oxford University Press.

Gittleman, J. L., and D. M. Decker. 1994. The phylogeny of behaviour. In *Behaviour and evolution*, edited by P. J. B. Slater and T. R. Halliday, 80–105. Cambridge: Cambridge University Press.

Gittleman, J. L., and P. H. Harvey. 1982. Carnivore home-range size, metabolic needs and ecology. *Behavioral Ecology and Sociobiology* 10:57–63.

Gittleman, J. L., K. E. Jones, and S. A. Price. 2004. Supertrees: Using complete phylogenies in comparative biology. In Bininda-Emonds 2004a, 439–60.

Gittleman, J. L., and M. Kot. 1990. Adaptation: Statistics and a null model for estimating phylogenetic effects. *Systematic Zoology* 39:227–41.

Gittleman, J. L., and H. K. Luh. 1992. On comparing comparative methods. *Annual Review of Ecology and Systematics* 23:383–404.

Gittleman, J. L., and A. Purvis. 1998. Body size and species-richness in carnivores and primates. *Proceedings of the Royal Society B: Biological Sciences* 265:113–19.

Goldman, N., J. Anderson, and A. Rodrigo. 2000. Likelihood-based tests of topologies in phylogenetics. *Systematic Biology* 49:652–70.

Goodman, M., W. J. Bailey, K. Hayasaka, M. J. Stanhope, J. Slightom, and J. Czelusniak. 1994. Molecular evidence on primate phylogeny from DNA sequences. *American Journal of Physical Anthropology* 94:3–24.

Goodman, M., C. Porter, J. Czelusniak, et al. 1998. Toward a phylogenetic classification of primates based on DNA evidence complemented by fossil evidence. *Molecular Phylogenetics and Evolution* 9:585–98.

Gordon, R. G. 2005. Ethnologue: Languages of the World. Dallas, TX: SIL International. http://www.ethnologue.com/.

Goss-Custard, J. D., R. I. M. Dunbar, and F. P. G. Aldrich-Blake. 1972. Survival, mating, and rearing in the evolution of primate social structure. *Folia Primatologica* 17:1–19.

Gould, S. J. 1966. Allometry and size in ontogeny and phylogeny. *Biological Reviews* 41: 587–640.

———. 1975a. Allometry in primates, with emphasis on scaling and the evolution of the brain. *Contributions to Primatology* 5:244–92.

———. 1975b. On the scaling of tooth size in mammals. *American Zoologist* 15:351–62.

Gould, S. J., and R. C. Lewontin. 1979. The spandrels of San Marco and the Panglossian paradigm: a critique of the adaptationist programme. *Proceedings of the Royal Society B: Biological Sciences* 205:581–98.

Gould, S. J., D. M. Raup, J. J. J. Sepkoski, T. J. M. Schopf, and D. S. Simberloff. 1977. The shape of evolution: A comparison of real and random clades. *Paleobiology* 3:23–40.

Gould, S. J., and E. S. Vrba. 1982. Exaptation: A missing term in the science of form. *Paleobiology* 8:4–15.

Grafen, A. 1989. The phylogenetic regression. *Philosophical Transactions of the Royal Society B: Biological Sciences* 326:119–57.

Grant, J., C. Chapman, and K. Richardson. 1992. Defended versus undefended home range size of carnivores, ungulates and primates. *Behavioral Ecology and Sociobiology* 31: 149–61.

Grant, P. R. 1986. *Ecology and evolution of Darwin's finches*. Princeton, NJ: Princeton University Press.

Gray, R. 2005. Pushing the time barrier in the quest for language roots. *Science* 309:2007–8.

Gray, R. D., and Q. D. Atkinson. 2003. Language-tree divergence times support the Anatolian theory of Indo-European origin. *Nature* 426:435–39.

Gray, R. D., A. J. Drummond, and S. J. Greenhill. 2009. Language phylogenies reveal expansion pulses and pauses in Pacific settlement. *Science* 323:479–83.

Gray, R. D., D. Bryant and S. J. Greenhill. 2010. On the shape and fabric of human history. *Philisophical Transactions of the Royal Society B: Biological Sciences* 365:3923–33.

Gray, R., S. Greenhill, and R. Ross. 2007. The pleasures and perils of Darwinizing culture (with phylogenies). *Biological Theory* 2:360–75.

Gray, R. D., and F. M. Jordan. 2000. Language trees support the express-train sequence of Austronesian expansion. *Nature* 405:1052–55.

Greene, H. W. 1994. Homology and behavioral repertoires. In *Homology*, edited by B. K. Hall, 369–91. San Diego, CA: Academic Press.

———. 1999. Natural history and behavioural homology.In *Novartis Foundation Symposium 222: Homology*, edited by G. R. Bock and G. Cardew, 173–88. New York: Wiley.

Greenhill, S., T. Currie, and R. Gray. 2009. Does horizontal transmission invalidate cultural phylogenies? *Proceedings of the Royal Society B: Biological Sciences* 276:2299–306.

Groves, C. P. 2001a. Why taxonomic stability is a bad idea, or why are there so few species of primates (or are there?). *Evolutionary Anthropology* 10:192–98.

————. 2001b. *Primate taxonomy*. Washington DC: Smithsonian Institution Press.

————. 2004. The what, why and how of primate taxonomy. *International Journal of Primatology* 25:1105–26.

Gudschinsky, S. C. 1956. The ABC's of lexicostatistics (glottochronology). *Word* 12:175–210.

Guégan, J. F., and A. T. Teriokhin. 2000. Human life history traits on a parasitic landscape. In *Evolutionary biology of host-parasite relationships: Theory meets reality*, edited by R. Poulin, S. Morand, and A. Skorping, 143–61. Amsterdam: Elsevier Science.

Guégan, J. F., F. Thomas, M. E. Hochberg, T. de Meeus, and F. Renaud. 2001. Disease diversity and human fertility. *Evolution* 55:1308–14.

Guernier, V., M. E. Hochberg, and J. F. Guégan. 2004. Ecology drives the worldwide distribution of human diseases. *PLoS Biology* 2:740–46.

Guglielmino, C. R., C. Viganotti, B. Hewlett, and L. L. Cavallisforza. 1995. Cultural variation in Africa—role of mechanisms of transmission and adaptation. *Proceedings of the National Academy of Sciences of the United States of America* 92:7585–89.

Gusfield, D. 1997. *Algorithms on strings, trees, and sequences*. Cambridge: Cambridge University Press.

Haldane, J. B. S. 1949. Suggestions as to quantitative measurement of rates of evolution. *Evolution* 3:51–56.

Hall, B. G. 2008. *Phylogenetic trees made easy: a how-to manual*. Sunderland, MA: Sinauer Associates.

Hansen, T. F. 1997. Stabilizing selection and the comparative analysis of adaptation. *Evolution* 51:1341–51.

Hansen, T. F., and E. P. Martins. 1996. Translating between microevolutionary process and macroevolutionary patterns: The correlation structure of interspecific data. *Evolution* 50:1404–17.

Hansen, T. F., J. Pienaar, and S. H. Orzack. 2008. A comparative method for studying adaptation to a randomly evolving environment. *Evolution* 62:1965–77.

Harcourt, A. H., S. A. Coppeto, and S. A. Parks. 2002. Rarity, specialization and extinction in primates. *Journal of Biogeography* 29:445–56.

Harcourt, A. H., P. H. Harvey, S. G. Larson, and R. V. Short. 1981. Testis weight, body weight and breeding system in primates. *Nature* 293:55–57.

Harcourt, A. H., and S. A. Parks. 2003. Threatened primates experience high human densities: Adding an index of threat to the IUCN Red List criteria. *Biological Conservation* 109:137–49.

Harcourt, A. H., A. Purvis, and L. Liles. 1995. Sperm competition: Mating system, not breeding season, affects testes size of primates. *Functional Ecology* 9:468–76.

Harmon, L., and R. Glor. 2010. Poor statistical performance of the mantel test in phylogenetic comparative analyses. *Evolution* 64:2173–78.

Harmon, L. J., and J. B. Losos. 2005. The effect of intraspecific sample size on type I and type II error rates in comparative studies. *Evolution* 59:2705–10.

Harmon, L. J., J. A. Schulte, A. Larson, and J. B. Losos. 2003. Tempo and mode of evolutionary radiation in iguanian lizards. *Science* 301:961–64.

Harmon, M. J., T. L. VanPool, R. D. Leonard, C. S. VanPool, and L. A. Salter. 2006. Reconstructing the flow of information across time and space: a phylogenetic analysis of ceramic traditions from Prehispanic Western and northern Mexico and the American southwest. In Lipo et al. 2006b, 209–30.

Hartung, J. 1982. Polygyny and the inheritance of wealth. *Current Anthropology* 23:1–12.

Harvey, P. H. 1996. Phylogenies for ecologists. *Journal of Animal Ecology* 65:255–63.

Harvey, P. H., A. J. L. Brown, J. M. Smith, and S. Nee. 1995. *New uses for new phylogenies.* Oxford: Oxford University Press.

Harvey, P. H., and T. H. Clutton-Brock. 1985. Life history variation in primates. *Evolution* 39:559–81.

Harvey, P. H., and J. R. Krebs. 1990. Comparing brains. *Science* 249:140–46.

Harvey, P. H., and G. M. Mace. 1982. Comparisons between taxa and adaptive trends: problems of methodology.In *Current problems in sociobiology*, ed. King's College Sociobiology Group, 343–61. Cambridge: Cambride University Press.

Harvey, P. H., R. M. May, and S. Nee. 1994. Phylogenies without fossils. *Evolution* 48:523–29.

Harvey, P. H., and M. D. Pagel. 1991. *The comparative method in evolutionary biology.* Oxford: Oxford University Press.

Harvey, P. H., and A. Purvis. 1991. Comparative methods for explaining adaptations. *Nature* 351:619–24.

Harvey, P. H., and A. Rambaut. 1998. Phylogenetic extinction rates and comparative methodology. *Proceedings of the Royal Society B: Biological Sciences* 265:1691–96.

———. 2000. Comparative analyses for adaptive radiations. *Proceedings of the Royal Society B: Biological Sciences* 355:1–7.

Haun, D. B. M., J. Call, G. Janzen, and S. C. Levinson. 2006a. Evolutionary psychology of spatial representations in the hominidae. *Current Biology* 16:1736–40.

Haun, D. B. M., C. J. Rapold, J. Call, G. Janzen, and S. C. Levinson. 2006b. Cognitive cladistics and cultural override in hominid spatial cognition. *Proceedings of the National Academy of Sciences of the United States of America* 103:17568–73.

Hauser, D. L., and G. Boyajian. 1997. Proportional change and patterns of homoplasy: Sanderson and Donoghue revisited. *Cladistics* 13:97–100.

Hauser, M. D. 1993. The evolution of nonhuman primate vocalizations: Effects of phylogeny, body weight, and social context. *American Naturalist* 142:528–42.

Hayes, J., and J. Shonkwiler. 2006. Allometry, antilog transformations, and the perils of prediction on the original scale. *Physiological and Biochemical Zoology* 79:665–74.

Healy, S. D., and C. Rowe. 2007. A critique of comparative studies of brain size. *Proceedings of the Royal Society B: Biological Sciences* 274:453–64.

Heard, S. B. 1992. Patterns in tree balance among cladistic, phenetic, and randomly generated phylogenetic trees. *Evolution* 46:1818–26.

Heard, S. B., and G. H. Cox. 2007. The shapes of phylogenetic trees of clades, faunas, and local assemblages: Exploring spatial pattern in differential diversification. *American Naturalist* 169:E107–18.

Heard, S.B., and D. L. Hauser. 1995. Key evolutionary innovations and their ecological mechanisms. *Historical Biology* 10:151–73.

Heard, S. B., and A. O. Mooers. 2000. Phylogenetically patterned speciation rates and extinction risks change the loss of evolutionary history during extinctions. *Proceedings of the Royal Society B: Biological Sciences* 267:613–20.

Heinsohn, R., and C. Packer. 1995. Complex cooperative strategies in group-territorial lions. *Science* 269:1260–62.

Hennig, W. 1966. *Phylogenetic systematics.* Urbana: University of Illinois Press.

Henrich, J. 2004. Demography and cultural evolution: How adaptive cultural processes can produce maladaptive losses: The Tasmanian case. *American Antiquity* 69:197–214.

Herrmann, E., J. Call, M. V. Hernandez-Lloreda, B. Hare, and M. Tomasello. 2007. Humans have evolved specialized skills of social cognition: The cultural intelligence hypothesis. *Science* 317:1360–66.

Hewlett, B. S., A. de Silvestri, and C. R. Guglielmino. 2002. Semes and genes in Africa. *Current Anthropology* 43:313–21.

Hey, J. 1992. Using phylogenetic trees to study speciation and extinction. *Evolution* 46: 627–40.

Hill, J. H. 2001. Proto-Uto-Aztecan: A community of cultivators in central Mexico? *American Anthropologist* 103:913–34.

Hillis, D. M. 1999. SINEs of the perfect character. *Proceedings of the National Academy of Sciences of the United States of America* 96:9979–81.

Hilton-Taylor, C. 2002. IUCN red list of threatened species. Morges: International Union for Conservation of Nature. http://www.iucnredlist.org.

Hocking, R. R. 1976. Analysis and selection of variables in linear-regression. *Biometrics* 32:1–49.

Hoenigswald, H. 1987. Language family trees, topological and metrical. In Hoenigswald and Wiener 1987, 257–67.

Hoenigswald, H. M., and L. F. Wiener, eds. 1987. *Biological metaphor and cladistic classification: An interdisciplinary perspective.* Philadelphia: University of Pennsylvania Press.

Hofreiter, M., D. Serre, H. N. Poinar, M. Kuch, and S. Paabo. 2001. Ancient DNA. *Nature Reviews Genetics* 2:353–59.

Hohenlohe, P. A., and S. J. Arnold. 2008. MIPoD: A hypothesis-testing framework for micro-evolutionary inference from patterns of divergence. *American Naturalist* 171:366–85.

Holden, C. J. 2002. Bantu language trees reflect the spread of farming across sub-Saharan Africa: a maximum-parsimony analysis. *Proceedings of the Royal Society B: Biological Sciences* 269:793–99.

———. 2006. The spread of Bantu languages, farming, and pastoralism in sub-equatorial Africa. In Lipo et al. 2006b, 249–67.

Holden, C. J., and R. Gray. 2006. Rapid radiation, borrowing and dialect continua in the Bantu languages. In *Phylogenetic Methods and the Prehistory of Languages,* edited by P. Forster and C. Renfrew, 33–42. Cambridge: McDonald Institute for Archaeological Research.

Holden, C. J., and R. Mace. 1997. Phylogenetic analysis of the evolution of lactose digestion in adults. *Human Biology* 69:605–28.

———. 1999. Sexual dimorphism in stature and women's work: a phylogenetic cross-cultural analysis. *American Journal of Physical Anthropology* 110:27–45.

———. 2003. Spread of cattle led to the loss of matrilineal descent in Africa: a coevolutionary analysis. *Proceedings of the Royal Society B: Biological Sciences* 270:2425–33.

Holder, M., and P. O. Lewis. 2003. Phylogeny estimation: Traditional and Bayesian approaches. *Nature Reviews Genetics* 4:275–84.

Holly Smith, B., T. Crummett, and K. L. Brandt. 1994. Ages of eruption of primate teeth: a compendium for aging individuals and comparing life histories. *Yearbook of Physical Anthropology* 37:177–231.

Holman, E. W., S. Wichmann, C. H. Brown, V. Velupillai, A. Müller, and D. Bakker. 2008. Explorations in automated language classification. *Folia Linguistica.*42: 331–54.

Hopkins, M. E., and C. L. Nunn. 2007. A global gap analysis of infectious agents in wild primates. *Diversity and Distributions* 13:561–72.

———. 2010. Gap analysis and the geographical distribution of parasites. In *The Biogeography of Host-Parasite Interactions,* edited by S. Morand and B. R. Krasnov, 129–42. Oxford: Oxford University Press.

Horner, V., and A. Whiten. 2005. Causal knowledge and imitation/emulation switching in chimpanzees (*Pan trogiodytes*) and children (*Homo sapiens*). *Animal Cognition* 8:164–81.

Houle, A. 1997. The role of phylogeny and behavioral competition in the evolution of co-existence among primates. *Canadian Journal of Zoology* 75:827–46.

Housworth, E. A., E. P. Martins, and M. Lynch. 2004. The phylogenetic mixed model. *American Naturalist* 163:84–96.

Howe, C., A. Barbrook, M. Spencer, P. Robinson, B. Bordalejo, and L. Mooney. 2001. Manuscript evolution. *Trends in Genetics* 17:147–52.

Howells, W. 1973. *Cranial variation in man: A study by multivariate analysis of patterns of difference among recent human populations.* Cambridge, MA: Peabody Museum of Archaeology and Ethnology, Harvard University.

———. 1989. *Skull shapes and the map: Craniometric analyses in the dispersion of modern Homo.* Cambridge, MA: Peabody Museum of Archaeology and Ethnology, Harvard University.

Hrdy, S. B. 1974. Male-male competition and infanticide among the langurs (*Presbytis entellus*) of Abu, Rajasthan. *Folia Primatologica* 22:19–58.

Hrdy, S. B., and P. L. Whitten. 1987. Patterning of sexual activity. In *Primate Societies*, edited by B. B. Smuts et al., 370–84. Chicago: University of Chicago Press.

Huelsenbeck, J. P., and J. P. Bollback. 2001. Empirical and hierarchical Bayesian estimation of ancestral states. *Systematic Biology* 50:351–66.

Huelsenbeck, J. P., and K. A. Crandall. 1997. Phylogeny estimation and hypothesis testing using maximum likelihood. *Annual Review of Ecology and Systematics* 28:437–66.

Huelsenbeck, J. P., B. Larget, R. E. Miller, and F. Ronquist. 2002. Potential applications and pitfalls of Bayesian inference of phylogeny. *Systematic Biology* 51:673–88.

Huelsenbeck, J. P., R. Nielsen, and J. P. Bollback. 2003. Stochastic mapping of morphological characters. *Systematic Biology* 52:131–58.

Huelsenbeck, J. P., and B. Rannala. 2003. Detecting correlation between characters in a comparative analysis with uncertain phylogeny. *Evolution* 57:1237–47.

Huelsenbeck, J. P., B. Rannala, and J. P. Masly. 2000. Accommodating phylogenetic uncertainty in evolutionary studies. *Science* 288:2349–50.

Huelsenbeck, J. P., F. Ronquist, R. Nielsen, and J. P. Bollback. 2001. Bayesian inference of phylogeny and its impact on evolutionary biology. *Science* 294:2310–14.

Huey, R. B. 1987. Phylogeny, history, and the comparative method. In *New directions in ecological physiology*, edited by M. E. Feder, A. F. Bennett, W. W. Burggren, and R. B. Huey, 76–98. Cambridge: Cambridge University Press.

Huey, R. B., and A. F. Bennett. 1987. Phylogenetic studies of coadaptation: preferred temperatures versus optimal performance temperatures of lizards. *Evolution* 41:1098–115.

Hugall, A. F., and M. S. Y. Lee. 2007. The likelihood node density effect and consequences for evolutionary studies of molecular rates. *Evolution* 61:2293–307.

Hugot, J. P. 1999. Primates and their pinworm parasites: The Cameron hypothesis revisited. *Systematic Biology* 48:523–46.

Hunt, G. 2007. The relative importance of directional change, random walks, and stasis in the evolution of fossil lineages. *Proceedings of the National Academy of Sciences of the United States of America* 104:18404.

Hurles, M. E., E. Matisoo-Smith, R. D. Gray, and D. Penny. 2003. Untangling Oceanic settlement: The edge of the knowable. *Trends in Ecology & Evolution* 18:531–40.

Huson, D. H. 1998. SplitsTree: Analyzing and visualizing evolutionary data. *Bioinformatics* 14:68–73.

Huson, D. H., and D. Bryant. 2006. Application of phylogenetic networks in evolutionary studies. *Molecular Biology and Evololution* 23:254–67.

Hutchinson, G. E., and R. H. MacArthur. 1959. A theoretical ecological model of size distributions among species of animals. *American Naturalist* 93:117.

Huxley, J. S. 1932. *Problems of relative growth.* London: Methuen.

Ingman, M., H. Kaessmann, S. Paabo, and U. Gyllensten. 2000. Mitochondrial genome variation and the origin of modern humans. *Nature* 408:708–13.

Isaac, N. J. B., P.-M. Agapow, P. H. Harvey, and A. Purvis. 2003. Phylogenetically nested comparisons for testing correlates of species richness: A simulation study of continuous variables. *Evolution* 57:18–26.

Isaac, N. J. B., and G. Cowlishaw. 2004. How species respond to multiple extinction threats. *Proceedings of the Royal Society B: Biological Sciences* 271:1135–41.

Isaac, N. J. B., K. E. Jones, J. L. Gittleman, and A. Purvis. 2005. Correlates of species richness in mammals: Body size, life history, and ecology. *American Naturalist* 165:600–7.

Isaac, N. J. B., J. Mallet, and G. M. Mace. 2004. Taxonomic inflation: its influence on macroecology and conservation. *Trends in Ecology & Evolution* 19:464–69.

Isler, K., A. D. Barbour, and R. D. Martin. 2002. Line-fitting by rotation: A nonparametric method for bivariate allometric analysis. *Biometrical Journal* 44:289–304.

Isler, K., and C. P. van Schaik. 2006. Metabolic costs of brain size evolution. *Biology Letters* 2:557–60.

Ives, A., and T. Garland 2010. Phylogenetic logistic regression for binary dependent variables. *Systematic Biology* 59:9–26.

Ives, A. R., P. E. Midford, and T. Garland. 2007. Within-species variation and measurement error in phylogenetic comparative methods. *Systematic Biology* 56:252–70.

Iwaniuk, A. N., J. E. Nelson, and S. M. Pellis. 2001. Do big-brained animals play more? Comparative analyses of play and relative brain size in mammals. *Journal of Comparative Psychology* 115:29–41.

Jablonski, N. G. 1993. *Theropithecus: The rise and fall of a primate genus.* Cambridge: Cambridge University Press.

Jablonski, N. G., and G. Chaplin. 2000. The evolution of human skin coloration. *Journal of Human Evolution* 39:57–106.

Jenkins, F. A. 1974. *Primate locomotion.* New York: Academic Press.

Jenner, R. A. 2004. Accepting partnership by submission? Morphological phylogenetics in a molecular millennium. *Systematic Biology* 53:333–42.

Jensen, J. S. 1990. Plausibility and testability: Assessing the consequences of evolutionary innovation. In *Evolutionary innovations,* edited by M. H. Nitecki and D. V. Nitecki, 171–190. Chicago: University of Chicago Press.

Jernvall, J., and P. C. Wright. 1998. Diversity components of impending primate extinctions. *Proceedings of the National Academy of Sciences of the United States of America* 95:11279–83.

Jolly, C. J. 1970. The seed-eaters: A new model of hominid differentiation based on a baboon analogy. *Man* 5:5–26.

Jolly, C. J. 2001. A proper study for mankind: Analogies from the papionin monkeys and their implications for human evolution. *Yearbook of Physical Anthropology* 44:177–204.

Jones, K., W. Sechrest, and J. Gittleman. 2005. Age and area revisited: identifying global patterns and implications for conservation. In Purvis, Gittleman, and Brooks 2005b, 141–67.

Jones, K. E., and E. C. Teeling. 2009. Phylogenetic tools for examining character and clade evolution in bats. In *Ecological and behavioral methods for the study of bats,* edited by T. H. Kunz and S. Parsons 715–38. Baltimore, MD: John Hopkins University Press.

Jordan, F., R. Gray, S. Greenhill, and R. Mace. 2009. Matrilocal residence is ancestral in Austronesian societies. *Proceedings of the Royal Society B: Biological Sciences* 276:1957–64.

Jordan, P., and T. Mace. 2006. Tracking culture-historical lineages: Can "descent with modification" be linked to "association by descent"? In Lipo et al. 2006b, 149–68.

Jordan, P., and S. J. Shennan. 2003. Cultural transmission, language, and basketry traditions amongst the California Indians. *Journal of Anthropological Archaeology* 22:43–74.

———. 2005. Cultural transmission in indigenous California. In Mace, Holden, and Shennan 2005, 133–64.

———. 2009. Diversity in hunter-gatherer technological traditions: Mapping trajectories of cultural "descent with modification" in northeast California. *Journal of Anthropological Archaeology* 28:342–65.

Jungers, W. L. 1978. Functional significance of skeletal allometry in *Megaladapis* in comparison to living prosimians. *American Journal of Physical Anthropology* 49:303–14.

———. 1984. Aspects of size and scaling in primate biology with special reference to the locomotor skeleton. *Yearbook of Physical Anthropology* 27:73–97.

———. 1985. Body size and scaling of limb proportions in primates*Size and scaling in primate biology*, edited by W. L. Jungers, 345–81. New York: Plenum.

Jungers, W. L., A. B. Falsetti, and C. E. Wall. 1995. Shape, relative size, and size-adjustments in morphometrics. *Yearbook of Physical Anthropology* 38:137–61.

Jungers, W. L., L. R. Godfrey, E. L. Simons, and P. S. Chatrath. 1997. Phalangeal curvature and positional behavior in extinct sloth lemurs (primates, Palaeopropithecidae). *Proceedings of the National Academy of Sciences of the United States of America* 94:1998–2001.

Kamilar, J. M. 2006. Geographic variation in savanna baboon (*Papio*) ecology and its taxonomic and evolutionary implications. *Primate Biogeography*, edited by S. M. Lehman and J. G. Fleagle, 169–200.

Kamilar, J. M., and L. M. Paciulli. 2008. Examining the extinction risk of specialized folivores: A comparative study of colobine monkeys. *American Journal of Primatology* 70:816–27.

Kappeler, P. M. 1997. Intrasexual selection and testis size in strepsirhine primates. *Behavioral Ecology* 8:10–19.

———. 1998. Nests, tree holes and the evolution of primate life histories. *American Journal of Primatology* 46:7–33.

———, ed. 2000. *Primate males*. Cambridge: Cambridge University Press.

Kappeler, P. M., and J. B. Silk. 2009. *Mind the gap: tracing the origins of human universals*. Berlin: Springer.

Kay, R. F. 1984. On the use of anatomical features to infer foraging behavior in extinct primates. *Adaptations for foraging in nonhuman primates*, edited by P. Rodman and J. Cant, 21–53. New York: Columbia University Press.

Kay, R. F., and E. L. Simons. 1980. The ecology of oligocene African anthropoidea. *International Journal of Primatology* 1:21–37.

Kelley, J., and W. Swanson. 2008. Dietary change and adaptive evolution of enamelin in humans and among primates. *Genetics* 178:1595.

Kendall, D. G. 1948. On the generalized "birth-and-death" process. *Annals of Mathematical Statistics* 19:1–15.

Kimbel, W., and L. Martin. 1993. Species, species concepts, and primate evolution. New York: Plenum.

Kinzey, W. G., ed. 1987. The evolution of human behavior: Primate models. Albany: State University of New York Press.

Kirkpatrick, M., and M. Slatkin. 1993. Searching For evolutionary patterns in the shape of a phylogenetic tree. *Evolution* 47:1171–81.

Kishino, H., and M. Hasegawa. 1989. Evaluation of the maximum likelihood estimate of the evolutionary tree topologies from DNA sequence data, and the branching order in Hominoidea. *Journal of Molecular Evolution* 29:170–79.

Kittler, R., M. Kayser, and M. Stoneking. 2003. Molecular evolution of *Pediculus humanus* and the origin of clothing. *Current Biology* 13:1414–17.

———. 2004. Molecular evolution of *Pediculus humanus* and the origin of clothing. *Current Biology* 14:2309.

Kleiber, M. 1961. The fire of life: An introduction to animal energetics. New York: Wiley.

Köbben, A. J. F. 1967. Why exceptions? Logic of cross-cultural analysis. *Current Anthropology* 8:3–19.

Koendgen, S., H. Kuhl, P. K. N'Goran, et al. 2008. Pandemic human viruses cause decline of endangered great apes. *Current Biology* 18:260–64.

Kolaczkowski, B., and J. W. Thornton. 2009. Long-branch attraction bias and inconsistency in Bayesian phylogenetics. *PLoS ONE* 4:e7891.

Kolokotrones, T., S. Van, E. J. Deeds, and W. Fontana. 2010. Curvature in metabolic scaling. *Nature* 464:753–56.

Krause, J., Q. Fu, J. Good, B. Viola, M. Shunkov, A. Derevianko, and S. Paabo. 2010. The complete mitochondrial DNA genome of an unknown hominin from southern Siberia. *Nature* 464:894–97.

Krings, M., A. Stone, R. W. Schmitz, H. Krainitzki, M. Stoneking, and S. Paabo. 1997. Neandertal DNA sequences and the origin of modern humans. *Cell* 90:19–30.

Krishnan, N. M., H. Seligmann, C. B. Stewart, A. P. J. de Koning, and D. D. Pollock. 2004. Ancestral sequence reconstruction in primate mitochondrial DNA: Compositional bias and effect on functional inference. *Molecular Biology and Evolution* 21:1871–83.

Kroeber, A. L. 1931. Historical reconstruction of culture growths and organic evolution. *American Anthropologist* 33:149–56.

Kruskal, J., I. Dyer, and P. Black. 1971. The vocabulary method of reconstructing family trees: innovations and large scale applications. *Mathematics in the archaeological and historical sciences*, edited by F. R. Hodson, D. G. Kendall, and P. Tantu, 361–80. Edinburgh: Edinburgh University Press.

Kubo, T., and Y. Iwasa. 1995. Inferring the rates of branching and extinction from molecular phylogenies. *Evolution* 49:694–704.

Kuris, A. M., A. R. Blaustein, and J. Javier Alio. 1980. Hosts as islands. *American Naturalist* 116:570–86.

Kutsukake, N., and C. L. Nunn. 2006. Comparative tests of reproductive skew in male primates: The roles of female mating behavior and incomplete control. *Behavioral Ecology and Sociobiology* 60:695–707.

LaBarbera, M. 1989. Analyzing body size as a factor in ecology and evolution. *Annual Review of Ecology and Systematics* 20:97–117.

Lajeunesse, M. 2009. Meta analysis and the comparative phylogenetic method. *American Naturalist* 174:369–81.

Laland, K. N., J. Odling-Smee, and M. W. Feldman. 2000. Niche construction, biological evolution, and cultural change. *Behavioral And Brain Sciences* 23:131–46.

Lande, R. 1981. Models of speciation by sexual selection on polygenic traits. *Proceedings of the National Academy of Sciences of the United States of America* 78:3721–25.

Larget, B., and D. L. Simon. 1999. Markov Chasin Monte Carlo algorithms for the Bayesian analysis of phylogenetic trees. *Molecular Biology and Evolution* 16:750–59.

Lauder, G. V. 1986. Homology, analogy, and the evolution of behavior. *Evolution of animal behavior: Paleontological and field approaches*, edited by M. Nitecki and J. Kitchell, 9–40. New York: Oxford University Press.

Laurin, M. 2004. The evolution of body size, Cope's rule and the origin of amniotes. *Systematic Biology* 53:594–622.

———. 2010. Assessment of the relative merits of a few methods to detect evolutionary trends. *Systematic Biology* 59:689-704.Lavin, S., W. Karasov, A. Ives, K. Middleton, and T. Garland. 2008. Morphometrics of the avian small intestine compared with that of nonflying mammals: A phylogenetic approach. *Physiological and Biochemical Zoology* 81:526–50.

Lee, P., ed. 1999. *Comparative primate socioecology*. Cambridge: Cambridge University Press.

Lees, R. B. 1953. The basis of glottochronology. *Language* 29:113–27.

Legras, J. L., D. Merdinoglu, J. M. Cornuet, and F. Karst. 2007. Bread, beer and wine: *Saccharomyces cerevisiae* diversity reflects human history. *Molecular Ecology* 16:2091–102.

Lehman, S. M. 2006. Conservation biology of Malagasy strepsirhines: A phylogenetic approach. *American Journal of Physical Anthropology* 130:238–53.

Leroi, A. M., M. R. Rose, and G. V. Lauder. 1994. What does the comparative method reveal about adaptation? *American Naturalist* 143:381–402.

Lesku, J. A., T. C. Roth II, C. J. Amlaner, and S. L. Lima. 2006. A phylogenetic analysis of sleep architecture in mammals: The intergration of anatomy, physiology, and ecology. *American Naturalist* 168:1–13.

Lewis, K. P. 2000. A comparative study of primate play behaviour: Implications for the study of cognition. *Folia Primatologica* 71:417–21.

Lewis, P. O. 2001. A likelihood approach to estimating phylogeny from discrete morphological character data. *Systematic Biology* 50:913–25.

Li, J. Z., D. M. Absher, H. Tang, et al. 2008. Worldwide human relationships inferred from genome-wide patterns of variation. *Science* 319:1100–4.

Li, W.-H., and A. Zharkikh. 1994. What is the bootstrap technique? *Systematic Biology* 43:424–30.

Lieberman, D. E. 1999. Homology and hominid phylogeny: Problems and potential solutions. *Evolutionary Anthropology* 7:142–51.

———. 2005. Further fossil finds from Flores. *Nature* 437:957–58.

Lieberman, D. E., B. A. Wood, and D. R. Pilbeam. 1996. Homoplasy and early *Homo*: An analysis of the evolutionary relationships of *H. habilis* sensu stricto and *H. rudolfensis*. *Journal of Human Evolution* 30:97–120.

Lieberman, E., J. B. Michel, J. Jackson, T. Tang, and M. A. Nowak. 2007. Quantifying the evolutionary dynamics of language. *Nature* 449:713–16.

Lindenfors, P. 2005. Neocortex evolution in primates: The "social brain" is for females. *Biology Letters* 1:407–10.

———. 2006. A method for calculating means and variances of comparative data for use in a phylogenetic analysis of variance. *Evolutionary Ecology Research* 8:975–95.

Lindenfors, P., L. Froberg, and C. L. Nunn. 2004. Females drive primate social evolution. *Proceedings of the Royal Society B: Biological Sciences* 271:S101–3.

Lindenfors, P., C. L. Nunn, and R. A. Barton. 2007. Primate brain architecture and selection in relation to sex. *BMC Biology* 5:20.

Lindenfors, P., L. J. Revell, and C. L. Nunn. 2010. Sexual dimorphism in primate aerobic capacity: A phylogenetic test. *Journal of Evolutionary Biology*, 23:1183–94.

Lindenfors, P., and B. S. Tullberg. 1998. Phylogenetic analyses of primate size evolution: the consequences of sexual selection. *Biological Journal of the Linnean Society* 64:413–47.

Lipo, C. P. 2006. The resolution of cultural phylogenies using graphs. In Lipo et al. 2006b, 89–107.

Lipo, C. P., M. J. O'Brien, M. Collard, and S. Shennan. 2006a. Cultural phylogenies and explanation: why historical methods matter. In Lipo et al. 2006b, 3–16.

———. 2006b. *Mapping our ancestors*. New Brunswick, NJ: Aldine.

Lloyd, S. 1995. Environmental influences on host immunity. In *Ecology of infectious diseases in natural populations*, edited by B. T. Grenfell and A. P. Dobson, 327–61. Cambridge: Cambridge University Press.

Lockwood, C. A., and J. G. Fleagle. 1999. The recognition and evaluation of homoplasy in primate and human evolution. *Yearbook of Physical Anthropology* 42:189–232.

Lorenz, K. 1941. Comparative studies of the motor patterns of Anatinae. Translated by R. Martin 1971. *Studies in Animal and Human Behaviour* 2:14–113.

Losos, J. B. 1990. A phylogenetic analysis of character displacement in Caribbean *Anolis* lizards. *Evolution* 44:58–569.

———. 1992. The evolution of convergent structure in Caribbean *Anolis* communities. *Systematic Biology* 41:403–20.

———. 1995. Community evolution in Greater Antillean anolis lizards: Phylogenetic patterns and experimental tests. *Philosophical Transactions of the Royal Society B: Biological Sciences* 349:69–75.

———. 1996. Phylogenetic perspectives on community ecology. *Ecology* 77:1344–54.

Losos, J. B., and F. R. Adler. 1995. Stumped by trees—A generalized null model for patterns of organismal diversity. *American Naturalist* 145:329–42.

Lovejoy, C. 2009. Reexamining human origins in light of *Ardipithecus ramidus*. *Science* 326:74.

Low, B. S. 1987. Pathogen stress and polygyny in humans. In *Human reproductive behaviour: A Darwinian perspective*, edited by L. Betzig, M. Borgerhoff Mulder, and P. W. Turke, 115–27. Cambridge: Cambridge University Press.

———. 1990. Marriage systems and pathogen stress in human societies. *American Zoologist* 30:325–39.

Lutzoni, F., and M. Pagel. 1997. Accelerated evolution as a consequence of transitions to mutualism. *Proceedings of the National Academy of Sciences* 94:11422–27.

Lutzoni, F., M. Pagel, and V. Reeb. 2001. Major fungal lineages are derived from lichen symbiotic ancestors. *Nature* 411:937–40.

Lycett, S. 2008. Acheulean variation and selection: does handaxe symmetry fit neutral expectations? *Journal of Archaeological Science* 35:2640–48.

———. 2009. Are Victoria West cores "proto-Levallois"? A phylogenetic assessment. *Journal of Human Evolution* 56:175–91.

Lyles, A. M., and A. P. Dobson. 1993. Infectious disease and intensive management: Population dynamics, threatened hosts, and their parasites. *Journal of Zoo and Wildlife Medicine* 24:315–26.

Lyman, R. L., and M. J. O'Brien. 2000. Measuring and explaining change in artifact variation with clade-diversity diagrams. *Journal of Anthropological Archaeology* 19:39–74.

Lynch, M. 1991. Methods for the analysis of comparative data in evolutionary biology. *Evolution* 45:1065–80.

Mace, R., and C. J. Holden. 2005. A phylogenetic approach to cultural evolution. *Trends in Ecology & Evolution* 20:116–21.

Mace, R., C. J. Holden, and S. Shennan. 2005. *The evolution of cultural diversity: A phylogenetic approach*. London: UCL Press.

Mace, R., F. Jordan, and C. Holden. 2003. Testing evolutionary hypotheses about human biological adaptation using cross-cultural comparison. *Comparative Biochemistry and Physiology A—Molecular & Integrative Physiology* 136:85–94.

Mace, R., and M. Pagel. 1994. The comparative method in anthropology. *Current Anthropology* 35:549–64.

———. 1995. A latitudinal gradient in the density of human languages in North America. *Proceedings of the Royal Society B: Biological Sciences* 261:117–21.

Macedonia, J. M., and K. F. Stanger. 1994. Phylogeny of the Lemuridae revisited—Evidence from communication signals. *Folia Primatologica* 63:1–43.

Maddison, W. P. 1990. A method for testing the correlated evolution of two binary characters: are gains or losses concentrated on certain branches of a phylogenetic tree. *Evolution* 44:539–557.

———. 1991. Squared-change parsimony reconstructions of ancestral states for continuous-valued characters on a phylogenetic tree. *Systematic Zoology* 40:304–14.

———. 1997. Gene trees in species trees. *Systematic Biology* 46:523–36.

———. 2000. Testing character correlation using pairwise comparisons on a phylogeny. *Journal of Theoretical Biology* 202:195–204.

———. 2006. Confounding asymmetries in evolutionary diversification and character change. *Evolution* 60:1743–46.

Maddison, W. P., and D. R. Maddison. 2000. *MacClade: Analysis of phylogeny and character evolution*. Sunderland, MA: Sinauer Associates, Inc.

———. 2006. Mesquite: a modular system for evolutionary analysis, version 2.5.http://mesquiteproject.org.

Maddison, W. P., P. E. Midford, and S. P. Otto. 2007. Estimating a binary character's effect on speciation and extinction. *Systematic Biology* 56:701–10.

Maddison, W. P., and M. Slatkin. 1991. Null models for the number of evolutionary steps in a character on a phylogenetic tree. *Evolution* 45:1184–97.

Maffi, L. 2005. Linguistic, cultural, and biological diversity. *Annual Review of Anthropology* 34:599–617.

Makarenkov, V., and P. Legendre. 2004. From a phylogenetic tree to a reticulated network. *Journal of Computational Biology* 11:195–212.

Makwana, S. C. 1978. Field ecology and behaviour of the rhesus macaque (*Macaca mulatta*): I. group compositition, home range, roosting sites, and foraging routes in the Asarori Forest. *Primates* 19:483–92.

Mallory, J. 1989. *In search of the Indo-Europeans: Language, archaeology and myth*. London: Thames and Hudson.

Mantel, N. 1967. The detection of disease clustering and a generalized regression approach. *Cancer Research* 27:209–20.

Marris, E. 2008. The language barrier. *Nature* 453:446–48.

Martin, R. D. 1981. Field studies of primate behaviour. *Symposia of the Zoological Society of London* 46:287–336.

———. 1990. *Primate origins and evolution*. London: Chapman and Hall.

———. 1996. Scaling of the mammalian brain: The maternal energy hypothesis. *News in Physiological Sciences* 11:149–53.

———. 1998. Comparative aspects of human brain evolution: scaling, energy costs and confounding variables. In *The origin and diversification of language*, edited by N. G. Jablonski and L. C. Aiello, 35–68. San Francisco: Memoirs of the California Academy of Sciences.

———. 2002. Primatology as an essential basis for biological anthropology. *Evolutionary Anthropology* 11:3–6.

Martin, R. D., D. J. Chivers, A. M. MacLarnon, and C. M. Hladik. 1985. Gastrointestinal allometry in primates and other mammals. In*Size and scaling in primate biology*, edited by W. L. Jungers, 61–89. New York: Plenum.

Martin, R. D., M. Genoud, and C. K. Hemelrijk. 2005. Problems of allometric scaling analysis: Examples from mammalian reproductive biology. *Journal of Experimental Biology* 208:1731–47.

Martins, E. P. 1994. Estimating the rate of phenotypic evolution from comparative data. *American Naturalist* 144:193–209.

———. 1996a. *Phylogenies and the comparative method in animal behavior*. New York: Oxford University Press.

———. 1996b. Phylogenies, spatial autoregression, and the comparative method: A computer simulation test. *Evolution* 50:1750–65.

———. 1999. Estimation of ancestral states of continuous characters: A computer simulation study. *Systematic Biology* 48:642–50.

———. 2000. Adaptation and the comparative method. *Trends in Ecology & Evolution* 15:296–99.

Martins, E. P., J. A. F. Diniz, and E. A. Housworth. 2002. Adaptive constraints and the phylogenetic comparative method: A computer simulation test. *Evolution* 56:1–13.

Martins, E. P., and T. Garland. 1991. Phylogenetic analyses of the correlated evolution of continuous characters: a simulation study. *Evolution* 45:534–57.

Martins, E. P., and T. F. Hansen. 1996. The statistical analysis of interspecific data: a review and evaluation of phylogenetic comparative methods. *Phylogenies and the comparative method in animal behavior*, edited by E. P. Martins, 22–75. New York: Oxford University Press.

———. 1997. Phylogenies and the comparative method: A general approach to incorporating phylogenetic information into the analysis of interspecific data. *American Naturalist* 149:646–67.

Martins, E. P., and J. Lamont. 1998. Estimating ancestral states of a communicative display: a comparative study of *Cyclura* rock iguanas. *Animal Behaviour* 55:1685–706.

Matthews, L. J., C. Arnold, Z. Machanda, and C. L. Nunn. 2010. Primate extinction risk and historical patterns of speciation and extinction in relation to body mass. *Proceedings of the Royal Society B: Biological Sciences*. doi:10.1098/rspb.2010.1489.

Matthews, L. J., C. Arnold, and C. L. Nunn. n.d. Niche construction influences the evolution of sex-biased dispersal in primates. Unpublished manuscript.

Matthews, L. J., J. Tehrani, F. M. Jordan, M. Collard, and C. L. Nunn. Forthcoming. Testing for cultural cores and components in Iranian textile assemblages: a Bayesian phylogenetic approach. *PLoS ONE*.

Mayden, R. L. 1997. A hierarchy of species concepts: the denouement in the saga of the species problem. *Systematics Association Special Volume* 54:381–424.

Mayr, E. 1963. *Animal species and evolution*. Cambridge MA: Harvard University Press.

McArdle, B. H. 1988. The structural relationship: regression in biology. *Canadian Journal of Zoology* 66:2329–39.

McCollum, M. A. 1999. The robust australopithecine face: A morphogenetic perspective. *Science* 284:301–5.

McElreath, R. 1997. Iterated parsimony: A method for reconstructing cultural histories. MA thesis, University of California at Los Angeles.

McGoogan, K., T. Kivell, M. Hutchison, et al. 2007. Phylogenetic diversity and the conservation biogeography of African primates. *Journal of Biogeography* 34:1962–74.

McGrew, W. 1981. The female chimpanzee as a human evolutionary prototype. In *Woman the gatherer*, edited by F. Dahlberg, 35–73. New Haven, CT: Yale University Press.

McHenry, H. M. 1975. Fossil hominid body weight and brain size. *Nature* 254:686–88.

McMahon, A., and R. McMahon. 2003. Finding families: Quantitative methods in language classification. *Transactions of the Philological Society* 101:7–55.

———. 2005. *Language classification by numbers.* New York: Oxford University Press.

McNab, B. K. 1963. Bioenergetics and the determination of home range size. *American Naturalist* 97:133–40.

———. 1986. The influence of food habits on the energetics of Eutherian mammals. *Ecological Monographs* 56:1–19.

McPeek, M. A. 1995. Testing hypotheses about evolutionary change on single branches of a phylogeny using evolutionary contrasts. *American Naturalist* 145:686–703.

McPeek, M. A., and T. E. Miller. 1996. Evolutionary biology and community ecology. *Ecology* 77:1319–20.

McShea, D. W. 1998. Possible largest-scale trends in organismal evolution: Eight "live hypotheses." *Annual Review of Ecology and Systematics* 29:293–318.

Mesoudi, A., A. Whiten, and K. N. Laland. 2004. Is human cultural evolution Darwinian? Evidence reviewed from the perspective of *The Origin of Species. Evolution* 58:1–11.

Milton, K., and M. L. May. 1976. Body weight, diet and home range area in primates. *Nature* 259:459–62.

Mitani, J. C., J. Gros-Louis, and A. F. Richards. 1996a. Sexual dimorphism, the operational sex ratio, and the intensity of male competition in polygynous primates. *American Naturalist* 147:966–80.

Mitani, J. C., J. Gros-Louis, and J. H. Manson. 1996b. Number of males in primate groups: Comparative tests of competing hypotheses. *American Journal of Primatology* 38:315–32.

Mitani, J. C., and P. S. Rodman. 1979. Territoriality: the relation of ranging pattern and home range size to defendability, with an analysis of territoriality among primate species. *Behavioral Ecology and Sociobiology* 5:241–51.

Mitani, J. C., and D. Watts. 1997. The evolution of non-maternal caretaking among anthropoid primates: do helpers help? *Behavioral Ecology and Sociobiology* 40:213–20.

Mitter, C., B. Farrell, and B. Wiegmann. 1988. The phylogenetic study of adaptive zones— Has phytophagy promoted insect diversification? *American Naturalist* 132:107–28.

Moen, D. S. 2006. Cope's rule in cryptodiran turtles: do the body sizes of extant species reflect a trend of phyletic size increase? *Journal of Evolutionary Biology* 19:1210–21.

Moller, A. P., and T. R. Birkhead. 1992. A pairwise comparative method as illustrated by copulation frequency in birds. *American Naturalist* 139:644–56.

Moodley, Y., B. Linz, Y. Yamaoka, et al. 2009. The peopling of the Pacific from a bacterial perspective. *Science* 323:527.

Mooers, A. O., and S. B. Heard. 1997. Inferring evolutionary process from phylogenetic tree shape. *Quarterly Review of Biology* 72:31–54.

———. 2002. Using tree shape. *Systematic Biology* 51:833–34.

Mooers, A. O., and D. Schluter. 1998. Fitting macroevolutionary models to phylogenies: An example using vertebrate body sizes. *Contributions to Zoology* 68:3–18.

———. 1999. Reconstructing ancestor states with maximum likelihood: Support for one- and two-rate models. *Systematic Biology* 48:623–33.

Mooers, A. O., S. M. Vamosi, and D. Schluter. 1999. Using phylogenies to test macroevolutionary hypotheses of trait evolution in cranes (Gruinae). *American Naturalist* 154: 249–59.

Moore, B. R., K. M. A. Chan, and M. J. Donoghue. 2004. Detecting diversification rate variation in supertrees. In Bininda-Emonds 2004a, 487–533.

Moore, J. H. 1994. Putting anthropology back together again: The ethnogenetic critique of cladistic theory. *American Anthropologist* 96:925–48.

Moore, J. L., L. Manne, T. Brooks, et al. 2002. The distribution of cultural and biological diversity in Africa. *Proceedings of the Royal Society B: Biological Sciences* 269:1645–53.

Moran, P. A. P. 1958. Random processes in genetics. *Proceedings of the Cambridge Philosophical Society* 54:60–71.

Morrow, E., and C. Fricke. 2004. Sexual selection and the risk of extinction in mammals. *Proceedings of the Royal Society B: Biological Sciences* 271:2395–401.

Morrow, E. H., and T. E. Pitcher. 2003. Sexual selection and the risk of extinction in birds. *Proceedings of the Royal Society B: Biological Sciences* 270:1793–99.

Moyà-Solà, S., and M. Köhler. 1996. A *Dryopithecus* skeleton and the origins of great-ape locomotion. *Nature* 379:156–59.

Moylan, J. W., M. Borgerhoff Mulder, C. M. Graham, C. L. Nunn, and T. Hakansson. 2006. Cultural traits and linguistic trees: Phylogenetic signal in east Africa. In Lipo et al. 2006b, 33–52.

Murdock, G. P. 1967. *Ethnographic atlas.* Pittsburgh: University of Pittsburgh Press.

Murdock, G. P., and D. White. 1969. Standard cross-cultural sample. *Ethnology* 8:329–69.

Nakhleh, L., D. Ringe, and T. Warnow. 2005. Perfect phylogenetic networks: A new methodology for reconstructing the evolutionary history of natural languages. *Language* 81: 382–420.

Napier, J. R. 1970. *The roots of mankind.* Washington DC: Smithsonian Institution Press.

Napier, J. R., and A. C. Walker. 1967. Vertical clinging and leaping—A newly recognized category of locomotor behaviour of primates. *Folia Primatologica* 6:204–19.

Naroll, R. 1961. Two solutions to galton problem. *Philosophy of Science* 28:15–39.

———. 1965. Galton problem—the logic of cross-cultural-analysis. *Social Research* 32: 428–51.

Naroll, R., and R. Cohen. 1970. *A handbook of method in cultural anthropology.* New York: Columbia University Press.

Nee, S. 2001. Inferring speciation rates from phylogenies. *Evolution* 55:661–68.

———. 2004. Extinct meets extant: simple models in paleontology and molecular phylogenetics. *Paleobiology* 30:172–78.

———. 2006. Birth-death models in macroevolution. *Annual Review of Ecology Evolution and Systematics* 37:1–17.

Nee, S., E. C. Holmes, R. M. May, and P. H. Harvey. 1994a. Extinction rates can be estimated from molecular phylogenies. *Philosophical Transactions of the Royal Society B: Biological Sciences* 344:77–82.

Nee, S., and R. M. May. 1997. Extinction and the loss of evolutionary history. *Science* 278: 692–94.

Nee, S., R. M. May, and P. H. Harvey. 1994b. The reconstructed evolutionary process. *Philosophical Transactions of the Royal Society B: Biological Sciences* 344:305–11.

Nee, S., A. O. Mooers, and P. H. Harvey. 1992. Tempo and mode of evolution revealed from molecular phylogenies. *Proceedings of the National Academy of Sciences of the United States of America* 89:8322–26.

Neff, H. 1992. Ceramics and evolution. *Archaeological Method and Theory* 4:141–93.

Nei, M., and S. Kumar. 2000. *Molecular evolution and phylogenetics*. New York: Oxford University Press.

Nettle, D. 1998. Explaining global patterns of language diversity. *Journal of Anthropological Archaeology* 17:354–74.

———. 1999a. Is the rate of linguistic change constant? *Lingua* 108:119–36.

———. 1999b. *Linguistic diversity*. Oxford: Oxford University Press.

———. 1999c. Linguistic diversity of the Americas can be reconciled with a recent colonization. *Proceedings of the National Academy of Sciences of the United States of America* 96:3325–29.

Newman, T. K., C. J. Jolly, and J. Rogers. 2004. Mitochondrial phylogeny and systematics of baboons (*Papio*). *American Journal of Physical Anthropology* 124:17–27.

Nicholls, G., and R. Gray. 2006. Quantifying uncertainty in a stochastic model of vocabulary evolution. In *Phylogenetic Methods and the Prehistory of Languages*, edited by P. Forster and C. Refrew, 161–72. Cambridge: McDonald Institute for Achaeological Research.

Nichols, J. 1990. Linguistic diversity and the first settlement of the new world. *Language* 66:475–521.

———. 1992. *Linguistic diversity in space and time*. Chicago: University of Chicago Press.

———. 1994. The spread of language around the Pacific rim. *Evolutionary Anthropology* 3:206–15.

———. 1997. Modeling ancient population structures and movement in linguistics. *Annual Review of Anthropology* 26:359–84.

Niklas, K. 1994. Plant allometry: The scaling of form and process. Chicago: University of Chicago Press.

Nunn, C. L. 1995. A simulation test of Smith's "degrees of freedom" correction for comparative studies. *American Journal of Physical Anthropology* 98:355–67.

———. 1999a. The evolution of exaggerated sexual swellings in primates and the graded signal hypothesis. *Animal Behaviour* 58:229–46.

———. 1999b. The number of males in primate social groups: a comparative test of the socioecological model. *Behavioral Ecology and Sociobiology* 46:1–13.

———. 2000. Collective action, "free-riders," and male extragroup conflict. In Kappeler 2000, 192–204.

———. 2002a. A comparative study of leukocyte counts and disease risk in primates. *Evolution* 56:177–90.

———. 2002b. Spleen size, disease risk and sexual selection: A comparative study in primates. *Evolutionary Ecology Research* 4:91–107.

Nunn, C. L., and S. Altizer. 2005. The *Global Mammal Parasite Database*: An online resource for infectious disease records in wild primates. *Evolutionary Anthroplogy* 14:1–2.

———. 2006. *Infectious diseases in primates: Behavior, ecology and evolution*. Oxford: Oxford University Press.

Nunn, C. L., S. Altizer, K. E. Jones, and W. Sechrest. 2003. Comparative tests of parasite species richness in primates. *American Naturalist* 162:597–614.

Nunn, C. L., S. M. Altizer, W. Sechrest, and A. Cunningham. 2005. Latitudinal gradients of disease risk in primates. *Diversity and Distributions* 11:249–56.

Nunn, C. L., S. Altizer, W. Sechrest, K. E. Jones, R. A. Barton, and J. L. Gittleman. 2004. Parasites and the evolutionary diversification of primate clades. *American Naturalist* 164:S90–103.

Nunn, C. L., C. Arnold, L. Matthews, and M. Borgerhoff Mulder. 2010. Simulating trait evolution for cross-cultural comparison. *Philisophical Transactions of the Royal Society of London B: Biological Sciences* 364:61–9.

Nunn, C. L., and R. A. Barton. 2000. Allometric slopes and independent contrasts: A comparative test of Kleiber's law in primate ranging patterns. *American Naturalist* 156: 519–33.

———. 2001. Comparative methods for studying primate adaptation and allometry. *Evolutionary Anthroplogy* 10:81–98.

Nunn, C. L., M. Borgerhoff Mulder, and S. Langley. 2006. Comparative methods for studying cultural trait evolution: A simulation study. *Cross-Cultural Research* 40:177–209.

Nunn, C. L., J. L. Gittleman, and J. Antonovics. 2000. Promiscuity and the primate immune system. *Science* 290:1168–70.

Nunn, C. L., and E. W. Heymann. 2005. Malaria infection and host behaviour: A comparative study of neotropical primates. *Behavioral Ecology and Sociobiology* 59:30–37.

Nunn, C. L., and R. J. Lewis. 2001. Cooperation and collective action in animal behaviour. In *Economics in nature: The evolutionary biology of economic behaviour*, edited by R. Noë, J. A. R. A. M. van Hooff, and P. Hammerstein, 42–66. Cambridge: Cambridge University Press.

Nunn, C. L., and C. P. van Schaik. 2002. Reconstructing the behavioral ecology of extinct primates. In *Reconstructing behavior in the fossil record*, edited by J. M. Plavcan, R. F. Kay, W. L. Jungers, and C. P. van Schaik, 159–216. New York: Kluwer Academic/Plenum.

O'Brien, M. J., J. Darwent, and R. L. Lyman. 2001. Cladistics is useful for reconstructing archaeological phylogenies: Palaeoindian points from the southeastern United States. *Journal of Archaeological Science* 28:1115–36.

O'Brien, M. J., T. D. Holland, R. J. Hoard, and G. L. Fox. 1994. Evolutionary implications of design and performance characteristics of prehistoric pottery. *Journal of Archeological Method and Theory* 1:259–304.

O'Brien, M. J., and R. L. Lyman. 2000. *Applying evolutionary archaeology*. New York: Kluwer Academic.

———. 2002. Evolutionary archeology: Current status and future prospects. *Evolutionary Anthropology* 11:26–36.

———. 2003. *Cladistics and archaeology*. Salt Lake City: University of Utah Press.

O'Brien, R. M. 2007. A caution regarding rules of thumb for variance inflation factors. *Quality and Quantity* 41:673–90.

O'Connor, M. P., S. J. Agosta, F. Hansen, et al. 2007. Phylogeny, regression, and the allometry of physiological traits. *American Naturalist* 170:431–42.

O'Meara, B. C., C. Ane, M. J. Sanderson, and P. C. Wainwright. 2006. Testing for different rates of continuous trait evolution using likelihood. *Evolution* 60:922–33.

O'Neill, M., and S. Dobson. 2008. The degree and pattern of phylogenetic signal in primate long-bone structure. *Journal of Human Evolution* 54:309–22.

Oakley, T. H., and C. W. Cunningham. 2000. Independent contrasts succeed where ancestor reconstruction fails in a known bacteriophage phylogeny. *Evolution* 54:397–405.

———. 2002. Molecular phylogenetic evidence for the independent evolutionary origin of an arthropod compound eye. *Proceedings of the National Academy of Sciences of the United States of America* 99:1426–30.

Omland, K. E. 1997a. Correlated rates of molecular and morphological evolution. *Evolution* 51:1381–93.

———. 1997b. Examining two standard assumptions of ancestral reconstructions: repeated loss of dichromatism in dabbling ducks (Anatini). *Evolution* 51:1636–46.

———. 1999. The assumptions and challenges of ancestral state reconstructions. *Systematic Biology* 48:604–11.

Organ, C. L., C. L. Nunn, Z. Machanda, and R. W. Wrangham. n.d. Phylogenetic Rate Shifts in Chewing Time and Molar Size During Human Evolution.

Organ, C. L., A. M. Shedlock, A. Meade, M. Pagel, and S. V. Edwards. 2007. Origin of avian genome size and structure in non-avian dinosaurs. *Nature* 446:180–84.

Ossi, K., and J. M. Kamilar. 2006. Environmental and phylogenetic correlates of *Eulemur* behavior and ecology (primates: Lemuridae). *Behavioral Ecology and Sociobiology* 61: 53–64.

Ostner, J., C. Nunn, and O. Schülke. 2008. Female reproductive synchrony predicts skewed paternity across primates. *Behavioral Ecology* 19:1150.

Packard, G. C., and T. J. Boardman. 1999. The use of percentages and size-specific indices to normalize physiological data for variation in body size: Wasted time, wasted effort? *Comparative Biochemistry and Physiology A—Molecular and Integrative Physiology* 122:37–44.

Page, R. D. M. 2003. Tangled trees: Phylogenies, cospeciation and coevolution. Chicago: University of Chicago Press.

Page, R. D. M., and M. A. Charleston. 1998. Trees within trees: Phylogeny and historical associations. *Trends in Ecology & Evolution* 13:356–59.

Page, R. D. M., and E. C. Holmes. 1998. *Molecular evolution: A phylogenetic approach*. Oxford:, Blackwell.

Pagel, M. D. 1992. A method for the analysis of comparative data. *Journal of Theoretical Biology* 156:431–42.

———. 1993. Seeking the evolutionary regression coefficient: An analysis of what comparative methods measure. *Journal of Theoretical Biology* 164:191–205.

———. 1994a. Detecting correlated evolution on phylogenies: A general method for the comparative analysis of discrete characters. *Proceedings of the Royal Society B: Biological Sciences* 255:37–45.

———. 1994b. The adaptationist wager.In *Phylogenetics and ecology*, edited by P. Eggleton and R. I. Vane-Wright, 29–51. London: Academic Press.

———. 1997. Inferring evolutionary processes from phylogenies. *Zoologica Scripta* 26: 331–48.

———. 1999a. Inferring the historical patterns of biological evolution. Nature 401:877–884.

———. 1999b. The maximum likelihood approach to reconstructing ancestral character states of discrete characters on phylogenies. Systematic Biology 48:612–622.

———. 2000a. Maximum-likelihood models for glottochronology and for reconstructing linguistic phylogenies. In Renfrew, McMahon, and Trask 2000, 189–207.

———. 2000b. The history, rate and pattern of world linguistic evolution. In *The evolutionary emergence of language*, edited by C. Knight, M. Studdert-Kennedy, and J. R. Hurford, 391–416. Cambridge: Cambridge University Press.

———. 2002. Modelling the evolution of continuously varying characters on phylogenetic trees: The case of hominid cranial capacity. In *Morphology, shape and phylogeny*, edited by N. MacLeod, and P. L. Forey, 269–286. London: Taylor and Francis.

Pagel, M. D., Q. D. Atkinson, and A. Meade. 2007. Frequency of word-use predicts rates of lexical evolution throughout Indo-European history. *Nature* 449:717–20.

Pagel, M. D., and P. H. Harvey. 1988. The taxon-level problem in the evolution of mammalian brain size: Facts and artifacts. *American Naturalist* 132:344–59.

———. 1989. Taxonomic differences in the scaling of brain on body weight among mammals. *Science* 244:1589–93.

———. 1992. On solving the correct problem: wishing does not make it so. *Journal of Theoretical Biology* 156:425–30.

Pagel, M. D., and F. Lutzoni. 2002. Accounting for phylogenetic uncertainty in comparative studies of evolution and adaptation. In *Biological evolution and statistical physics*, edited by M. Lässig and A. Valleriani, 148–61. Berlin: Springer-Verlag.

Pagel, M. D., and R. Mace. 2004. The cultural wealth of nations. *Nature* 428:275–78.

Pagel, M. D., R. M. May, and A. R. Collie. 1991. Ecological aspects of the geographical-distribution and diversity of mammalian-species. *American Naturalist* 137:791–815.

Pagel, M. D., and A. Meade. 2004. A phylogenetic mixture model for detecting pattern-heterogeneity in gene sequence or character-state data. *Systematic Biology* 53:571–81.

———. 2005. Bayesian estimation of correlated evolution across cultures: A case study of marriage systems and wealth transfer at marriage. In Mace, Holden, and Shennan 2005, 235–56.

———. 2006a. Bayesian analysis of correlated evolution of discrete characters by reversible-jump Markov chain Monte Carlo. *American Naturalist* 167:808–25.

———. 2006b. Estimating rates of lexical replacement on phylogenetic trees of languages. *Phylogenetic Methods and the Prehistory of Languages* 1:173.

Pagel, M. D., A. Meade, and D. Barker. 2004. Bayesian estimation of ancestral character states on phylogenies. *Systematic Biology* 53:673–84.

Paradis, E. 1998. Detecting shifts in diversification rates without fossils. *American Naturalist* 152:176–87.

———. 2004. Can extinction rates be estimated without fossils? *Journal of Theoretical Biology* 229:19–30.

———. 2005. Statistical analysis of diversification with species traits. *Evolution* 59:1–12.

Paradis, E., and J. Claude. 2002. Analysis of comparative data using generalized estimating equations. *Journal of Theoretical Biology* 218:175–85.

Parker, G. A. 1970. Sperm competition and its evolutionary consequences in insects. *Biological Reviews of the Cambridge Philosophical Society* 45:525–67.

Paul, A. 1997. Breeding seasonality affects the association between dominance and reproductive success in nonhuman male primates. *Folia Primatologica* 68:344–49.

Pedersen, A., and T. Davies. 2009. Cross-species pathogen transmission and disease emergence in primates. *EcoHealth* 6:496–508.

Pedersen, A. B., M. Poss, S. Altizer, A. Cunningham, and C. Nunn. 2005. Patterns of host specificity and transmission among parasites of wild primates. *International Journal for Parasitology* 35:647–57.

Pereira, M. E. 1991. Asynchrony within estrous synchrony among ringtailed lemurs (primates: Lemuridae). *Physiology and Behavior* 49:47–52.

Peters, R. H. 1983. *The ecological implications of body size*. Cambridge: Cambridge University Press.

Petraitis, P. S., A. E. Dunham, and P. H. Niewlarowski. 1996. Inferring multiple causality: The limitataions of path analysis. *Functional Ecology* 10:421–31.

Platnick, N. I., and H. D. Cameron. 1977. Cladistic methods in textual, linguistic, and phylogenetic analysis. *Systematic Zoology* 26:380–85.

Plavcan, J. M. 2001. Sexual dimorphism in primate evolution. *Yearbook of Physical Anthropology* 44:25–53

———. *Reconstructing behavior in the primate fossil record.* New York: Kluwer Academic/Plenum.

Plavcan, J. M., and D. Daegling. 2006. Interspecific and intraspecific relationships between tooth size and jaw size in primates. *Journal of Human Evolution* 51:171–84.

Plavcan, J. M., and C. Ruff. 2008. Canine size, shape, and bending strength in primates and carnivores. *Yearbook of Physical Anthropology* 136:65–84.

Plavcan, J. M., and C. P. van Schaik. 1992. Intrasexual competition and canine dimorphism in anthropoid primates. *American Journal of Physical Anthropology* 87:461–77.

———. 1997a. Interpreting hominid behavior on the basis of sexual dimorphism. *American Journal of Physical Anthropology* 32:345–74.

———. 1997b. Intrasexual competition and body weight dimorphism in anthropoid primates. *American Journal of Physical Anthropology* 103:37–68.

Plavcan, J. M., C. P. van schaik, and P. M. Kappeler. 1995. Competition, coalitions and canine size in primates. *Journal of Human Evolution* 28:245–76.

Pocklington, R. 2006. What is a culturally transmitted unit, and how do we find one? In Lipo et al. 2006b, 19–31.

Polly, P. D. 2001. Paleontology and the comparative method: Ancestral node reconstructions versus observed node values. *American Naturalist* 157:596–609.

Posada, D., and T. R. Buckley. 2004. Model selection and model averaging in phylogenetics: Advantages of akaike information criterion and Bayesian approaches over likelihood ratio tests. *Systematic Biology* 53:793–808.

Posada, D., and K. A. Crandall. 2001a. Intraspecific gene genealogies: Trees grafting into networks. *Trends in Ecology & Evolution* 16:37–45.

———. 2001b. Selecting the best-fit model of nucleotide substitution. *Systematic Biology* 50:580–601.

Posadas, P., D. R. M. Esquivel, and J. V. Crisci. 2001. Using phylogenetic diversity measures to set priorities in conservation: An example from southern South America. *Conservation Biology* 15:1325–34.

Price, T. 1997. Correlated evolution and independent contrasts. *Philosophical Transactions of the Royal Society B: Biological Sciences* 352:519–29.

Price, T., I. J. Lovette, E. Bermingham, H. L. Gibbs, and A. D. Richman. 2000. The imprint of history on communities of North American and Asian warblers. *American Naturalist* 156:354–67.

Purvis, A. 1995. A composite estimate of primate phylogeny. *Philosophical Transactions of the Royal Society B: Biological Sciences* 348:405–21.

Purvis, A., P. M. Agapow, J. L. Gittleman, and G. M. Mace. 2000a. Nonrandom extinction and the loss of evolutionary history. *Science* 288:328–30.

Purvis, A., M. Cardillo, R. Grenyer, and B. Collen. 2005a. Correlates of extinction risk: Phylogeny, biology, threat and scale. In Purvis, Gittleman, and Brooks 2005b, 295–316.

Purvis, A., and T. Garland. 1993. Polytomies in comparative analyses of continuous characters. *Systematic Biology* 42:569–75.

Purvis, A., J. L. Gittleman, and T. Brooks, ed. 2005b. *Phylogeny and conservation.* Cambridge: Cambridge University Press.

Purvis, A., J. L. Gittleman, G. Cowlishaw, and G. M. Mace. 2000b. Predicting extinction risk in declining species. *Proceedings of the Royal Society B: Biological Sciences* 267:1947–52.

Purvis, A., J. L. Gittleman, and H. Luh. 1994. Truth or consequences: Effects of phylogenetic accuracy on two comparative methods. *Journal of Theoretical Biology* 167:293–300.

Purvis, A., S. Nee, and P. H. Harvey. 1995. Macroevolutionary inferences from primate phylogeny. *Proceedings of the Royal Society B: Biological Sciences* 260:329–33.

Purvis, A., C. D. L. Orme, and K. Dolphin. 2003. Why are most species small-bodied? A phylogenetic view. *Macroecology: Concepts and Consequences*, edited by T. M. Blackburn and K. J. Gaston, 155–173. Oxford: Blackwell Science Ltd.

Purvis, A., and A. Rambaut. 1995. Comparative analysis by independent contrasts (CAIC): An Apple Macintosh application for analysing comparative data. *Computer Applications in the Biosciences* 11:247–51.

Purvis, A., and A. J. Webster. 1999. Phylogenetically independent contrasts and primate phylogeny. In Lee 1999, 44–68.

Pybus, O. G., and P. H. Harvey. 2000. Testing macro-evolutionary models using incomplete molecular phylogenies. *Proceedings of the Royal Society of London Series B: Biological Sciences* 267:2267–72.

R Development Core Team. 2009. R: *A language and environment for statistical computing.* Vienna, Austria: R Foundation for Statistical Computing.

Rabosky, D. 2010. Extinction rates should not be estimated from molecular phylogenies. *Evolution* 64:1816–24.

Rabosky, D., and M. Alfaro. 2010. Evolutionary bangs and whimpers: Methodological advances and conceptual frameworks for studying exceptional diversification. *Systematic Biology* 59:615–18.

Rambaut, A., N. C. Grassly, S. Nee, and P. H. Harvey. 1996. Bi-De: An application for simulating phylogenetic processes. *Computer Applications in the Biosciences* 12:469–71.

Rapoport, E. H. 1982. *Areography: Geographical strategies of species.* Oxford: Pergamon.

Raup, D. M., S. J. Gould, T. J. M. Schopf, and D.S. Simberloff. 1973. Stochastic models of phylogeny and the evolution of diversity. *Journal of Geology* 81:525–42.

Ravosa, M. J. 1996. Jaw morphology and function in living and fossil old world monkeys. *International Journal of Primatology* 17:909–32.

Ravosa, M. J., and W. Hylander. 1994. Function and fusion of the mandibular symphysis in primates: Stiffness or strength. In *Anthropoid origins*, edited by J.G. Fleagle and R.F. Kay, 447–68. New York: Plenum.

Rayner, J. 1985. Linear relations in biomechanics: the statistics of scaling functions. *Journal of Zoology London* 206:415–39.

Read, A. F., and S. Nee. 1995. Inference from binary comparative data. *Journal of Theoretical Biology* 173:99–108.

Reader, S. M., and K. N. Laland. 2001. Primate innovation: Sex, age and social rank differences. *International Journal of Primatology* 22:787–805.

———. 2002. Social intelligence, innovation, and enhanced brain size in primates. *Proceedings of the National Academy of Sciences of the United States of America* 99: 4436–41.

Ree, R. H. 2005. Detecting the historical signature of key innovations using stochastic models of character evolution and cladogenesis. *Evolution* 59:257–65.

Ree, R. H., and M. J. Donoghue. 1998. Step matrices and the Interpretation of homoplasy. *Systematic Biology* 47:582–88.

———. 1999. Inferring rates of change in flower symmetry in asterid angiosperms. *Systematic Biology* 48:633–41.

Reeve, H., and P. Sherman. 1993. Adaptation and the goals of evolutionary research. *Quarterly Review of Biology* 68:1–32.

Relethford, J. H. 1994. Craniometric variation among modern human populations. *American Journal of Physical Anthropology* 95:53–62.

Rendall, D., and A. Di Fiore. 2007. Homoplasy, homology, and the perceived special status of behavior in evolution. *Journal of Human Evolution* 52:504–521.

Renfrew, C. 1987. *Archaeology and language: the puzzle of Indo-European origins*. Cambridge: Cambridge University Press.

———. 1992. Archaeology, genetics and linguistic diversity. *Man*: 27:445–78.

———. 2000a. *At the edge of knowability: Towards a prehistory of language*. Cambridge Archeological Journal 10:7–34.

———. 2000b. The problem of time depth. In Renfrew, McMahon, and Trask 2000, ix–vix.

Renfrew, C., A. McMahon, and L. Trask, ed. 2000. *Time depth in historical linguistics*. Cambridge: McDonald Institute for Archaeological Research.

Revell, L. J. 2008. On the analysis of evolutionary change along single branches in a phylogeny. *American Naturalist* 172:140–47.

———. 2009. Size-correction and principal components for interspecific comparative studies. *Evolution* 63:3258–68.

———. 2010. Phylogenetic signal and linear regression on species data. *Methods in Ecology and Evolution*. In press.

Revell, L. J., and D. C. Collar. 2009. Phylogenetic analyses of the evolutionary correlation using likelihood. *Evolution* 63:1090–100.

Revell, L. J., and L. J. Harmon. 2008. Testing quantitative genetic hypotheses about the evolutionary rate matrix for continuous characters. *Evolutionary Ecology Research* 10: 311–31.

Revell, L. J., L. J. Harmon, and D. C. Collar. 2008. Phylogenetic signal, evolutionary process, and rate. *Systematic Biology* 57:591–601.

Rexová, K., D. Frynta, and J. Zrzavy. 2003. Cladistic analysis of languages: Indo-European classification based on lexicostatistical data. *Cladistics* 19:120–27.

Reznick, D. N., F. H. Shaw, F. H. Rodd, and R. G. Shaw. 1997. Evaluation of the rate of evolution in natural populations of guppies (*Poecilia reticulata*). *Science* 275:1934–37.

Richerson, P. J., and R. Boyd. 2005. *Not by genes alone: How culture transformed human evolution*. Chicago: University of Chicago Press.

Ricklefs, R. E., and J. M. Starck. 1996. Applications of phylogenetically independent contrasts: A mixed progress report. *OIKOS* 77:167–72.

Ridley, M. 1983. *The explanation of organic diversity: the comparative method and adaptations of mating*. Oxford: Clarendon.

———. 1986. The number of males in a primate troop. *Animal Behaviour* 34:1848–58.

Riede, F. 2009. Tangled trees: Modeling material culture evolution as host-associate cospeciation. In Shennan 2009, 85–99.

Ringe, D., T. Warnow, and A. Taylor. 2002. Indo-European and Computational Cladistics. *Transactions of the Philological Society* 100:59–129.

Robson-Brown, K. 1999. Cladistics as a tool in comparative analysis. In Lee 1999, 23–43.

Roca, A., N. Georgiadis, and S. O'Brien. 2005. Cytonuclear genomic dissociation in African elephant species. *Nature Genetics* 37:96–100.

Roch, S. 2010. Toward extracting all phylogenetic information from matrices of evolutionary distances. *Science* 327:1376–79.

Rodseth, L., R. W. Wrangham, A. M. Harrigan, et al. 1991. The human community as a primate society. *Current Anthropology* 32:221–54.

Rohlf, F. J. 2001. Comparative methods for the analysis of continuous variables: Geometric interpretations. *Evolution* 55:2143–60.

———. 2006. A comment on phylogenetic correction. *Evolution* 60:1509–15.

Rohlf, F. J., W. S. Chang, R. R. Sokal, and J. Y. Kim. 1990. Accuracy of estimated phylogenies—Effects of tree topology and evolutionary model. *Evolution* 44:1671–84.

Ronquist, F. 2004. Bayesian inference of character evolution. *Trends in Ecology & Evolution* 19:475–81.

Ronquist, F., and J. P. Huelsenbeck. 2003. MrBayes 3: Bayesian phylogenetic inference under mixed models. *Bioinformatics* 19:1572–74.

Rose, M. R., and G. V. Lauder. 1996. *Adaptation.* San Diego, CA: Academic Press.

Rosenzweig, M. L. 1995. *Species diversity in space and time.* Cambridge: Cambridge University Press.

Ross, C., M. Henneberg, M. Ravosa, and S. Richard. 2004. Curvilinear, geometric and phylogenetic modeling of basicranial flexion: Is it adaptive, is it constrained? *Journal of Human Evolution* 46:185–213.

Ross, C., and K. E. Jones. 1999. Socioecology and the evolution of primate reproductive rates. In Lee 1999, 73–110.

Rowe, N. 1996. *The pictorial guide to the living primates.* East Hampton, NY: Pogonias Press.

Ruvolo, M. 1987. Reconstructing genetic and linguistic trees: Phenetic and cladistic approaches. In Hoenigswald and Wiener 1987, 193–216.

———. 1997. Molecular phylogeny of the hominoids: Inferences from multiple independent DNA sequence data sets. *Molecular Biology and Evolution* 14:248–65.

Ryan, M. J., and A. S. Rand. 1995. Female responses to ancestral advertisement calls in Túngara frogs. *Science* 269:390–92.

Saitou, N., and M. Nei. 1987. The neighbor-joining method: A new method for reconstructing phylogenetic trees. *Molecular Biology and Evolution* 4:406–25.

Salisbury, B. A., and J. H. Kim. 2001. Ancestral state estimation and taxon sampling density. *Systematic Biology* 50:557–64.

Sanderson, M. J. 1990. Estimating rates of speciation and evolution: a bias due to homoplasy. *Cladistics* 6:387–91.

———. 1993. Reversibility in evolution—A maximum-likelihood approach to character gain loss bias in phylogenies. *Evolution* 47:236–52.

———. 1995. Objections to bootstrapping phylogenies. *Systematic Biology* 44:299–320.

———. 2002. Estimating absolute rates of molecular evolution and divergence times: A penalized likelihood approach. *Molecular Biology and Evolution* 19:101–9.

Sanderson, M. J., and G. Bharathan. 1993. Does cladistic information affect inferences about branching rates? *Systematic Biology* 42:1–17.

Sanderson, M. J., and M. J. Donoghue. 1989. Patterns of variation in levels of homoplasy. *Evolution* 43:1781–95.

———. 1994. Shifts in diversification rate with the origin of angiosperms. *Science* 264: 1590–93.

————. 1996. Reconstructing shifts in diversification rates on phylogenetic trees. *Trends in Ecology & Evolution* 11:15–20.

Sanderson, M. J., A. Purvis, and C. Henze. 1998. Phylogenetic supertrees: Assembling the trees of life. *Trends in Ecology & Evolution* 13:105–9.

Sawaguchi, T., and H. Kudo. 1990. Neocortical development and social structure in primates. *Primates* 31:283–89.

Schluter, D. 2000. *The ecology of adaptive radiation.* Oxford: Oxford University Press.

Schluter, D., T. Price, A. O. Mooers, and D. Ludwig. 1997. Likelihood of ancestor states in adaptive radiation. *Evolution* 51:1699–711.

Schmidt-Nielsen, K. 1975. Scaling in biology: The consequences of size. *Journal of Experimental Zoology* 194:287–307.

————. 1984. *Scaling: Why is animal size so important?* Cambridge: Cambridge University Press.

Schultz, A. H. 1930. The skeleton of the trunk and limbs of higher primates. *Human Biology* 2:303–438.

Schultz, T. R., and G. A. Churchill. 1999. The role of subjectivity in reconstructing ancestral character states: A Bayesian approach to unknown rates, states, and transformation asymmetries. *Systematic Biology* 48:651–64.

Scotland, R. W., R. G. Olmstead, and J. R. Bennett. 2003. Phylogeny reconstruction: The role of morphology. *Systematic Biology* 52:539–48.

Sechrest, W., T. M. Brooks, G. A. B. da Fonseca, et al. 2002. Hotspots and the conservation of evolutionary history. *Proceedings of the National Academy of Sciences of the United States of America* 99:2067–71.

Serre, D., and S. P. Pääbo. 2004. Evidence for gradients of human genetic diversity within and among continents. *Genome Research* 14:1679–85.

Shao, K.-T., and R. R. Sokal. 1990. Tree balance. *Systematic Zoology* 39:266–76.

Shennan, S. 2002. *Genes, memes and human history: Darwinian archaeology and cultural evolution.* London: Thames and Hudson.

————. 2009. *Pattern and process in cultural evolution.* Berkeley: University of California Press.

Shettleworth, S. 1998. *Cognition, evolution, and behavior.* New York: Oxford University Press.

Short, R. V. 1979. Sexual selection and its component parts, somatic and genital selection, as illustrated by man and the great apes. *Advances in the Study of Behavior* 9:131–58.

Shoshani, J., C. P. Groves, E. L. Simons, and G. F. Gunnell. 1996. Primate phylogeny: Morphological vs. molecular results. *Molecular Phylogenetics and Evolution* 5:102–54.

Sieg, A., M. O'Connor, J. McNair, B. Grant, S. Agosta, and A. Dunham. 2009. Mammalian metabolic allometry: Do intraspecific variation, phylogeny, and regression models matter? *American Naturalist* 174: 720–33.

Silcox, M. T., J. I. Bloch, D. M. Boyer, et al. 2009. Semicircular canal system in early primates. *Journal of Human Evolution* 56:315–27.

Sillén-Tullberg, B., and A. P. Møller. 1993. The relationship between concealed ovulation and mating systems in anthropoid primates: A phylogenetic analysis. *American Naturalist* 141:1–25.

Simons, E. 1995. Egyptian oligocene primates: A review. *American Journal of Physical Anthropology* 38:199–238.

Simpson, G. G. 1944. *Tempo and mode in evolution.* New York: Columbia University Press.

Slowinski, J. B., and C. Guyer. 1989. Testing null models in questions of evolutionary success. *Systematic Zoology* 38:189–91.

———. 1993. Testing whether certain traits have caused amplified diversification—An improved method based on a model of random speciation and extinction. *American Naturalist* 142:1019–24.

Smith, E. A. 2001. On the coevolution of cultural, linguistic, and biological diversity. *On biocultural diversity: Linking language, knowledge, and the environment,* edited by L. Maffi, 95–117. Washington DC: Smithsonian Books.

Smith, R. J. 1980. Rethinking allometry. *Journal of Theoretical Biology* 87:97–111.

———. 1983. The mandibular corpus of female primates: Taxonomic, dietary, and allometric correlates of interspecific variations in size and shape. *American Journal of Physical Anthropology* 61:315–30.

———. 1984a. Determination of relative size: The "criterion of subtraction" problem in allometry. *Journal of Theoretical Biology* 108:131–42.

———. 1993. Logarithmic transformation bias in allometry. *American Journal of Physical Anthropology* 90:215–28.

———. 1994. Degrees of freedom in interspecific allometry: an adjustment for the effects of phylogenetic constraint. *American Journal of Physical Anthropology* 93:95–107.

———. 1996. Biology and body size in human evolution. *Current Anthropology* 37:451–81.

———. 2005. Relative size versus controlling for size. *Current Anthropology* 46:249–73.

Smith, R. J., and J. M. Cheverud. 2002. Scaling of sexual dimorphism in body mass: A phylogenetic analysis of Rensch's rule in primates. *International Journal of Primatology* 23:1095–135.

Smith, R. J., and W. L. Jungers. 1997. Body mass in comparative primatology. *Journal of Human Evolution* 32:523–59.

Smith, R. L. 1984b. *Sperm competition and the evolution of animal mating strategies.* Orlando, FL: Academic Press.

Smouse, P. E., and J. C. Long. 1992. Matrix correlation-analysis in anthropology and genetics. *Yearbook of Physical Anthropology* 35:187–213.

Smouse, P. E., J. C. Long, and R. R. Sokal. 1986. Multiple regression and correlation extensions of the mantel test of matrix correspondence. *Systematic Zoology* 35:627–32.

Sokal, R. R., N. L. Oden, and C. Wilson. 1991. Genetic evidence for the spread of agriculture in Europe by demic diffusion. *Nature* 351:143–45.

Sokal, R. R., and F. J. Rohlf. 1995. *Biometry.* New York: W. H. Freeman and Company.

Soltis, J., R. Boyd, and P. J. Richerson. 1995. Can group-functional behaviors evolve by cultural group selection? An empirical test. *Current Anthropology* 36:473–94.

Spathelf, M., and T. A. Waite. 2007. Will hotspots conserve extra primate and carnivore evolutionary history? *Diversity and Distributions* 13:746–51.

Spence, A. 2009. Scaling in biology. *Current Biology* 19:57–61.

Spocter, M. A., and P. R. Manger. 2007. The use of cranial variables for the estimation of body mass in fossil hominins. *American Journal of Physical Anthropology* 134:92–105.

Spoor, F., T. Garland Jr., G. Krovitz, T. M. Ryan, M. T. Silcox, and A. Walker. 2007. The primate semicircular canal system and locomotion. *Proceedings of the National Academy of Sciences of the United States of America* 104:10808–12.

Srivastava, A., and R. I. M. Dunbar. 1996. The mating system of Hanuman langurs: A problem in optimal foraging. *Behavioral Ecology and Sociobiology* 39:219–26.

Stahl, W. R. 1965. Organ weights in primates and other mammals. *Science* 150:1039–42.

Stanford, C. B., and J. S. Allen. 1991. On strategic storytelling—Current models of human behavioral evolution. *Current Anthropology* 32:58–61.

Stanley, S. M. 1973. An explanation for cope's rule. *Evolution* 27:1–26.

————. 1979. *Macroevolution, pattern and process*. San Francisco: Freeman.

Sterck, E. H. M., D. P. Watts, and C. P. van Schaik. 1997. The evolution of female social relationships in nonhuman primates. *Behavioral Ecology and Sociobiology* 41:291–309.

Stern, J., and R. Susman. 1983 The locomotor anatomy of *Australopithecus afarensis*. *American Journal of Physical Anthropology* 60:279–317.

Stevens, G. C. 1989. The latitudinal gradient in geographical range—How so many species coexist in the tropics. *American Naturalist* 133:240–56.

Stewart, C. B. 1993. The powers and pitfalls of parsimony. *Nature* 361:603–7.

Strait, D. S., and F. E. Grine. 2004. Inferring hominoid and early hominid phylogeny using craniodental characters: The role of fossil taxa. *Journal of Human Evolution* 47:399–452.

Strait, D., F. E. Grine, and J. G. Fleagle. 2007. Analyzing hominid phylogeny. In *Handbook of paleoanthropology*, edited by W. Henke and I. Tattersall, 1781–806. Berlin: Springer-Verlag.

Strait, D. S., F. E. Grine, and M. A. Moniz. 1997. A reappraisal of early hominid phylogeny. *Journal of Human Evolution* 32:17–82.

Strier, K. B. 2003. Primatology comes of age: 2002 AAPA luncheon address. *Yearbook of Physical Anthropology* 46:2–13.

Struhsaker, T. T. 2000. Variation in adult sex ratios of red colobus monkey social groups: implications for interspecific comparisons. In Kappeler 2000, 108–19.

Supriatna, J., B. O. Manullang, and E. Soekara. 1986. Group composition, home range, and diet of the Maroon leaf monkey (*Presbytis rubicunda*) at Tanjung Puting reserve, Central Kalimantan, Indonesia. *Primates* 27:185–90.

Susman, R. L. 1987. Pygmy chimpanzees and common chimpanzees: Models for the behavioral ecology of the earliest hominids. In Kinzey 1987, 72–86.

Sutherland, W. J. 2003. Parallel extinction risk and global distribution of languages and species. *Nature* 423:276–79.

Swadesh, M. 1952. Lexico-statistic dating of prehistoric ethnic contacts: With special reference to North American Indians and Eskimos. *Proceedings of the American Philosophical Society* 96:452–63.

————. 1955. Towards greater accuracy in lexicostatistic dating. *International Journal of American Linguistics* 21:121–37.

Sweet, S. S. 1980. Allometric inference in morphology. *American Zoologist* 20:643–52.

Swofford, D. L., and W. P. Maddison. 1992. Parsimony, character-state reconstructions, and evolutionary inferences. *Systematics, historical ecology, and North American freshwater fishes*, edited by R. Mayden, 186–223. Palo Alto, CA: Stanford University Press.

Symonds, M. R. E. 2002. The effects of topological inaccuracy in evolutionary trees on the phylogenetic comparative method of independent contrasts. *Systematic Biology* 51: 541–53.

Symonds, M. R. E., and M. A. Elgar. 2002. Phylogeny affects estimation of metabolic scaling in mammals. *Evolution* 56:2330–33.

Tattersall, I. 1987. Cathemeral activity in primates: A definition. *Folia Primatologica* 49: 200–2.

Taub, D. M. 1980. Female choice and mating strategies among wild Barbary macaques (*Macaca sylvanus* L.).*The macaques: Studies in ecology, behavior and evolution*, edited by D. G. Lindburg, 287–344. New York: van Nostrand.

Tehrani, J., and M. Collard. 2002. Investigating cultural evolution through biological phylogenetic analyses of Turkmen textiles. *Journal of Anthropological Archaeology* 21:443–63.

————. 2009. On the relationship between interindividual cultural transmission and population-level cultural diversity: A case study of weaving in Iranian tribal populations. *Evolution and Human Behavior* 30:286–300.e281.

Tehrani, J. J., M. Collard, and S. J. Shennan. 2010. Tracking traditions thorugh trees and jungles: a cophylogenetic approach to cultural inheritance. *Philisophical Transactions of the Royal Society B: Biological Sciences* 365:3865–74.

Tëmkin, I., and N. Eldredge. 2007. Phylogenetics and material cultural evolution. *Current Anthropology* 48:146–54.

Tennie, C., J. Call, and M. Tomasello. 2006. Push or pull: Imitation vs. emulation in great apes and human children. *Ethology* 112:1159–69.

———. 2009. Ratcheting up the ratchet: On the evolution of cumulative culture. *Philosophical Transactions of the Royal Society B: Biological Sciences* 364:2405.

Thierry, B., F. Aureli, C. L. Nunn, O. Petit, C. Abegg, and F. B. de Waal. 2008. A comparative study of conflict resolution in macaques: Insights into the nature of trait co-variation. *Animal Behaviour* 75:847–60.

Thomas, G. H., R. P. Freckleton, and T. Székely. 2006. Comparative analyses of the influence of developmental mode on phenotypic diversification rates in shorebirds. *Proceedings of the Royal Society B: Biological Sciences* 273:1619–24.

Thorne, J., and H. Kishino. 2002. Divergence time and evolutionary rate estimation with multilocus data. *Systematic Biology* 51:689–702.

Tilson, R. L., and R. R. Tenaza. 1976. Monogamy and duetting in an old world monkey. *Nature* 263:320–21.

Tishkoff, S. A., F. A. Reed, A. Ranciaro, et al. 2006. Convergent adaptation of human lactase persistence in Africa and Europe. *Nature Genetics* 39:31–40.

Tocheri, M. W., C. M. Orr, S. G. Larson, et al. 2007. The primitive wrist of *Homo floresiensis* and its implications for hominin evolution. *Science* 317:1743–45.

Tooby, J., and I. DeVore. 1987. The reconstruction of hominid behavioral evolution through strategic modeling. In Kinzey 1987, 183–237.

Tosi, A., J. Morales, and D. Melnick. 2002. Y-chromosome and mitochondrial markers in *Macaca fascicularis* indicate introgression with Indochinese *M. mulatta* and a biogeographic barrier in the Isthmus of Kra. *International Journal of Primatology* 23:161–78.

Turelli, M., N. H. Barton, and J. A. Coyne. 2001. Theory and speciation. *Trends in Ecology & Evolution* 16:330–43.

Tuttle, R. 1972. *The functional and evolutionary biology of primates.* Chicago: Aldine-Atherton.

Tylor, E. B. 1889. On a method of investigating the development of institutions applied to the law of marriage and descent. *Journal of the Royal Anthropological Institute* 18:245–72.

Valkenburgh, B. V. 1988. Trophic diversity in past and present guilds of large predatory mammals. *Paleobiology* 14:155–73.

Vamosi, S., S. Heard, J. Vamosi, and C. Webb. 2009. Emerging patterns in the comparative analysis of phylogenetic community structure. *Molecular Ecology* 18:572–92.

Van Riper, C., S. G. Van Riper, M. L. Goff, and M. Laird. 1986. The epizootiology and ecological significance of malaria in Hawaiian land birds. *Ecological Monographs* 56:327–44.

Van Schaik, C. P. 1989. The ecology of social relationships amongst female primates. In *Comparative socioecology*, edited by V. Standen, and R. A. Foley, 195–218. Oxford: Blackwell.

———. 1996. Social evolution in primates: The role of ecological factors and male behaviour. *Proceedings of the British Academy* 88:9–31.

Van Schaik, C. P., and C. Janson. 2000. Infanticide by Males and Its Implications. Cambridge, Cambridge University Press.

Van Schaik, C. P., and P. M. Kappeler. 1997. Infanticide risk and the evolution of male-female association in primates. *Proceedings of the Royal Society B: Biological Sciences* 264: 1687–94.

Van Schaik, C. P., M. A. van Noordwijk, and C. L. Nunn. 1999. Sex and social evolution in primates. In Lee 1999, 204–40.

Vane-Wright, R. I., C. J. Humphries, and P. H. Williams. 1991. What to protect? Systematics and the agony of choice. *Biological Conservation* 55:235–54.

Venditti, C., A. Meade, and M. Pagel. 2006. Detecting the node-density artifact in phylogeny reconstruction. *Systematic Biology* 55:637–43.

Venditti, C., and M. Pagel. 2010. Speciation as an active force in promoting genetic evolution. *Trends in Ecology & Evolution* 25:14–20.

Vigilant, L., M. Stoneking, H. Harpending, K. Hawkes, and A. C. Wilson. 1991. African populations and the evolution of human mitochondrial-DNA. *Science* 253:1503–7.

Vinyard, C., and J. Hanna. 2005. Molar scaling in strepsirrhine primates. *Journal of Human Evolution* 49:241–69.

Vitone, N. D., S. M. Altizer, and C. L. Nunn. 2004. Body size, diet and sociality influence the species richness of parasitic worms in anthropoid primates. *Evolutionary Ecology Research* 6:1–17.

Wanntorp, H. E., D. R. Brooks, T. Nilsson, et al. 1990. Phylogenetic approaches in ecology. *OIKOS* 57:119–32.

Warnow, T. 1997. Mathematical approaches to comparative linguistics. *Proceedings of the National Academy of Sciences of the United States of America* 94:6585–90.

Warton, D. I., I. J. Wright, D. S. Falster, and M. Westoby. 2006. Bivariate line-fitting methods for allometry. *Biological Reviews* 81:259–91.

Webb, C. O. 2000. Exploring the phylogenetic structure of ecological communities: An example for rain forest trees. *American Naturalist* 156:145–55.

Webb, C. O., D. D. Ackerly, M. A. McPeek, and M. J. Donoghue. 2002. Phylogenies and community ecology. *Annual Review of Ecology and Systematics* 33:475–505.

Webster, A. J., R. J. H. Payne, and M. Pagel. 2003. Molecular phylogenies link rates of evolution and speciation. *Science* 301:478.

Webster, A. J., and A. Purvis. 2002a. Ancestral states and evolutionary rates of continous characters. In *Morphology, shape and phylogeny*, edited by N. MacLeod and P. L. Forey 247–68. London: Taylor and Francis.

Webster, A. J., and A. Purvis. 2002b. Testing the accuracy of methods for reconstructing ancestral states of continuous characters. *Proceedings of the Royal Society B: Biological Sciences* 269:143–49.

Wenzel, J. W. 1992. Behavioral homology and phylogeny. *Annual Review of Ecology and Systematics* 23:361–81.

White, C., T. Blackburn, and R. Seymour. 2009a. Phylogenetically informed analysis of the allometry of mammalian basal metabolic rate supports neither geometric nor quarter-power scaling. *Evolution* 63:2658–67.

White, C., P. Cassey, and T. Blackburn. 2007. Allometric exponents do not support a universal metabolic allometry. *Ecology* 88:315–23.

White, D. R., M. L. Burton, and M. M. Dow. 1981. Sexual division of labor in African agriculture—A network auto-correlation analysis. *American Anthropologist* 83:824–49.

White, T., B. Asfaw, Y. Beyene, et al. 2009b. *Ardipithecus ramidus* and the paleobiology of early hominids. *Science* 326:64.

Whiten, A., D. M. Custance, J. C. Gomez, P. Teixidor, and K. A. Bard. 1996. Imitative learning of artificial fruit processing in children (*Homo sapiens*) and chimpanzees (*Pan troglodytes*). *Journal of Comparative Psychology* 110:3–14.

Whiten, A., I. Horner, C. A. Litchfield, and S. Marshall-Pescini. 2004. How do apes ape? *Learning & Behavior* 32:36–52.

Whiten, A., W. McGrew, L. Aiello, et al. 2010. Studying extant species to model our past. *Science* 327:410.

Wichmann, S. 2008. The emerging field of language dynamics. *Language and Linguistics Compass* 2:442–55.

Wichmann, S., and E. W. Holman. 2009. *Assessing temporal stability for linguistic typological features.*Munich: Lincom Europa.

Wiens, J. J. 2004. The role of morphological data in phylogeny reconstruction. *Systematic Biology* 53:653–61.

Wildman, D. E., T. J. Bergman, A. al-Aghbari, et al. 2004. Mitochondrial evidence for the origin of hamadryas baboons. *Molecular Phylogenetics and Evolution* 32:287–96.

Williams, G. C. 1966. *Adaptation and natural selection*. Princeton, NJ: Princeton University Press.

Wilson, D. E., and D. M. Reeder. 2005. *Mammal species of the world*. Baltimore, MD: Johns Hopkins University Press.

Wlasiuk, G., and M. Nachman. 2010. Promiscuity and the rate of molecular evolution at primate immunity genes. *Evolution* 64:2204–20.

Wrangham, R., and D. Pilbeam. 2001. African apes as time machines In *All apes great and small*, edited by B. M. F. Galdikas, N. E. Briggs, L. K. Sheeran et al., 5–17. New York: Kluwer Academics/Plenum Publishers.

Wrangham, R. W. 1987. The significance of African apes for reconstructing human social evolution. In Kinzey 1987, 51–71

Wurm, S. A., I. Heyward, and UNESCO. 2001. *Atlas of the world's languages in danger of disappearing*. Paris: UNESCO Publishing.

Yang, Z., and B. Rannala. 2005. Branch-length prior influences Bayesian posterior probability of phylogeny. *Systematic Biology* 54:455.

Yoder, A. D. 1992. The applications and limitations of ontogenetic comparisons for phylogeny reconstruction: The case of the strepsirhine internal carotid artery. *Journal of Human Evolution* 23:183–95.

Yoder, A. D., B. Rakotosamimanana, and T. J. Parsons. 1999. Ancient DNA in subfossil lemurs: Methodological challenges and their solutions. *New directions in lemur studies*, edited by B. Rakotosamimanana, S. M. Goodman, and J. U. Ganzhorn, 1–17. New York: Plenum.

Yule, G. U. 1925. A mathematical theory of evolution, based on the conclusions of Dr. J. C. Willis, F.R.S. *Philosophical Transactions of the Royal Society B: Biological Sciences* 213:21–87.

Zhang, J. Z., and M. Nei. 1997. Accuracies of ancestral amino acid sequences inferred by the parsimony, likelihood, and distance methods. *Journal of Molecular Evolution* 44:S139–46.

Zihlman, A. L., J. E. Cronin, D. L. Cramer, and V. M. Sarich. 1978. Pygmy chimpanzee as a possible prototype for the common ancestor of humans, chimpanzees and gorillas. *Nature* 275:744–46.

Zink, R. M., and J. B. Slowinski. 1995. Evidence from molecular systematics for decreased avian diversification in the Pleistocene epoch. *Proceedings of the National Academy of Sciences of the United States of America* 92:5832–35.

Index

Page numbers in italics indicate figures; page numbers followed by "t" indicate tables.

acceleration-deceleration (ACDC) model, 112t, 121

acquisition bias, with morphological data, 34

adaptation: ancestral state reconstruction and, 56, 75; Brownian motion and, 101–2; vs. chance in evolutionary history, 180; clade shifts and, 143, 215; controlling for body mass and, 215–16; controversial aspects of, 126, 147; convergence approach to, 130, 131–33, 134, 280, 296, 297; cultural hypotheses of, 228–31; directional models in studies of, 165; examples of comparative findings, 127–31, 128, 130; experimental approaches to, 135; intraspecific variation and, 275; limitations of comparative methods for, 134–39, 147; in Ornstein-Uhlenbeck process, 166; phylogenetic context of, 127, 132–33, 133; role of comparative methods for, 126–27, 146–47, 176; singular traits and, 282–83; stabilizing selection and, 161; statistical nonindependence and, 130–31, 139–40, 148, 275. *See also* correlated evolution

adaptationist trap, 147

adaptive radiation, 181; of human populations, 11

adaptive radiation model. *See* ecological niche-filling model

Aegyptopithecus zeuxis, 222

aerobic function, in primates, 276–77

African societies: cultural traits in, 110, 238; linguistic diversity in, 249, 250; origin of clothing in, 232

agriculture: adaptive radiations and, 181; human dispersal and, 11, 46; Middle Eastern and European crops, 232; technological advances in, 124

agriculture and language, 11, 38, 46; Indo-European, 41, 44, 46, 247; linguistic diversity and, 253; proto-Indo-European, 95

Akaike information criterion (AIC), and diversification rates, 195

allometric coefficient, defined, 207

allometric exponent, defined, 207

allometry, 207–15; clade shifts and, 210, 210–11, 215–21, 216, 218, 221; evolutionary model used in, 214–15, 226; independent contrasts and, 209–10, 210, 211, 214; line-fitting methods in, 212–14, 226; logarithmic transformation bias in, 215; logarithmic transformation in, 208, 215. *See also* Kleiber's law

alternative hypotheses, 135–36, 147; phylogenetic targeting and, 314, 314–15. *See also* confounding variables

American Indians: basketry of, 236, 239–40; Kishino-Hasegawa test applied to, 239; language diversity in, 9–11, 10, 196, 251; migration to New World, 247, 249; need for language data on, 319; projectile points of, 232, 233, 249; tree congruence in studies of, 241

amino acid sequence reconstruction: computer simulation of, 77; parsimony approach to, 58

analysis of covariance (ANCOVA), 216–19, 296

analysis of variance (ANOVA), nested, 110, 176

ancestral state reconstruction: adaptation and, 56, 75; background principles of, 53–56; computer simulations of, 77, 84, 89, 93, 96, 319; diversification rates and, 77, 96; evolutionary singularities and, 282, 291–92; extinct lineages and, 54, 54–55, 222; of hominins, 298; horizontal transmission and, 319; with imperfect knowledge, 54–56, 55; intraspecific variation and, 77, 86; new methods for, 77–78; nodal estimates in, 56, 64, 77; phylogenetic signal and, 109, 117; questions addressed by, 52–53; for referential model, 282, 284, 285; root-level uncertainty in, 76; sampling bias in, 77, 80–81, 96; taxon sampling and, 77

ancestral state reconstruction for continuous traits, 79–96; with Bayesian methods, 87, 88; branch lengths in, 95–96, 113; confidence limits on (*see* confidence limits on continuous trait reconstructions); evolutionary trends and, 86, 89–93, 90, 91, 92, ancestral state reconstruction for continuous traits (*continued*) 93, 293; in extinct species, 222; future of, 96; with generalized least squares, 79, 86–87, 92, 96; with independent contrasts, 80, 83, 87; introduction to, 79–81; limited to selected nodes, 81; with linear parsimony, 80; linguistic, 93–95; with maximum likelihood, 79, 81–84, 82, 83, 85, 96; phylogenetic mean in, 80, 88–89; with squared-change parsimony, 80, 84–86, 85, 89, 90, 91, 96; summary of, 95–96. *See also* ancestral state reconstruction

ancestral state reconstruction for discrete traits, 52–78; background principles of, 53–56; with Bayesian methods, 71–74, 73, 75, 76; confidence limits on, 56, 298; introduction to, 57–58; with maximum likelihood, 63–71, 65, 66t, 68, 70, 76, 77; with parsimony, 58–63, 60, 61t, 62, 76, 77; questions addressed by, 52–53; summary of, 75–78. *See also* ancestral state reconstruction

ANCOVA. *See* analysis of covariance (ANCOVA)

Anolis lizards, 260

ANOVA. *See* analysis of variance (ANOVA)

anthropology. *See* evolutionary anthropology

AnthroTree, 3, 16

apes. *See* primates

archaeological artifacts: cladistics of, 232, 233; niche-filling model of evolution in, 124–25. *See also* culturally transmitted objects

Ardipithecus ramidus, 291

artiodactyls: genetic evolution in, 73, 73–74, 76

Asian languages, 319

ASPM gene, 129

asymmetric model, maximum likelihood, 67, 68–69, 70

Australian languages, 46

Australian possums, 261

Australopithecus: *A. afarensis*, 129, 130; *A. robustus*, 222; cranial variation in, 294; modeling evolutionary change in, 292

Austronesian languages: borrowing among, 241; settlement of Polynesia and, 196, 248, 248–49

Austronesian societies, residence patterns in, 74

autapomorphic traits, 281

autocorrelation: clade shifts and, 219; cultural borrowing and, 246; phylogenetic signal and, 110, 178; statistical nonindependence and, 178

baboons: nonindependence of populations, 140; as referential models, 283

bacteriophage phylogeny, 89, 90

Bantu expansion, 46, 247

Bantu languages: cattle-keeping and, 67, 120, 172; matrilineal descent and, 67, 120, 172; origin of agriculture and, 46

Barbary macaque mating behavior, 6

basketry of American Indians, 236, 239–40

Bayesian analysis of correlated evolution: genome size in vertebrates and, 223, 224, 225; with independent contrasts, 158, 178

Bayesian analysis of languages: borrowing and, 47, 245; diversification rates in, 196; Indo-European, 41, 43, 44; rate of word change in, 46; settlement of Polynesia and, 249

Bayesian methods for ancestral state reconstruction: advantages of, 76; continuous traits, 87, 88; discrete traits, 71–74, 73, 75, 76; evolutionary singularities and, 282, 291, 292

Bayesian methods for inferring phylogenies, 35–38, 37; of fossil hominins, 298; of lichenization among fungi, 67; node-density artifact in, 200; partitioning data in, 242; of primates, 36, 48–49; strengths and weaknesses of, 36–38

Bayesian methods with birth-death model, 186, 189

beetles: diversity of, 196; mandible length in tiger beetles, 114

behavioral traits: in hominins, 129, 283–90, 285; homologous, 256, 277–78; for inferring phylogeny, 26; phylogenetic signal of, 112, 113t, 278; in primates, 112, 113t, 127, 128, 138, 278; rates of evolution in, 124

binary traits. *See* discrete traits

biogeography, 256, 261; of islands, 259–60

biological diversity: comparative studies and, 1; documenting before extinction, 319, 321; language diversity and, 251, 252; latitude and, 9; phylogenetic aspects of, 260; quantification of, 265–69; undiscovered, 272

biological species concept, 265–66

biomechanics: in *Megaladapis*, 205; scaling in, 202–3. *See also* locomotor behavior

birds: evolutionary models for, 118–19, 119; genome size of, 223, 224, 225; Hawaiian, malaria in, 270; nonindependence of populations of, 140; plumage coloration in, 118; sexual dichromatism in, 61, 63, 196; testes mass in, 138–39

birth-death model, 183–89; applications to primates, 189, 195t; for languages, 251

bivariate correlations, 159

body mass: brain size and, 166, 202; controlling for, 215–21, 216, 218, 221, 226; diversification rates and, 182, 198; extinction risk and, 270; home range size and, 202, 203, 217–18, 218; intermembral index and, 205,

206, 207; metabolic rate and, 149, *150*, 166, 202, 211–12, *212*; phylogenetic mean of, 88–89; phylogenetic signal of, 112, 112t, 209; rate of evolution for, 121, 124; reconstructing in extinct species, 79, 222–23, 225, 226; scaling with, 202, 203 (*see also* allometry); semicircular canals in mammals and, 166, *167*; trend of increase in, 89, 92

body mass in primates: brain size and, 166; dimorphism of (*see* sexual dimorphism in primates); diversification and, 198; group size and, 255; home range and, *151*, 151–56, 152t, *155*, *156*, 158, 214, 255; intermembral index and, 205, *206*, 207; metabolic rate and, 149, *150*, 211, *212*; phylogenetic signal of, 112, 112t, 121; rate of evolution for, 121; small intestine surface area and, 211; spleen mass and, 207–8

body size, vertebrate, Brownian motion model of, 118

bonobos, as referential models, 283

bootstrap support indices, 32, 33, 34–35

borrowing, cultural: vs. common descent, 234–42, *240*; comparative methods and, 234, 242–46, *244*, *245*, 320. *See also* horizontal transmission

borrowing, linguistic, 46–47, *47*, 95, 241, 319; statistical nonindependence and, 141–42; tree inference and, 245. *See also* horizontal transmission; loanwords

brain evolution in hominins, 293

brain size: body mass and, 166, 202; ecological and social correlates of, 131–32, *133*; energetic costs of, 163; play and, 163; in primates, 131–32, 143, *144*, 166; sleep patterns in mammals and, 138. *See also* cranial capacity; neocortex size

brain structures: activity period categories and, 173; allometry with body mass and, 213; dietary categories and, 173; dimorphism in corpus callosum, 203. *See also* cognition; neocortex size

branches of phylogenetic tree, 21, *21*; ancestral state reconstruction and, *54*, 54–55, *55*; dated, *23*, 24, 100; long-branch attraction and, 32, 38; two different representations of, 100–1; undated, 22, *23*, 24. *See also* dated phylogenies; polytomies

branch lengths, *23*, 24; in Bayesian approach, 35, 36, 38, 71, 76; Brownian motion and, 81, 101, 102, 118; in continuous-trait reconstructions, 95–96, 113; in distance-matrix methods, 30; evolutionary models and, 100–1, 108–9, 118–20, *119*, 121–23; evolutionary trends and, 91–92; in independent contrasts method, 150, 152, 153, 154, 156, 157, *158*, 158–59; in linguistic reconstructions, 95, 120, 196; in maximum likeli-

hood methods, 33, 64, *65*, 65–66, 69, *70*, 71, 76; node-density artifact and, 200; in Pagel's discrete method, 168; in parsimony analysis, 32, 59, 63; phylogenetic diversity and, 266–67, *267*; in primate phylogeny, 320; in squared-change parsimony, 84, 85; in supermatrix approach, 49; variance-covariance matrix and, 164

branch length transformations, 113–16, *115*, 117; adaptive radiation and, 121–22, 123; allometric analyses using, 214–15; Brownian motion and, 117–19, *119*; for independent contrasts, 159, 161, 167; speciational model and, 119–20. *See also* κ (kappa); λ (lambda)

bride-price: in Indo-European societies, 74, *75*, 173; in Kipsigis of Kenya, 74; polygyny and, 173, 229; statistical nonindependence of data, *141*

Brownian motion model, 80, 81, 96, 98, 100t, 101–3, 106; diagnostics for, 117–19; evolutionary singularities and, 294; future research and, 318; generalized least squares with, 96, 165, 166; *K* value under, 111; λ (lambda) parameter and, 113–14, *115*, 117–18; maximum likelihood and, 81, 85, 96; niche-filling model and, 106, 122; Ornstein-Uhlenbeck model and, 103, 120; in simulation of horizontal transmission, 243, *244*; squared-change parsimony and, 96; for standardization of independent contrasts, 156–57, 157t, *158*; statistical nonindependence and, 140

Brunch algorithm, 175, 219

burn-in, for Bayesian analysis, 36, *37*, 72

CAIC computer program, 175, 219

canine teeth: height dimorphism in extinct hominins, 129, *130*; size of, in primates, 203

Canterbury Tales (Chaucer), 49, 241

carnivores: extinction risk in, 267, 268, 269, 270, 272; home range size vs. body mass, 217–18, *218*; molar teeth in, 90, *91*

cathemerality, 57, 60

cattle-keeping: in Bantu-speaking people, 67, 120, 172; lactose tolerance and, 163–64, 171; patrilineal descent and, 172, 229

causation: vs. correlation, 134–35; of diversification, 199; Pagel's discrete method and, 171

cerebral cortex: *ASPM* gene and, 129. *See also* neocortex size

chance in evolutionary history: vs. adaptation, 180; push of the past and, 186

character states. *See* continuous traits; discrete traits

chimpanzees: in ape phylogeny, 22, *23*, 24, 26; cognition in, 1, 8; parasites of, 264; as referential models, 283. *See also* primates

chronogram, 24
CI. *See* consistency index (CI)
clade-diversity diagrams, in archaeology, 232
clade shifts, 142–43, *144*; allometry and, *210*, 210–11, 215–21, *216*, *218*, *221*; regression through the origin and, 160
cladogenesis, 181. *See also* speciation
cladogram, 24
clothing: in Australian Aborigines, 231, *231*; lice and, 232
Clovis points, 249
coevolution: gene-culture, 171; host-parasite, 11–13, *12*, 241
cognates, linguistic, 40, 41; glottochronology and, 45; of Indo-European languages, 41, *42*, 232; need for more data on, 319; protolanguage reconstruction and, 94
cognition: comparative psychology and, 8, 162–63; phylogenetic targeting for studies of, 312–16; in primates, 1, 3, 7–8, 9; transitions in hominins, 290
collinearity, unstable regression models with, 221
community ecology, phylogenetic, 256, 259–65, *262*, *263*
comparative linguistics, 39, 247
comparative models, 282, 286–90, *287*, *288*
comparative studies: of adaptation, 126–27, 146–47, 176; defined, 2; foundational work on, 146, 317; future directions for, 317–21; overview of, 16–19; role of, 1–5; without phylogenetic context, 20, 226 (*see also* nonphylogenetic tests)
computer programs, 3, 16; to calculate contrasts, 160, 175, 219; *R*, 3, 16, 124, 321. *See also* databases, comparative
computer simulations. *See* simulations
conceptual (strategic) modeling, 282, 283–84, 286, 288
confidence limits on continuous trait reconstructions, 81; with evolutionary trends, 90–91, *91*; for extinct species, 223, 225, 289, 290, 291, 292; with generalized least squares, 86; with independent contrasts, 83; for intermembral index, *83*, 83–84, 85–86, 87, *88*, 95; with maximum likelihood, *83*, 83–84; with squared-change parsimony, 85–86; summary of, 96
confidence limits on discrete trait reconstructions, 56, 298
confidence limits on phylogenetic mean, 89
confidence limits on tree balance statistics, 191
confounding variables, 135–36; phylogenetic, 218; reducing effects of, 142; selecting species to study and, 311, 315; size as, 202, 203. *See also* alternative hypotheses
congruence of phylogenetic trees, 241, 319–20
consensus tree, 35, 36

conservation of biodiversity, 256, 265; impact of phylogeny on, 265–72, *267*, *268*
consistency index (CI), 31, 236–37, 238; Austronesian languages and, 248; with behavioral traits, 277
constraint, evolutionary, 107–8
continuous traits, 57; in analysis of covariance, 218–19; controlling for body mass with, 220–21, *221*; correlated evolution of, 148; in datasets including discrete traits, 148, 173, 175–76, 178; discrete coding of, 57; diversification related to, 182; evolutionary lag in, 135; evolutionary models for, 100t (*see also specific models*); phylogenetic signal in, 110–16, 112t, 113t, *114*, *115*; phylogenetic targeting with, 315; speciation and extinction rates in, 198–99; statistical power and, 258. *See also* ancestral state reconstruction for continuous traits
contrasts analysis of diversification, 198. *See also* independent contrasts
convergence approach to adaptation, 130, 131–33, 134, 280, 296, 297. *See also* homoplasy
convergent evolution: analysis of covariance and, 218; distance-matrix methods and, 30
Cope's rule, 89
core vocabulary, 40–41, 95
correlated evolution: of continuous traits, 148, 149, 159–60, 164–68, *167*, 178; cultural borrowing and, 234, 242, 244–45, 246, 319; of discrete traits, 148, 168–73, 169t, *170*, *174*, *175*, 176; as evidence for adaptation, 126, 127, 129, 132, 176; generalized least squares studies of, 148, 149, 164–68, *167*, 168; independent contrasts and, 132, *133*, 148, 159–60; of mixed continuous and discrete traits, 173, 175, 176, 178; niche-filling model and, 106; Ornstein-Uhlenbeck (OU) model and, 215; phylogenetic signal and, 109, 117; phylogenetic targeting and, 316; reconstructions in extinct species and, 222, 223, 225, 226; scaling relationships as, 208; singularities and, 282, 296–97; stabilizing selection with, 161; statistical nonindependence and, 142, *143*; statistical power and, 243; summary of, 176, 178. *See also* adaptation; statistical nonindependence
correlation vs. causation, 134–35
cost structures, in parsimony analysis, 32–33, 60–63, *62*, 76. *See also* gains and losses
covariance, evolutionary, 113, *114*
cranial capacity, in hominins, 86, 93, *93*, 120, 121–22, 289–90, 293
cranial characters: diet in mammals and, 202; of hominins, 28, 293–94; phylogenetic signal of, 112
criterion of subtraction, 220
cross-cultural studies: ambiguity of data in,

305; borrowing and comparative methods in, 234, 242–46, *244*, *245*, 320; borrowing vs. common descent in, 234–42, *240*; data coding in, 137; of discretely varying traits, 245, 297; examples of, 228–34; future directions for, 319–20; georeferenced data in, 252, 253; of lactose tolerance, 163–64, 171; of migration patterns, 58–59, 234, 247–49, *248*, 319; questions addressed by, 227–28, 234; simulations in, 320; statistical nonindependence in, 140–42, *141*, 178, 228, 320; statistical validity as challenge in, 142, 228; summary of, 253–54
cultural diversity, 1, 9–11. *See also* cross-cultural studies; language diversity
cultural evolution, parsimony analyses of, 31
cultural group extinction: in computer simulation, 243, *244*, *245*; documentation prior to, 319, 321; rates of, 201
culturally transmitted objects, 49. *See also* archaeological artifacts
cultural traits: Bayesian approaches to, 74, *75*; community ecology approach to, 264–65; comparative approach to, 128, 146; family organization and, 58; maximum likelihood reconstruction, 58, 67; phylogenetic signal and, 110, 116, 125, 237–38, 319; rates of evolution in, 124–25; similarities to genetic traits, 234–35; stabilizing selection and, 104; tree congruence of, 241, 320. *See also* cross-cultural studies; evolutionary anthropology; languages
culture, defined, 234

d (death rate), 183, *186*, 187–89, 195t
d (scaling parameter), 112, 112t, 118, 120–21
Darwin, Charles, 38
darwin, unit of, 123
Darwin's finches, diet in, 64, *65*, 65–66
data, phylogenetic, 25–26. *See also* behavioral traits; fossil data; genetic data; morphological data
databases, comparative: intraspecific variation in, 276; introduction to, 299–300; sacrifice of detail in, 137. *See also* informatics revolution; relational databases
data collection, 299, 300; data model for, 301; future needs for, 318–19, 321. *See also* phylogenetic targeting
data model, 301
data quality in comparative studies, 136–39
dated phylogenies, 23, 24, 100; for estimating speciation and extinction rates, 25; evolutionary rates and, 123; with linguistic data, 44–46; reconstructing evolution on, 183–87, *185*, *186*; signature of extinction on, 182, *182*, 183–84, *184*; variance-covariance matrix for, *114*

degrees of freedom: for analysis of covariance, 218; independent contrasts and, 145–46, 150–51, 154, 160; in node-averaging approach, 176; in Pagel's discrete method, 171; statistical nonindependence and, 145–46
δ (delta), 121–22, 166
delta score, horizontal transmission and, 241
demic expansion, 46, 247
dental traits: diet in mammals and, 202; in fossil horses, 292; in hominins, 28; in primates, 52, 203–4, 222, 225, 320. *See also* canine teeth; molar teeth
dependent model. *See* Pagel's discrete method
derived characters. *See* shared derived characters (synapomorphies)
diagnostics for evolutionary models, 106, 117–23, *119*; comparative analyses based on, 161; with cultural traits, 238; statistical nonindependence and, 130
diet: cranial/dental morphology in mammals and, 202; in Darwin's finches, 64, *65*, 65–66; *enamelin* gene and, 129; home range size and, 255; human, spicy foods in, 229–31, *230*; jaw morphology in Old World monkeys and, 215–16, *216*; of primates, 52, 320
dimorphism. *See* sexual dimorphism in primates
dinosaurs, genome size in, 223, *224*, 225, 291–92
directional models: in ancestral state reconstruction, 91–93, *92*, 93, 120; for brain evolution in hominins, 293; node-density artifact and, 200; in studies of adaptation, 165
directional selection, 96. *See also* evolutionary trends
discrete traits, 57; continuous distribution underlying, 219; continuous traits in datasets with, 148, 173, 175–76, 178; correlated evolution of, 148, 176 (*see also* Pagel's discrete method); in cross-cultural datasets, 245, 297; intraspecific variation in, 77; order of events for, 135; phylogenetic signal with, 110, 116, 238; phylogenetic targeting with, 315; as probability distributions at nodes, *75*; speciation and extinction rates in, 197–98; statistical power and, 258. *See also* ancestral state reconstruction for discrete traits
disease risk: latitude and, 128–29, 135, 229. *See also* infectious diseases
distance-matrix methods: for ancestral state reconstruction, 77; for inferring phylogenies, 29–30; Mantel tests as, 239–40, *240*
diversification: assumptions in studies of, 186–88; background concepts of, 181–87, *182*; defined, 181; null models of, 180, 182,

diversification (*continued*)
191, 192, 201; of primates (*see* primate diversification); questions about, 180–81, 182; reconstruction on dated phylogeny and, 183–87, *185, 186*; summary of, 200–1; unifying extant and fossil data on, 180. *See also* dated phylogenies; extinction; extinction rates; speciation; speciation rates
diversification rates: ancestral state reconstruction and, 77, 96; in birth-death model, 187–88, 189; body mass and, 182, 198; evolutionary rates and, 182, 188, 199–200; factors influencing, 196–99; quantifying variation in, 193–96, 195t; shifts in, 194–95; variation in, 190–96, *191*. *See also* evolutionary rates; speciation rates
diversity. *See* biological diversity; cultural diversity; ecological diversity; language diversity; phylogenetic diversity (PD)
diversity skewness, in primate communities, 262
DNA: from fossil material, 26, 52; nuclear ribosomal, mutualism and, 200; sequence alignment by researcher, 28; sequence changes on phylogram, 24; sequence reconstruction with parsimony, 58; substitution models of, 34; transitions vs. transversions in, 32, 34. *See also* genetic data; molecular evolution
Dollo parsimony, 61
domestication of animals, 11, 181; horses and Indo-European populations, 44, 247
dowry: female-female competition and, 128; in Indo-European societies, 74, *75*; statistical nonindependence of data, *141*
ducks, sexual dichromatism in, 61, 63

ecological conditions: cultural transmission and, 238, 239, 246; in referential models, 283; in strategic models, 285. *See also* environmental variables
ecological diversity: biological variation and, 321; phylogenetic divergence and, 262
ecological niche-filling model, 100t, 104–6, *105*; branch length transformation for, 115; cultural traits and, 124–25; diagnostics for, 121–23; generalized least squares with, 165; tests of phylogenetic signal and, 112
ecological risk, and language diversity, 252
ecological traits, phylogenetic signal of, 114–15
enamelin gene, 129
ensemble indices of homoplasy, 31
entangled bank hypothesis, 248
enumerated text, 303, 305
environmental variables: causation and, 135; diversification and, 196–97; language diversity and, 251–52, *253*. *See also* ecological conditions

estrus synchrony in primates, 258–59, 275
ethnogenesis, 234, 235, 236
Ethnographic Atlas (Murdock), 128, 178
European populations, gene distribution in, 247
evolutionary anthropology: examples of comparison in, 5–16; future of comparative methods in, 147, 321; horizontal transmission in, 321; maximum likelihood reconstructions in, 66–67; models of evolution and, 124–25; niche-filling model and, 106; role of comparison in, 1–5; scope of, 2. *See also* hominins; humans; primates
evolutionary archaeology, 232, *233*
evolutionary change: under Brownian motion, 118; correlated (*see* correlated evolution); independent contrasts and, 149; inertia concept and, 107; on single branch of a tree, 294–96, *295*; time proportionality of, 118, 120. *See also* models of evolutionary change
evolutionary constraint, 107–8
evolutionary lag, and independent contrasts, 162
evolutionary patterns: behavioral shaping of, 279; convergence approach and, 280; models of evolutionary change and, 101, 106, 108, 117
evolutionary rates: of behavioral traits, 124; of body mass, 121, 124; in Brownian motion model, 101; of cultural traits, 124–25; detecting heterogeneity in, 296; diversification rates and, 182, 188, 199–200; in generalized least squares studies, 166; increased by evolution of a trait, 200; independent contrasts and, 123, 150, 162; linguistic, 45–46; method of reconstruction and, 77; in niche-filling model, 106; node-density artifact and, 200; of primate locomotion, 82, 84; in primates, 124, 298; quantification of, 123–24; temporal variation of, for trait on tree, 121–23; uncertainty of reconstruction and, 71; variable, in phylogram, 100–1. *See also* diversification rates
evolutionary singularities, 280–98; confidence measures and, 318; methodological approaches to, 282, 291–97, *295*; models for, 282, 283–90, *285*; overview of, 280–83; summary of, 297–98
evolutionary steps, in parsimony analysis, 31, 32, 59, 60, 61, 61t, 63, 76
evolutionary trends: ancestral state reconstruction and, 86, 89–93, *90, 91, 92, 93*, 293; of increasing body mass, 89, 92
executive brain ratio, and social learning in primates, 8, *9*
express train hypothesis, 59, 248, *248*
extinction: diversification and, 181–82, *182*;

macroevolutionary patterns and, 180, 181; mass extinctions, 188, 189–90, *190*; signature of, on dated phylogeny, 182, *182*, 183–84, *184*. *See also* cultural group extinction

extinction rates, 182; in birth-death model, 183, 184, 187–89; of cultural groups, 201; estimation of, 25, 201; factors influencing, 196–99; varying, 187. *See also* diversification rates

extinction risk: comparative studies of, 269–72, *271*; of languages, 273–74; latent, 271, *271*; phylogenetic diversity and, 267–68, *268*; sexual selection and, 173

extinct languages, 251, 252

extinct lineages, 25; ancestral state reconstruction and, *54*, 54–55, 222; inferring behavior in, 283; reconstructing body mass and other traits in, 79, 222–23, *224*, 225, 226, 291–92; sampling biases and, 137; speciational model and, 102–3. *See also* ancestral state reconstruction; fossil data; hominins

family organization, parsimony reconstruction of, 58

farming. *See* agriculture

feature diversity, 266–67, 268

female philopatry, in Old World monkeys, 58

fieldwork, need for, 278–79, 318–19

fish, swordtail, female choice in, 79

fitness, 124, 126

flat files. *See* spreadsheets

Flores, 1

flower morphology, 68–69

foraging: group composition and, 257; ranging patterns and, 134

Foraminifera, fossil, 90, 93

foreign key, in database, 301, 302, *302*, 306

fossil data: in ancestral state reconstruction, 52, 90–91, *91*, 93, *93*; in building phylogenies, 25, 26, 27, 28; diversification and, 180; intermembral index and, 85–86, 91, 205; Pagel's discrete method and, 171; on semicircular canals, 166. *See also* extinct lineages; hominins

founder effects, 147; in linguistic evolution, 233

free model, 118, *119*

frog vocalizations, 79

fungi: lichenization of, 67–68, *68*; mutualism among, 200

g parameter, 112, 112t, 118, 121

gains and losses: adaptation and, 127; in Bayesian analysis, 74; in maximum likelihood, 64, 67–69, *68*, *70*, 71, 98; in parsimony analysis, 61–63, *62*. *See also* cost structures

Galapagos finches, diet in, 64, *65*, 65–66

Galton's problem, 141, 228

gene-culture coevolution, 171

generalized least squares (GLS): allometric slopes estimated with, 209; for ancestral state reconstruction, 79, 86–87, 92, 96; assumptions checks with, 167; clade shifts and, 219; correlated evolution and, 148, 149, 164–68, *167*, 178; independent contrasts and, 149, 164–65, 167–68; line-fitting methods with, 226; with mixed discrete and continuous variables, 175–76, 178; multiple sources of nonindependence and, 246; phylogenetic signal and, 111, 161, 166, 168; phylogenetic targeting with, 312, 315

gene sequences. *See* DNA

genetic change: evolutionary model based on, *119*; horizontal transmission in, *321*

genetic data: independence of, 25; primate phylogeny and, 50, 317–18. *See also* DNA

genetic drift: hominin cranial remains and, 293–94; Ornstein-Uhlenbeck model and, 103

genetics of human populations: gene distribution in European populations, 247; languages and, 13–16, *14*

genome size, in dinosaurs, 223, *224*, 225, 291–92

geographic distance, and cultural transmission, 239–40, *240*, 246

geographic information systems (GIS), 252, 279, 300

geographic range: extinction risk and, 270; in primates, evolution in, 122. *See also* home range size and body mass; home range size in primates

geographic signal, 238

ghost lineages, 25

glottochronology, 44–45, 46

GLS. *See* generalized least squares (GLS)

Grafen transformation, 159, *206*

graphical user interfaces (GUIs), 303, 304

GRIN2A gene, 129

ground squirrels, North American, 261

group size and cognitive ability, *313*, 313–14

group size in primates: body mass and, 255; diversification rates and, 199; in hominin evolution, 289–90, *290*, 293; intraspecific variation in, 274; vs. neocortex size, 127, *128*, 132, *133*, 296

growing season, and language diversity, 251–52, *253*

Haldane, J. B. S., 196

Hanuman langurs, mating systems in, 275

hard polytomy, 22

Helicobacter pylori, 249

helminths, primate hosts of, 264

heteroscedasticity, 155–56

higher nodes approach, 176, *177*, 255

historical linguistics, 39; comparative method in, 39–41, 247; reconstruction in, 93–95

home range size and body mass, 202, 203, 217–18, *218*; in primates, *151*, 151–56, 152t, *155*, *156*, 158, 214, 255

home range size in primates: body mass and, *151*, 151–56, 152t, *155*, *156*, 158, 214, 255; day range relative to, 127, *128*, 134; diet and, 255; evolutionary rates and, 124; field observations of, 279; group composition and, 138; group metabolic needs and, 214, *215*, 275; number of males and, 257

home range size of carnivores vs. ungulates, 217–18, *218*

hominins: adaptive radiation of, 106; behavioral evolution in, 129, 283–84, *285*, 286–88, *287*, *288*, 289–90, *290*, 292; body mass estimation, 222–23; brain evolution in, 293; cognitive transitions in, 290; cranial capacity in, 86, 93, *93*, 120, 121–22, 289–90, 293; DNA from, 52; genetic drift in, 293–94; group size in, 289–90, *290*, 293; homoplasy in evolution of, 58; male competition in, 289; model of evolutionary change in, 292–94; phylogenetic reconstruction of, 1, 25, 28, *28*, 34, 50, 298; sexual dimorphism in, 129, *130*; trait reconstruction of, 291, 292–93, 298

Homo erectus, 290, *290*

homology, 26; adaptation and, 131; behavioral, 256, 277–78; linguistic equivalent of, 40, 94; referential models and, 283, 284

homoplasy, 26; adaptation and, 131; behavioral, 277; in cultural transmission, 236; genomic insertions and, 50; in hominin evolution, 58; indices of, 31, 236; in primate evolution, 58. *See also* convergence approach to adaptation

homoscedasticity, 155

horizontal transmission: in anthropology, 321; computer simulations of, 237, 240, 243–45, *244*, *245*, 319. *See also* borrowing, cultural; borrowing, linguistic

horses: domestication of, 44, 247; fossil tooth morphology, 292

host-parasite tree congruence, 11–13, *12*, 241, 319–20

humans: in ape phylogeny, 22, *23*, 24, 26, 285, *285*; bipedality of, 95, 285, 295; comparative method for, 146; comparative models of evolution in, 286–90, *287*, *288*, *290*, 297; corpus callosal dimorphism in, 203; disease risk and latitude in, 128–29, 135, 229; mating system and wealth transfer in, 128, 172–73; parasite pressure on, 129; parasites shared with other primates, 264; referential models of behavior in, 283; sex ratios in populations, 167; singularities in evolution

of, 280–82; social learning in, 7–8; statistical nonindependence of populations, 140–41, *141*; testes mass in, 6, *7*; variation in populations, 108. *See also* cultural diversity; cultural traits; evolutionary anthropology; evolutionary singularities; hominins; language diversity; linguistic data

humeral measurements, 121

iguana displays, 86

IMI. *See* intermembral index (IMI)

imitation in primates, 7–8

immunity: habitat shrinkage and, 272; promiscuity in primates and, 129, 173

independent characters, 25, 28. *See also* statistical nonindependence

independent contrasts, 148–64; allometry and, 209–10, *210*, 211, 214; for ancestral state reconstruction, 80, 83, 87; assumptions checks with, 156–59, 157t, *158*, 160, 165, 167; basic concept of, 149–51; change on single branch and, 294–96, *295*; citations of original paper on, 148–49, *149*; clade shifts and, 219; correlated evolution and, 132, *133*, 148, 159–60; degrees of freedom in, 145–46, 150–51, 154, 160; evolutionary rates and, 123, 150, 162; evolutionary singularities and, 296; examples of, 162–64, *163*; generalized least squares and, 149, 164–65, 167–68; heteroscedasticity in, 155–56; horizontal transmission and, 243–45, *244*, *245*, 246; of intermembral index and body mass, 205, *206*, 207; issues involved with, 160–62; log transformation of data for, 158–59, *206*; males in primate social groups and, 256; measurement error in, 157t, 161; with mixed discrete and continuous variables, 175; nodal estimates in, 153–54, *155*; node-density artifact with, 200; nonphylogenetic studies concurrent with, 161; nonphylogenetic tests vs., *133*, 158, 161; outliers in, 157t, 159, 160; phylogenetic signal and, 111, 162; phylogenetic targeting and, 312, *313*, 315; range of questions addressed by, 162; standardized, 152, 152t, 155–59, *156*, 157t, *158*; statistical nonindependence and, *145*, 145–46; unstandardized, *151*, 151–55, 152t, *155*, 162

independent model. *See* Pagel's discrete method

Indo-European languages: horizontal borrowing among, 241; inferred phylogeny of, 41, *42*, *43*, 44; mating system–wealth transfer relationship and, 173; origin of agriculture and, 46, 247; rate heterogeneity and, 45–46; word evolution in, 232

inertia, phylogenetic, 107–8

infanticide in primates, 257

infectious diseases: human female fertility and diversity of, 229, *230*; information gaps in primates, 279; latitude and, 128–29, 135, 229; in threatened species, 272. *See also* parasitic infections
informatics revolution, 4, 5, 136, 147, 276, 316. *See also* databases, comparative
information theoretic approaches, 124
ingroup, 29
innovations: adaptive radiation and, 181; agriculture as, 38, 44; comparative psychology of, 8; diversification rates and, 182, 196–97; language spread and, 46; large character change and, 294; words for, in ancestral language, 40. *See also* evolutionary singularities; technological advances
insectivores, neocortex size in, *144*
instinctive behavioral patterns, 277
interdisciplinary research, 4
intermembral index (IMI), 81–82, 203–7; Bayesian analysis of, 87, *88*; body mass and, 205, *206*, 207; branch lengths and, 85; fossil evidence and, 85–86, 91, 205; future directions for research on, 320; generalized least squares for, 87; independent contrasts method for, 87, 205, *206*, 207; maximum likelihood estimates for, *82*, 82–84, *83*, 85; McPeek's single-branch method and, 295; for *Megaladapis*, 205; squared-change parsimony for, 86; summary of reconstruction methods, 95
International Union for Conservation of Nature. *See* IUCN Red List of Threatened Species
intraspecific variation: adaptive hypotheses and, 275; allometric, 214, 226; in ancestral state reconstruction, 77, 86; Bayesian methods and, 178; controlling for, 256, 275–77, 320; critiques of comparative studies based on, 137, 138, 147; with independent contrasts, 157t, 160–61; in primates, 256, 274–77, 320
Iranian textiles, 236
irreversible traits, 61, 63
island biogeography, 259–60
isometric traits, 208, 220
IUCN Red List of Threatened Species, 267, 269, 270, 272, 273

jaw morphology, in Old World monkeys, 215–16, *216*

κ (kappa), 119–20, 124; allometric applications of, 214–15; evolutionary models and, 292
key innovations, 182, 196–97; large character change and, 294
kin bias, in macaques, 113t
Kipsigis of Kenya, bride-price in, 74

Kishino-Hasegawa (K-H) test, 239
Kleiber's law, 149, *150*, 166, 211–12, *212*; primate range size and, *215*
Kurgan expansion, 44
K value, 111–12, 112t, 113t

lactose tolerance, 163–64, 171
λ (lambda), 113–16; Brownian motion model and, 113–14, 115, 117–18; in generalized least squares method, 165, 166, 168; phylogenetic signal and, 122; as symbol for speciation rate, 183
language diversity, 249–53; biological methods applied to, 41, 250–51; drivers of, 251–53, *253*; global distribution of, 249–50, *250*, 251; habitat diversity and, 10, *10*; latitude and, 9, *10*, 251, 253; linguistic change associated with, 200; sampling bias and, 137; time since human settlement and, 10–11; unanswered questions about, 196, 201. *See also* linguistic evolution
languages: borrowing of words in (*see* borrowing, linguistic; loanwords); conservation of, 256; dating first appearance of, 290; extinction risk of, 273–74; farming and (*see* agriculture and language); genetics of populations and, 13–16, *14*; latitudinal extent of, 9–10, *10*, 11; phylogenetic data derived from, 26; phylogenetic methods and, 146; phylogenetic signal and, 108; tonal, genetics of, 15; two types of change in, 46–47
language trees: biological methods applied to, 41, *42*, *43*, 44, 46; building of, 28, 39, 40–41 (*see also* cognates); compared to biological phylogenies, 39; congruence of, 241; dating of, 44–46; evolutionary questions and, 181; geographical diffusion and, 46–47; growth of interest in, 38–39; horizontal transmission and, 245; linguistic signal with, 110; migration patterns and, 58–59, 234, 247–49, *248*, 319; pull of the present in, 274; time depth limits on, 44, 50–51; as tool for studying prehistory, 247
latitude: biological diversity and, 9; disease risk and, 128–29, 135, 229; language diversity and, 9, *10*, 251, 253; language extent and, 9–10, *10*, 11
least squares (LS) regression: in allometry, 213–14; taking residuals in, 220, *221*; two-stages, 246
lemurs: behavioral variation in, 275; conservation of, 269. *See also* strepsirrhines
lexicostatistics, 45, 46, 319
lice, clothing and divergence of, 232
lichenization, 67–68, *68*
likelihood: in Bayesian analysis, 36, 38. *See also* maximum likelihood

likelihood ratio test (LRT): to compare diversification rates, 193, 195; to compare models, 124, 164; of directional model, 92, *92*; λ (lambda) scaling and, 114, 117; in Pagel's discrete method, 171, 296–97
lineages, 21, *21*, 22. *See also* branches of phylogenetic tree; extinct lineages
lineages through time plots, 183–86, *184*, *185*, *186*, 187–88; with mass extinction, 189–90, *190*; with missing extant lineages, 187–88, *188*
linear parsimony, 80
line-fitting methods, 212–14, 226
linguistic data: human prehistory and, 247; statistical nonindependence of, 141–42
linguistic ecology, 250
linguistic evolution, 232–34; rates of, 45–46. *See also* language diversity
linguistic reconstructions, 93–95
linguistics, comparative method in, 1, 39–41, 94, 247; future directions for, 319
linguistic signal, 110, 238
lizards: character displacement in, 84; community ecology of, 260; correlated evolution in, 84; iguana displays, 86; variation of subclades in, 122
loanwords, 46–47, 94, 142, 246. *See also* borrowing, linguistic
locomotor behavior: digit morphology of extinct species and, 222; in primates, 52, 204, *204*; semicircular canals in mammals and, 166, *167*. *See also* biomechanics; intermembral index (IMI)
logistic regression, phylogenetic, 176
log transformation: in allometry, 208, 215; of independent contrasts data, 158–59, *206*
long-branch attraction, 32, 38
LRT. *See* likelihood ratio test (LRT)
LS. *See* least-squares (LS) regression
lumbar vertebra, scaling of, in hominins, 222

macaques: Barbary macaque mating behavior, 6; coexistence of different species, 261; loud calls and seasonality in, 172; phylogenetic signal in behavioral traits of, 112, 113t; rhesus group size, 274, 275. *See also* sexual swellings, exaggerated
major axis regression, 213
malaria: in Hawaiian birds, 270; human female fertility and, 229; phylogenetic signal in prevalence of, 109
mammals: artiodactyl genetic evolution, *73*, 73–74, 76; brain size and body mass in, 166; clade shifts in development of, 143; decelerating body mass evolution in, 121; diet and cranial/dental morphology in, 202; diversification rates in, 199; extinction risk in, 271, *271*; metabolic rate and body

mass in, 166, 211–12, *212*; phylogenetic dispersion in, 261, *262*; phylogenetic signal in, 114–15; rates of evolution in, 124; semicircular canals in, 166, *167*; sleep patterns in, 138; small intestine surface area in, 211; supertrees of, 48. *See also* carnivores; primates
Mantel tests, 239–40, *240*, 244, 262
manuscript evolution, 49, 241, 318
mapping uncertainty: in Bayesian reconstruction, 71, 87; in primate research, 320; for ribonuclease gene in artiodactyls, 73
Markov chain Monte Carlo (MCMC), 36, 72, 73, 74, 87; Pagel's discrete method with, 173
Markov model: of lineage accumulation, 192; in maximum likelihood reconstructions, 63–64; in Pagel's discrete method, 168
marmosets, forelimb morphology in, 202
maroon leaf monkeys, group size in, 274, 275
mass extinctions, 188, 189–90, *190*
mating systems in humans: wealth transfer and, 128, 172–73. *See also* polygyny
mating systems in primates, 6–7, *7*, 58, 256–59, *257*, 297; intraspecific variation in, 275
matrilineal descent, in Bantu-speaking people, 67, 120, 172
matrix representation with parsimony, 48, *49*
maximum likelihood: with birth-death model, 187, 189; to compare evolutionary models, 118–19, *119*; with diversification rates, 194, 195, 197, 198; to estimate evolutionary rates, 123; to estimate λ (lambda), 113, 114–15; with evolution on part of a phylogeny, 296; in generalized least squares method, 165; in Pagel's discrete method, 168–69, 171
maximum likelihood for ancestral state reconstruction: advantages over parsimony, 64, 65, 66, 69, 71, 76, 77; vs. Bayesian methods, 76; of continuous traits, 79, 81–84, *82*, *83*, 85, 96; of discrete traits, 63–71, *65*, 66t, *68*, *70*, 76, 77; in evolutionary anthropology, 66–67; one-parameter vs. two-parameter, 67, 68–69, *70*, 71
maximum likelihood for inferring phylogenies, 33–35, 37; node-density artifact in, 200
mean, phylogenetic, 80, 88–89; independent contrasts and, 209
mean square error, and phylogenetic signal, 111, 113t
measurement error: generalized least squares and, 86; intraspecific variation as, 276
Megaladapis, 205
meta-analysis, phylogenetic, 165
metabolic needs of group, and home range size, 214, *215*, 275
metabolic rate, and body mass, 149, *150*, 166, 202, 211–12, *212*

MHC evolution, and parasite richness, 129
migration, human: linguistic phylogeny and, 58–59, 234, 247–49, *248*, 319; yeast strains and, 13
mitochondrial Eve, 15, 247
models of evolutionary change: in allometry, 214–15, 226; in analysis of covariance, 217, 218; Bayesian methods and, 35, 36, 37, 38, 72, 76; branch lengths and, 100–1, 108–9, 118–20, *119*, 121–23; for continuously varying traits, 80, 96; defined, 100; generalized least squares and, 86, 164, 168; hominin evolution and, 292–94; hominin phylogeny and, 50; introduction to, 98, 100; maximum likelihood comparison of, 118–19, *119*; maximum likelihood methods and, 33, 34, 63, 67, 68–69, *70*, 71, 76; pattern of evolution and, 101, 106, 108, 117; phylogenetic signal and, 108; process of evolution and, 101, 106, 108, 117; rates of evolution and, 123–24; summary of, 124–25; three major examples of, 100t, 106. *See also* Brownian motion model; diagnostics for evolutionary models; ecological niche-filling model; Ornstein-Uhlenbeck (OU) model
molar teeth: body mass in apes and, 222, 225; in carnivores, 90, *91*
molecular evolution: alignment of sequences and, 27–28; diversification rates and, 199–200; maximum likelihood methods for, 34; in primates, 129; rate-smoothing techniques for, 46. *See also* DNA
monkeys. *See* macaques; New World monkeys; Old World monkeys; primates
monophyletic groups, 27, *27*
Monte Carlo simulation: in analysis of covariance, 217–18. *See also* Markov chain Monte Carlo (MCMC)
Moran process, 183
morphological data, 25, 26; acquisition bias with, 34; for hominins, 50; maximum likelihood methods with, 34; methods available for analyzing, 33; for primates, 50, 203–4; shape in, 202–3
"most recent common ancestor" method, 72–73, 74
multiple regression: to control for body mass, 220–21; correlated evolution and, 159–60; with cultural traits, 238
musical instruments, 241
mutualism, among fungi, 200
MySQL, 304

Native Americans. *See* American Indians
natural selection: Brownian motion and, 102; directional, 96. *See also* adaptation; sexual selection; stabilizing selection

Neanderthals: cranial capacity of, 122; DNA from, 52
Nee transformation, 159, *206*
negative allometry, 208
neighbor joining, 30
Neighbor Net, 47
neocortex size: allometry with body mass and, 213; clade shifts in, *144*; vs. group size in primates, 127, *128*, 132, *133*, 289
nested analysis of variance (ANOVA), 110, 176
neutral evolution: Brownian motion and, 101, 102; phenotypic divergence and, 124
New Guinea: cultural group extinction rates in, 201; diversity of languages in, 196, 249, *253*
New World monkeys: diversification in, 189, *194*, 194–95, 195t; host-parasite co-evolution in, 13. *See also* primates
niche conservatism, and statistical nonindependence, 140
niche-filling model. *See* ecological niche-filling model
niche space, 259–60; competition for, 261
node-averaging approach, 176, *177*
node-density artifact, 200
nodes, 21, *21*, 22, 24; ancestral state reconstruction and, *54*, 54–55, *55*, 56; diversification and number of, 198, 200; limiting reconstruction to subset of, 81; unrooted tree and, 29
nonindependence. *See* statistical nonindependence
nonphylogenetic tests, 110, 133, *133*; concurrent with independent contrasts, 161; vs. independent contrasts, *133*, 158, 161; of intermembral index vs. body mass, 205; intraspecific variation and, 138; problem of nonindependence and, 142–46, *143*; statistical power of, 223; type I error in, 210
numerical taxonomy, 30

Old World monkeys: diversification in, 189, 192, 193–94, *194*, 195, 195t; exaggerated sexual swellings in, 54, 59, 311; female philopatry in, 58; jaw morphology in, 215–16, *216*. *See also* macaques; primates
ordered character states, 60, 61t, 63
Ornstein-Uhlenbeck (OU) model, 100t, 103–4, *104*, 106; adaptation toward fitness peak in, 166; in correlated evolution studies, 215, 218; of dental evolution in fossil horses, 292; diagnostics for, 118, 120; generalized least squares with, 165; small intestine surface area and, 211; tests of phylogenetic signal and, 112, 112t
osmolarity, plasma, 84–85
osteocyte size, and genome size, 223, *224*, 225, 291

outgroup: to resolve ambiguous reconstructions, 59, *60*; rooting of tree based on, 29

Pagel's discrete method, 148, 168–73, 169t, *170*; in Bayesian framework, 172–73, *174, 175*; evolutionary singularities and, 296–97; examples of, 171–72
paleontology, statistical, 53, 71, 256
paraphyletic group, *27*
parasite-host tree congruence, 11–13, *12*, 241, 319–20
parasite richness in primates, 109–10, 128–29, 135, 199, 259
parasites of primates: communities of, 264; database of, 309; pinworms, 11–13, *12*; protozoan, 128–29, 135, 264
parasitic infections: in humans, 229, *230*, 249, 251; phylogenetic signal of, 114
parsimony for ancestral state reconstruction: cost structures in, 60–63, *62*, 76; of discrete traits, 58–63, *60*, 61t, *62*, 76, 77; linear, 80; vs. maximum likelihood, 64, *65*, 66, 69, 71, 76, 77; overall disadvantages of, 76; squared-change, 80, 84–86, *85*, 89, *90, 91, 96*; transformation types in, 60–61, *63*
parsimony for inferring language trees: Austronesian, 248; Indo-European, 41, *42*
parsimony for inferring phylogenies, 30–33; cost frameworks for, 32–33; matrix representation with, 48, *49*; node-density artifact in, 200; quantifying support for, 32
PD (phylogenetic diversity), 266–68, *267*, 269
phenetics, 30
PHP scripting language, 304, 309
phylogenesis, cultural, 234, 235, 236
phylogenetically independent contrasts. *See* independent contrasts
phylogenetic distance, 239, 240, *240*
phylogenetic diversity (PD), 266–68, *267*, 269
phylogenetic inertia, 107–8
phylogenetic mean, 80, 88–89; independent contrasts and, 209
phylogenetic reconstruction. *See* ancestral state reconstruction
phylogenetic referential models, 282, 283, 284–86, *285*
phylogenetic signal, 98, *99*, 100, 106, 107–17; autocorrelation and, 110, 178; of behavioral traits, 112, 113t, 278; of body mass, 112, 112t, 121, 209; cultural traits and, 110, 116, 125, 237–38, 319; *d* parameter and, 120–21; defined, 107; diagnostics for use of, 161; in discrete traits, 110, 116, 238; *g* parameter and, 121; general examples of, 109–10; horizontal transmission and, 319; independent contrasts and, 111, 162; introduction to, 107–9; language relationships and, 108, 245; logistic regression and, 176; in popu-

lations, 108; quantification of, 110–16, 112t, 113t, *114, 115*; rates of evolution and, 124; reasons for estimating, 109; star phylogeny and, *108*, 108–9, *115*, 116, 120; statistical nonindependence and, 130, 140; summary of, 117
phylogenetic species concept, 265–66
phylogenetic targeting, 300, 310–16, *312, 313, 314*, 321
phylogenetic tree building, 21, 25–38; Bayesian methods for, 35–38, *37*; data for, 25–28; distance-matrix methods for, 29–30; key concepts of, 25, *26*–29; large-scale, 47–49, *49*; maximum likelihood for, 33–35; parsimony for, 30–33. *See also* Bayesian methods for inferring phylogenies; language trees
phylogenetic trees: for ancestral state reconstruction, 52–53, 57–58; basic-level understanding of, 20; basic structural features of, *21*, 21–25; extinct lineages and, 25; number of possible topologies, 30; unrooted, 29, *29*. *See also* branches of phylogenetic tree; dated phylogenies; language trees; nodes; star phylogeny; tree congruence
phylogenetic uncertainty: Bayesian analysis and, 37, 38, 71–72, 73, 74, 87, 178, 186; future significance of, 318, 321; hominin phylogeny and, 50; maximum likelihood and, 50, 67; in primate research, 186, 320; squared-change parsimony and, 85; in supermatrix approach, 49, 50
phylogram, 24, 100–1
pinworms of primates, 11–13, *12*
plasma osmolarity, 84–85
polygyny: bride-price and, 173, 229; male bias in inheritance and, 228–29, 229t; phylogenetic signal with, 238
polymorphic character states, 77
polymorphism parsimony, 77
Polynesian ceramics, 232
Polynesian languages, 46, *47*, 196, 241
polyphyletic group, *27*
polytomies, 22, *23*; independent contrasts with, 152–53; in language trees, 46–47, 51; in supertree analysis, 48
population density: primate diversification rates and, 199; primate extinction risk and, 270
population-level data points, 275–76
population movements. *See* migration, human
populations: phylogenetic signal in, 108; statistical nonindependence of, 140–41, *141*. *See also* community ecology
positive allometry, 208
positivized contrasts, 149, 152
possums, Australian, 261

posterior probability distribution, 35, 36, 37, 72, 87
power. *See* statistical power
primary key, in database, 301–2, *302*
primate diversification, 320; body mass and, 198; descendent lineages per taxon, 193, *194*; major rate shifts in, 194–95; parasite richness and, 199; phylogenetic uncertainty and, 186, 320; rates of, 192, 193–95, 195t, 198–99; varying speciation or extinction rates and, 187
primate phylogenies: evolutionary questions and, 181; hominin evolution and, 298; inference of, 25–26, 36, 48–50, 76, 317–18; ρ (rho) transformation of, 115, *115*; uncertainty in, 320
primates: adaptation in, 127–28, *128*; aerobic function in, 276–77; behavioral and ecological research issues in, 255–56, 274–79; behavioral traits in, 112, 113t, 127, *128*, 138, 278; birth-death model of evolution in, 189, 195t; body mass dimorphism in (*see* sexual dimorphism in primates); body mass in (*see* body mass in primates); brain size in, 131–32, 143, *144*, 166; cognition in, 1, 3, 7–8, *9*; community ecology of, 259, 260, 261–65, *263*; community structure in, 106; competition in, 162, *163*, 261, 289; conservation of, 269; correlated evolution and, 178; day range in, 127, *128*, 134; diet of, 52, 320; dimorphism in (*see* sexual dimorphism in primates); discovery of species of, 272; disease risk and latitude in, 135; diversification of (*see* primate diversification); early comparative studies of, 146; estrus synchrony in, 258–59, 275; evolutionary rates in, 124, 298; exaggerated sexual swellings in (*see* sexual swellings, exaggerated); extinction risk in, 267, 268, 270; field studies of, 278–79; fossil taxa of, 34, 288–89; future directions for comparative research, 320; genetic data for, 50, 317–18; group size in (*see* group size in primates); home range size in (*see* home range size in primates); homoplasy in evolution of, 58; immunity and promiscuity in, 129, 173; infanticide in, 257; infectious disease information gaps in, 279; intraspecific variation in, 256, 274–77, 320; locomotor behavior in, 52, 204, *204* (*see also* intermembral index); malaria prevalence in, 109; male competition in, 162, *163*, 289; mating systems in, 6–7, *7*, 58, 256–59, *257*, 275, 297; molecular evolution in, 129; morphological studies of, 50, 203–4 (*see also* intermembral index); number of males in groups of, 256–59, *257*; parasite communities of, 264; parasite database for,

309; parasite richness in, 109–10, 128–29, 135, 199, 259; phylogenetic diversity in, 267, 268, 269; phylogenetic signal in, 109–10, 112, 112t, 113t, 120–21, 278; phylogenies of (*see* primate phylogenies); pinworms of, 11–13, *12*; range defense in, *128*, 134, 320; ranging patterns in, 143; referential models with, 283, 284–86, *285*; singularities in evolution of, 281–82; sleeping habits in, 58; small intestine surface area in, 211; sperm competition in, 6–7, *7*; territoriality in, 127, *128*, 134; testes mass in, 6–7, *7*, 112t, 121, 136; updated data sources on, 137; white blood cell counts in, 109. *See also* hominins; humans; New World monkeys; Old World monkeys
primatology, comparative method in, 1, 255–56
principal components analysis, 159
prior, Bayesian, 35, 36, 37, 72
Proconsul africanus, 222
projectile points, American Indian, 232, *233*, 249
prosimians: host-parasite coevolution in, 13; subfossil giant, 205
proto-Indo-European, 94–95
protolanguage, 40; reconstruction of, 93–95
protozoan parasites of primates, 128–29, 135, 264. *See also* malaria
pseudoreplication, 81
psychology, comparative, 8, 162–63
pull of the present, *184*, 184–85, *186*, 187; in language trees, 274
punctuational evolution, 103, 105
punctuational model, and κ (kappa) parameter, 292
pure birth process, 183, 187
push of the past, 185–86, *186*, 187

quantitative traits. *See* continuous traits
queries, database, 303

R, 3, 16, 124, 321
rainforest trees, community ecology of, 260–61
random walk model, 81; vs. directional model, 92, *92*, 93. *See also* Brownian motion model
range. *See* geographic range; home range size and body mass; home range size in primates
Rapoport's Rule, 10
rate heterogeneity: linguistic, 45–46; methods with biological data, 46
rates. *See* diversification rates; evolutionary rates; speciation rates
reconstruction. *See* ancestral state reconstruction
Red List of Threatened Species, 267, 269, 270, 272, 273

reduced major axis (RMA), 213, 214
referential models, 282, 283, 284–86, *285*
regression: through the origin, 159–60. *See also* least-squares (LS) regression; multiple regression
relational database management systems (RDBMSs), 300, 303, 304, 310, 316
relational databases, 300–310; errors and double-checking in, 307–8; fully referenced, 304–5; introduction to, 300–4, *302*; notes field of, 305, 308; sharing data of, 309; summary of, 310; taxonomic ambiguity and mismatches in, 305–7, *307*. *See also* databases, comparative
relational integrity, 302–3
Rensch's rule, 166
reproductive skew, 275–76
residence patterns: in Austronesian societies, 74; descent mechanisms and, 229
residuals: in least-squares regression, 220, *221*; in multiple regression, 221
retention index (RI), 31, 236–37, 238
reticulate data, 241
rhesus macaques, group size in, 274, 275
ρ (rho), *115*, 115–16, 123
ribonuclease gene, in artiodactyls, *73*, 73–74, 76
ridge regression, 221
RMA (reduced major axis), 213, 214
root, 21, *21*, 28–29

sample sizes, 137–38; in human evolution, 296; population-level data points and, 275–76; splitting of species and, 307. *See also* statistical power
sampling bias: in ancestral state reconstruction, 77, 80–81, 96; language diversity and, 137
scaling parameters. *See* branch length transformations
scaling with size, 202–3; as correlated trait evolution, 208; incorporating phylogeny with, 208–9; reconstructing traits in extinct species and, 222; summary of, 225–26. *See also* allometry; body mass
schema for data collection, 301
semicircular canals, in mammals, 166, *167*
sex ratios: in human populations, 167; operational, 162, *163*
sexual dichromatism in birds, 61, 63, 196
sexual dimorphism in primates, 6, 107, 121, 129, 162, *163*, 203; body mass and, 166–67, 220, 255; in extinct hominins, 129, *130*; male competition and, 289
sexual division of labor in humans, 238
sexual selection, 6–7; diversification and, 196; in extinct hominins, 129; extinction risk categories and, 173; infanticide and, 257; speciation rate and, 77, 181

sexual swellings, exaggerated, in primates, 53–54; Bayesian analysis of, 71–72; diversification and, 197; maximum likelihood analysis of, 69, *70*, 71; multi-male mating systems and, 173, *174*, *175*, 297; parsimony analysis of, 59, *60*, 62, 62–63; selection of species to study, 311
shape, scaling in studies of, 202–3
shared derived characters (synapomorphies), 26–27, *27*; distance-matrix methods and, 30; parsimony and, 31; retention index and, 31; as singular evolutionary events, 281
shared derived characters in linguistics, 39, 45
σ2 (sigma2) parameter, 118
simulations: analysis of covariance in, 217–18; of ancestral state reconstruction, 77, 84, 89, 93, 96, 319; with birth-death model, 188, 189; comparing λ and ρ, 116; controlling for phylogeny in, 133; in cross-cultural research, 320; with defined variance, 118; extinction risk in, 268; of horizontal transmission, 237, 240, 243–45, *244*, *245*, 319; ignoring phylogenetic information in, 143–44, *145*, 146; of independent contrasts method, 157; of macroevolutionary patterns, 180; of niche-filling model, 122; phylogenetic signal and, 108, 110, 116; of phylogenetic targeting, 316; of singular evolutionary changes, 296; speciation and extinction rates in, 197–98, 201; statistical performance measures studied in, 242–43; taxon sampling in, 137; tree balance in, 191, *192*
singularity problem, 281. *See also* evolutionary singularities
sister clade comparison, 190, 192–93, 197, 198–99
sister species, 24
size, scaling with, 202–3. *See also* allometry; body mass
sleeping habits in primates, 58
sleep patterns in mammals, 138
small intestine surface area, in primates, 211
social learning in primates, 7–8, *9*
social relationships. *See* behavioral traits
soft polytomy, 22
Southeast Asia, colonization of, 58–59
spatial memory in primates, 129
spatial reasoning in primates, 132
speciation: allopatric, statistical nonindependence and, 140; in bacteriophage, experimental, 89, *90*; community structure and, 260; diversification and, 181; macroevolutionary patterns and, 180; time period required for, 192; tree structure and, 21, 22, 25
speciational model, 102–3; allometric analysis using, 214–15; biodiversity under, 268;

diagnostics for, *119*, 119–20, 124; κ (kappa) parameter and, 292

speciation rates, 182; in birth-death model, 183, 187–89; character states that affect, 77; estimation of, 25; factors influencing, 196–99; sexual selection and, 77, 181; varying, 187. *See also* diversification rates

species concepts, 265–66

specific effects, vs. phylogenetic effects, 107

sperm competition: in birds, 138–39; in primates, 6–7, *7*

spleen mass, in primates, 207–8

split decomposition, 47, 240–41; Austronesian languages and, 249; incongruence in cultural trait histories and, 320

splits graph, 47, *47*, 241

spreadsheets, 300, 301, 303, 310

SQL (structured query language), 303

squared-change parsimony, 80, 84–86, *85*, 89, *90*, *91*, 96

stabilizing selection: adaptation and, 161; ancestral state reconstruction under, 96; diagnostics for, 120–21; generalized least squares with, 165; phylogenetic signal and, 117; statistical models for, 96. *See also* Ornstein-Uhlenbeck (OU) model

standard cross-cultural sample, 178

standard errors: in generalized least squares, 165; for reconstructed values, 84, 86, 96

star phylogeny: branch length transformations and, 114, *115*, 116; nonphylogenetic tests and, 146, 161; phylogenetic signal and, *108*, 108–9, 120

statistical nonindependence, 4–5, 139–46, 148; adaptive hypotheses and, 130–31, 139–40, 148, 275; autocorrelation methods and, 178; basic concept of, 139; in cross-cultural studies, 140–42, *141*, 178, 228, 320; generalized least squares and, 246; geographic proximity and, *253*; in higher nodes approach, 176, *177*, 255; of hominin characters, 50; horizontal transmission and, 242; independent contrasts and, *145*, 145–46; of intermembral index and body mass, 205, *206*, 207; phylogenetic signal and, 130, 140; phylogenies and, 142–46, *143*, *144*, *145*; of populations, 140; potential drivers of, 140

statistical paleontology, 53, 71, 256

statistical power: in computer simulations, 243, 296; with continuous vs. discrete values, 258; defined, 243; with directional model, 293; with evolutionary singularities, 218, 296, 297; horizontal transmission and, 243; of nonphylogenetic tests, 223; phylogenetic targeting and, 311, 315, 316. *See also* sample sizes

statistical support measures: with Bayesian analysis, 36, 37; bootstrap indices, 32, 33, 34–35; with maximum likelihood methods, 34–35, 37, 64

step matrix, in parsimony analysis, 61, 61t, 63, 76

stochastic character mapping, 77–78

strategic (conceptual) models, 282, 283–84, 285, 288

stratigraphic parsimony, 61

strepsirrhines, 189, 194, 195, 195t, 262, 281, 298. *See also* lemurs

structural relations model, 213–14

structured query language (SQL), 303

substitution models, 34; rate heterogeneity in, 46

supermatrix approach, 48–49, 50

supertrees, 26, 48, *49*; for primates, 194

support indices. *See* statistical support measures

Swadesh list, 40–41

swordtail fish, female choice in, 79

symmetric model: Bayesian, 74; maximum likelihood, 67, 68, 69, *70*, 71

symmetry of phylogenies. *See* tree balance

synapomorphies. *See* shared derived characters (synapomorphies)

Taiwan, and settlement of Polynesia, 248, *248*, 249

tamarins, forelimb morphology in, 202

targeting, phylogenetic, 300, 310–16, *312*, *313*, *314*, 321

taxonomic names, 19; translation of different classifications, 305–7, *307*; trend toward splitting and, 306–7

taxon sampling: in ancestral state reconstruction, 77; incomplete, 137; lineages through time plots and, 187–88, *188*

technological advances: rates of evolution in, 124–25; settlement of Polynesia and, 249. *See also* innovations

teeth. *See* dental traits

10kTrees website, 305

testes mass: in birds, 138–39; in primates, 6–7, *7*, 112t, 121, 136

textiles, 49; Iranian, 236

tiger beetles, mandible length in, 114

time depth in language relationships, 44, 50–51

tonal languages, 15

tool use, 8

topology, tree, 22, 24; phylogenetic uncertainty and, 71, 74; possible number of, 30

traits. *See* behavioral traits; continuous traits; data, phylogenetic; discrete traits

transformation types: as models, 98; in parsimony analysis, 60–61

transition matrix, 64–65, 66t; in Pagel's discrete method, 169, 169t

transition rates: in Bayesian reconstruction, 71, 72, 73, 74; for characters affecting diversification, 77; for evolutionary singularities, 296–97; in maximum likelihood analysis, 64–65, 66, 66t, 67–69, 76, 296–97, 298; in Pagel's discrete method, 169–71, 169t, 170
translation tables, 306, 307
tree balance, 190–92, 191, 201
tree congruence, 241, 319–20; of primates and pinworms, 11–13, 12
trees. See phylogenetic trees
Tree-Thinking Challenge, 20
two-stages least squares, 246
Tylor, Edward B., 141, 227–28
type I error, 139; in computer simulations, 243, 296; in cross-cultural studies, 141, 178; horizontal transmission and, 243, 245; by omitting phylogeny, 144, 145, 210, 223
type II error, 139

ultrametric trees, 23, 24, 100
uncertainty: in data values, 308. See also mapping uncertainty; phylogenetic uncertainty
ungulates, home range size in, 217–18, 218, 219
unordered character states, 60, 61t, 63

variance: of error in allometry, 213–14; heteroscedasticity and, 155, 156; partitioned among species, 146; of trait change, 118, 153
variance-covariance matrix, 113–14, 114, 164, 165
variance inflation factors, 221
variation: in archaeological artifacts, 232; behavior in extinct species and, 207; in human populations, 108; in Ornstein-Uhlenbeck model, 103; phylogenetic signal and, 107, 108, 110, 117; temporal, of rates on tree, 121. See also intraspecific variation
vertebrates: body size of, Brownian motion model, 118; genome size of, 223, 224, 225. See also birds; mammals; primates
viruses: bacteriophage, 89, 90; primate hosts of, 264

web-based database applications, 304, 309
websites: AnthroTree, 3, 16; databases on, 304, 309; for taxonomic translation, 305
wheeled vehicles, 95
white blood cell counts, in primates, 109
Woods Hole Group, 180

yeast strains, and human migrations, 13
Yule model, 182–83